New Aspects of
Human Ethology

New Aspects of Human Ethology

Edited by

Alain Schmitt
Ludwig-Boltzmann-Institute for Urban Ethology
Vienna, Austria

Klaus Atzwanger
Research Center for Human Ethology
Max-Planck-Society
Andechs, Germany

Karl Grammer
Ludwig-Boltzmann-Institute for Urban Ethology
Vienna, Austria

and

Katrin Schäfer
Institute for Human Biology
Vienna, Austria

Plenum Press • New York and London

Library of Congress Cataloging-in-Publication Data

New aspects of human ethology / edited by Alain Schmitt ... [et al.].
      p.   cm.
    "Proceedings of the 13th Conference of the International Society
  for Human Ethology, held August 5-10, 1996, in Vienna, Austria"-
  -T.p. verso.
    Includes bibliographical references and index.
    ISBN 0-306-45695-8
    1. Human behavior--Congresses. 2. Genetic psychology--Congresses.
  3. Psychobiology--Congresses. 4. Behavior evolution--Congresses.
  I. Schmitt, Alain. II. International Society for Human Ethology.
  Conference (13th : 1996 : Vienna, Austria)
  BF701.N48   1997
  155--dc21   97-30572
                                                              CIP

Proceedings of the 13th Conference of the International Society for Human Ethology,
held August 5–10, 1996, in Vienna, Austria

ISBN 0-306-45695-8

© 1997 Plenum Press, New York
A Division of Plenum Publishing Corporation
233 Spring Street, New York, N. Y. 10013

http://www.plenum.com

10 9 8 7 6 5 4 3 2 1

All rights reserved

No part of this book may be reproduced, stored in a retrieval system, or transmitted in any form or by any means, electronic, mechanical, photocopying, microfilming, recording, or otherwise, without written permission from the Publisher

Printed in the United States of America

THE NEW ASPECTS OF THIS VOLUME

The New never grows from nowhere, but is rooted in the Old and Unexpected. The latter is a function of curiosity and emotional not rational involvement in the subject. We hope that the following chapters will stimulate emotion and thought, and provoke new systematic and empirical research. The dilemma is that in one sense, we as humans abhor speculation and the unknown, but that in another sense, as scientists, we need the speculative as life needs water. Curiosity drives us and fills the gap. And fortunately, theories pass over, but the frog persists, as Jean Rostand expressed so succinctly the relation between conjecture and refutation ["Les théories passent, la grenouille reste." *Carnets d'un biologiste*].

Human ethology studies human behavior in a very interdisciplinary way. In this century, it started in the forties from comparative zoology and evolutionary theory, with a specialization in the direct observation of behavior in a "natural" context. These roots provided not only the drive to search for behavioral universals among cultures, but also to see humans before the background of the natural (animal) world, as part of the overwhelming diversity of life. Two factors were of outstanding importance in reconciling the wealth of "languages" spoken at the Babel of interdisciplinarity: evolutionary theory and the observation of undisturbed "natural" behavior, which clearly showed the limits of laboratory and questionnaire studies when confronted with social complexity, and thus induced the involved scientists to go well beyond the frontiers of the discipline they were trained in. This volume reflects the interdisciplinarity of human ethology. Our aim when organizing the conference was to bring together ideas some of which "normally" are not considered to be of immediate and burning interest to human ethology. The authors, zoologists, psychologists, a geneticist, a mathematician and a physiologist have found a transdisciplinary language to write on human social behavior without losing the specific world views of their branches. All papers contain many new aspects, and even some brand-new insights and (yet unpublished) data. In the following, we direct the reader's attention to some of these aspects. Obviously, there are more; there was never an editors' instruction to restrain to some pieces of news; the reader has to discover the other's by her/himself, and to judge their relevance.

Irenäus Eibl-Eibesfeldt has written a contribution which combines a very personal history of biographical entanglements of the proponents of modern ethology with a chronologically structured summary of the fundamental concepts needed by the founders of sociobiology and human ethology. Thus the process of developing a new science is presented as a profoundly social enterprise, not only as a brilliant logical exercise in re-combining old ideas and creating new insights. Eibl goes further and comments on some of the newest developments of human ethology. For him, the application of principles derived from classical (human) ethology to culture, e.g. art, law and urban environments, are the most interesting and promising future areas of investigation.

Glenn Weisfeld presents a new allegedly complete list of human basic emotions, which, he convincingly argues, may be used to systematically classify human behavior. Some of his emotions very seldom appear in psychologically inspired lists, e.g. esthetic feelings, humor appreciation, and pride and shame. The list is based on a phylogenetic perspective of man, and Weisfeld gives adaptive value hypotheses in each case. He argues that emotions are phylogenetically much older than cognitions and higher learning mechanisms (and notes that neurobiology says the same since some time), and are thus most probably heavily related to fundamental biological processes. In sum, it may be very productive to use emotions as a basis for the human ethogram. This leads Weisfeld to try to find criteria to identify a basic emotion. Specific adaptive value, involving the whole organism, eventually specific facial or bodily or visceral or hormonal associated patterns, distinct affect, to list only a few of them he uses to construct his promising "new complete inventory".

Peter K. Smith compares observational and interview data of real and play fighting. Play fighting or rough-and-tumble play (in short R&T) is one among those human behaviors whose existence as an element of the human ethogram is not questioned, and which is well analysed in classical ethological terms. As far as one can know from the existing data, it is an universal, that is, a behavior in which all youngsters all over the world are involved, boys more than girls. Most interesting is Smith's discussion of the developmental and evolutionary function of R&T. In Smith's opinion, it is not very probable that R&T serves social cohesion or enhances social skills, since these hypotheses are unable to explain the sex differences. R&T as a means to practice fighting may explain sex differences, but is rejected by both theoretical and empirical counter-evidence (see 9.3). The most solid grounds exist for invoking a dominance function, that is, R&T serves to establish or maintain dominance relations and to inform about strength of others (see 9.4). The comparison of observational and interview data, particularly interviews with teachers, has educational consequences. E.g., teachers overestimate the overall frequency of real fighting and, largely, the tendency of R&T to degrade into real fighting. Consequently, they intervene more often than necessary.

Karl Sigmund has the talent to simplify complex mathematical connections and to illustrate them by adequate and (ethologically) relevant examples. Here, he reviews the state-of-the-art in game theory applied to cooperation. One of the important insights of his chapter is that no single strategy can do it on its own, that its success depends both on the other strategies present in the population and on very subtle differences in historical and environmental contexts. For example, one must quite exactly know when the other player makes his move, and if he then will be informed about my preceeding move, or not. This makes meta-strategies necessary, that is, one has to know at which historical moment in the development of the population one stands and what exactly are the circumstances which surround the game. Moreover, it leads to states of punctuated equilibrium, where seemingly small fluctuations in the population may suddenly provoke abrupt transitions from e.g. a period of cooperation to one of defection. Sigmund goes as far as to propose, quasi *en passant* (end of §4), that emotions or feelings may be such meta-strategies, and that they may have evolved because of the fuzziness and complexities of strategic social iife.

Robin I. M. Dunbar has developed in the last years a new language evolution story, one never told before. Moreover, his insights are unique in that they for the first time allow empirical quantitative testing, which conventional theories do not. Classically, language is said to have evolved as a teaching and/or hunting device. Starting from the finding that neocortex volume increases with social group size, and not with ecological variables, Dunbar hypothesizes that thinking power required to handle the fuzziness and complexities of sociality and to reduce the centrifugal forces within social groups, may be the driving force of language evolution. After having collected non-human primate evidence, Dunbar asks where this leaves modern humans. If language is a sort of grooming-Ersatz or social kit as Dunbar's bonding mechanism hypothesis says, some quantitative and testable predictions may be derived from it. One is that human conversation cliques are proportionately larger than non-human primate grooming "cliques" (talking in that view is symbolic grooming, for example chatting or exchanging niceties and gossip). "Proportionately" means that clique size manageable through language should be about three times that manageable by real grooming, which reaches only ONE other individual. The other is that (social) gossip, not teaching e.g. technical matters or planning "hunts," is predominant in informal conversations. His data corroborate these hypotheses surprisingly well. Most interestingly, the above proportionality mirrors into group or band size: The ratio for humans and chimpanzees is 148 to 53.5, that is 2.8 to 1.

Karl Grammer, Valentina Filova and Martin Fieder think that there is now enough (negative) evidence to suggest that the classical methods used in behavior observation have to be completed by a series of methods able to automatically analyse the behavior of humans (and animals) unobstrusively filmed in their "natural" environment. They convincingly argue that the new methods must have a theoretical rationale. In fact, they suggest human everyday analysis of others' behavior is not only based on categorization and "intuitive ethograms" (from that perspective, ethologists currently do the same as the lay person, but they act more systematically). They propose that the lay person may also judge others' behavior globally, without categorizing, by evaluating the qualities (e.g. speed and complexity) of each and every movement and motion. Their idea is corroborated by physiological insights of the last 20 years, e.g. by Hubel and Livingstone who have shown that form, color and movement are processed parallelly in visual systems. They propose two new methods of digital image analysis based on neural networks and on a software developed by themselves.

Thomas J. Bouchard, Jr. reviews the insights one may obtain from twin studies, that is, by comparing the behavior, personality, intelligence, interests, social attitudes, psychopathologies, etc, observed in monozygotic and homozygotic twins reared apart or together. When examining the correlations in intelligence of genetically unrelated individuals reared together in order to evaluate environmental influences, the data (Bouchard, Figure 2), besides showing that the correlations are about .30, raise an interesting question. Indeed, the figure shows that the influence of the environment decreases with increasing age. From that observation, Bouchard hypothesizes that heritability of IQ may change over time, more exactly, that it may increase with age. A quite counter-intuitive hypothesis, which it is however easy to test. Bouchard's Figure 3 shows the result of the test: there is a strong increase in heritability and a decrease in common family environmental influence with age. This is a powerful indirect hint at the genetical basis of intelligence (note: intelligence as measured by psychological tests.). My second example is of direct interest for sociobiologists. They have constructed over the past thirty years an elaborate theory of dominance and social status, and on how it differs among the sexes, animal and human. Essentially, they say that status plays a much stronger role in male than in female life. Most interestingly, Bouchard (Table 11) finds that the genetic effects on socioeconomic status are large in both genders (as compared to those on e.g. personal interests), but that there are large differences between women and men, men being much more influenced (by a factor 2 to 5) by their genetic make-up. This is a surprising indirect corroboration of sociobiological theory.

C. Sue Carter introduces her chapter with a historical perspective, summarizing that "an awareness of a relationship between the testes and human behavior predates written history." Clearly then, the hormonal influences on human behavior are not a new insight. What is surprising however, is the very firm overview message on sex hormones, which usually, for example in medical textbooks, are put forward as THE paradigm of efficient and probative research on human hormones: "Although reproductive hormones have been studied in some detail, remarkably few strong conclusions are possible." When we say "surprising conclusion," we obviously mean surprising to us; there is an impression of overgeneralization of results and underestimation of flawed assumptions we got when trying to be up-to-date in that area of research; Carter eliminates this impression by a wealth of arguments on assumptions and on connections of the data to the general bio-logical background. E.g., she questions the appropriateness of laboratory rodents to advance understanding of human behavior, because they are as rodents a long phylogenetic way off our evolutionary road and, as white lab rats, artificially selected, that is, domesticated. Nobody knows what this means when the results are carried over from the animal model to humans. Carter makes a very interesting attempt to explain the observation that sex steroids seem to have very few effects on human behavior that can be identified reliably. She proposes that since humans rely heavily on social stimuli for the regulation of reproduction, the connection between steroid hormones and behavior has weakened in order to allow rapid behavioral, emotional and "vegetative" adaptation to the situation. The "fast" peptide hormones like vasopressin and oxytocin are candidates for this regulation.

R. Robin Baker has written what seems to be at first sight a quite dry state-of-the-art paper, plenty of numbers and equations. But a second look reveals many exciting hypotheses and tests of them. Not all were corroborated; for example, Baker hypothesized that bisexuality is an adaptation to sperm competition. His data eliminate that hypothesis. [Note that the following is in a sense purely descriptive, and that to analyse a behavior is neither to value nor to judge or recommend it.] Particularly strong is his empirical evidence for a three-phasic model of female choice. In late adolescence (phase I), women seem to meet a series of men and pair quite stably with a partner, but remain vigilant for a male who is better, either in terms of genes, resources offered, or both. Seeking or allowing copulation outside the established pair may be part of this strategy and sperm competition may often occur (17% of females generate sperm competition in their first 50 lifetime copulations). Phase II begins when the woman has found and attracted a male who is an acceptable compromise in terms of the genes and resources offered. Reproduction eventually begins. The woman's second child is the most likely of all her children to have been fathered by her partner, the first sometimes having been fathered by one of her previous partners towards the end of phase I. In the middle of phase II, fidelity is highest and the risk of sperm competition lowest. Phase III is one in which, from the springboard of her established relationship and existing children, the female renews her shopping around for genes and resources better than provided by her current partner. For infidelity to occur, the female needs to meet someone with genes and/or resources sufficiently better than her partner's to outweigh the costs of infidelity. The chances of infidelity and sperm competition slowly increase as the woman ages and each child after the second is less and less likely to be fathered by her partner.

THE EDITORS
Vienna, in October 1997

CONTENTS

1. Human Ethology: Origins and Prospects of a New Discipline 1
   Irenäus Eibl-Eibesfeldt

2. Research on Emotions and Future Developments in Human Ethology 25
   Glenn Weisfeld

3. Play Fighting and Real Fighting: Perspectives on Their Relationship 47
   Peter K. Smith

4. Games Evolution Plays 65
   Karl Sigmund

5. Groups, Gossip, and the Evolution of Language 77
   R. I. M. Dunbar

6. The Communication Paradox and Possible Solutions: Towards a Radical
   Empiricism 91
   Karl Grammer, Valentina Filova, and Martin Fieder

7. Twin Studies of Behavior: New and Old Findings 121
   Thomas J. Bouchard, Jr

8. Hormonal Influences on Human Behavior 141
   C. Sue Carter

9. Copulation, Masturbation, and Infidelity: State-of-the-Art 163
   R. Robin Baker

Abstracts of the 13th Biennial Conference of the International Society for Human
   Ethology (ISHE), Vienna 5–10 August, 1996 189

Contributors 225

Curricula Vitae of Editors 227

Name Index 229

Subject Index 235

New Aspects of
Human Ethology

HUMAN ETHOLOGY

Origins and Prospects of a New Discipline

Irenäus Eibl-Eibesfeldt

Max-Planck-Institut für Verhaltensphysiologie/Humanethologie
D-82346, Andechs, Germany

1. INTRODUCTION

With the discovery of the conditioned reflexes, I. P. Pavlov introduced into the behavioral sciences a promising experimental approach and the hope that even the most complex behaviors could be explained by stepwise associative integration starting from some elementary precursors. In the United States of America, the school of Behaviorism fostered stimulus-response psychology, focussing on the experimental investigation of learning processes. "Experience" became an all-encompassing explanatory principle for students of the ontogeny of behavior. Z. Y Kuo's (1932, 1967) speculations on the development of the pecking response of the chick within the egg were widely quoted as evidence for the experiental shaping of behavior during embryosis. That chaffinches which in their eggs are exposed to nearly identical experiences as the chick embryo, however, develop a gaping response instead, did not occur as a problem to the "environmentalists."[1]

How could it happen that genetic factors controlling the self differentiation processes of the nervous system could be considered unimportant in the behavioral realm? Two main reasons can be pointed out. At the time, when stimulus-response (S-R) psychology was born, there were also psychologists such as W. James (1890) and C. Lloyd Morgan (1894, 1900) who presented fairly modern conceptions of instinct. They agreed that animals are not only outfitted with organs, but also with an innate ability to use them in specific ways, the neuronal structures underlying these abilities being the result of evolu-

---

[1] The subject matter was discussed in detail in my text "Ethology, the Biology of Behavior" (^{2}1975, German ^{7}1987). In his 1967 publication, Kuo even speculated that if he was able to exchange a human brain against a chimpanzee brain and vice versa nothing much would be changed, since it are the bodies which dictate in a conditioning way the brain what to do. There was nothing to grotesque to be published at that time, if it were supporting the ideology in fashion, in this case environmentalism and the Stimulus-Response-Psychology. This situation must be remembered in order to understand the impact of Ethology and thus of the pioneering contributions in particular of Konrad Lorenz and Niko Tinbergen which we are going to discuss.

New Aspects of Human Ethology, edited by Schmitt *et al.*
Plenum Press, New York, 1997

tion. But there were many others, who saw in "instinct" a vitalistic nearly mystical force guiding an animal. It was against these, that the materialisitic S-R psychologists tried to define themselves. In addition, J. B. Watson (1919, 1930), the founder of behaviorism, was convinced that environmental conditions more than genetic heritage decided the outcome in the behavioral development of the individual. The Lockean conviction that human beings come into the world as blank slates and that education could achieve wonders, experienced a renaissance and became dogma, sustainable only by ignoring certain fact of life. It must have been clear of course even to a stubborn environmentalist that there could not be any learning program to teach a fellow human being to experience the feelings of love, hatred, fear or any other feelings accompanying emotions or any of the sensations. Theoretically, we actually could never be sure that other persons feel the way we do, but we all act "as if," and this hypothesis works. We "understand" each other and are touched by poetry, love stories and other creation of art, even when produced by persons from other cultures. But such recalcitrant phenomena were simply omitted from the S-R research agenda.

Eventually it became evident that a "comparative" psychology based upon the studies of pigeons and rats in labyrinths and Skinner boxes—even though objectivistic—was too narrow an approach to cover the multifaceted phenomena of behavior.

Konrad Lorenz and Niko Tinbergen must be credited with providing the basic concepts and layout of ethology as a new discipline, defined as the "biology of behavior." It stimulated research and basic concepts such as that of the template, the motor generator system, and self differentiation as well as phenomena such as imprinting which have been validated down to the neuronal and molecular levels.

Since we are meeting in Vienna I want to discuss in particular the contribution of Konrad Lorenz to ethology and follow with some comments on how as a consequence human ethology grew as a new subdiscipline[2].

I began with ethological studies in 1946 at the small biological station at Wilhelminenberg, a field station located at the border of the Wienerwald. When Konrad Lorenz returned to Austria in 1948, I became closely attached to his small research group which moved to Germany in 1951 and finally became established as the *Max-Planck-Institut für Verhaltensphysiologie* in Seewiesen. In 1970, I started my own enterprise which in 1975 became an independent Research Center of the Max-Planck-Society. Thus, I was lucky to experience and contribute to the decisive postwar development of ethology.

2. THE EARLY DAYS

Konrad Lorenz was born in Vienna on November 7, 1903. He was a late child, his brother being 18 years older. His father Prof. Dr. Adolf Lorenz was the founder of orthopedics in Vienna. Konrad grew up in a spacious villa within a magnificent park-like garden. From his early days he was interested in animal behavior, and soon he adhered to the principle of keeping most animals free, though this sometimes created problems for visitors. His brother Albert vividly describes the situation in his autobiography: "Since in freedom tamed animals lose their fear of man, but at the same time also as a consequence

---

[2] There are certainly important precursors such as Charles Darwin, Charles Otis Witman, Oskar Heinroth or Jacob von Uexküll, just to mention a few, but it was Konrad Lorenz who formulated the main theoretical concepts and who integrated the knowledge on animal behavior of his time for the creation of a new discipline.

lose their respect, it could happen to a visitor that a civet cat unexpectedly attacked his calves from a bush, or a cockatoo clawed his neck. Or when someone was taking his coffee outdoors it occurred that Konrad's two ravens—the birds of Odin—would swoop down on metallic objects such as the glittering silver spoons or the sugar tongs, which would never be seen again!" (A. Lorenz 1952, p. 187)

Konrad craved to study zoology, but his father did not consider this an appropriate profession and his son followed his advice to study medicine. In 1928 he won his promotion to medical doctor. In the same year he began studying zoology which he concluded in 1933. His zoological interest had never been interrupted during his medical studies. He had kept detailed diaries on his pets and in 1927 he published the first paper on jackdaws in the *Journal für Ornithologie*, a paper which aroused the interest of Oskar Heinroth, one of the leading ornithologists at his time. During his medical training Lorenz worked as assistant of the anatomist Ferdinand Hochstetter who familiarized him with the field of functional and comparative anatomy. In his introduction to the 1975 edition of his collected papers Lorenz wrote: "A particularly happy event in my life as scientist was that I was thoroughly trained by my teacher Ferdinand Hochstetter in the theory and methodology of comparative morphology and that I established close contact with Oskar Heinroth, who became my second great teacher. His classical paper *Beiträge zur Biologie, insbesondere Psychologie und Anatomie der Anatiden* became a guideline for my future."

Konrad Lorenz started his behavioral research by focussing on those behavior patterns which were, like morphological features, characteristic of a species[3] and which could, by applying the criteria of homology, as elaborated by morphologists, serve for taxonomic purposes, since comparable movement patterns could be found in related species, their graded resemblance indicating closer or more distant relationships. Lorenz termed them *Instinkthandlungen*. They increasingly became the focal point of his interest. Stepping stones in the development of his ideas were in particular two papers *Beiträge zur Ethologie sozialer Corviden* (1931) and *Betrachtungen über das Erkennen der arteigenen Triebhandlungen bei Vögeln* (1932).

In 1935, the *Kumpan in der Umwelt des Vogels* (The companion in the world of birds) came out. Here, Konrad Lorenz presented the first layout of the field which was later to be termed "ethology."

He introduced the concept of the "*angeborenes Schema*"[4] (innate template) which are thought to exist as neuronal reference patterns in which knowledge about specific stimulus situations is encoded in such a way that at the income of specific trigger stimuli (*Auslöser* – social releasers) specific behaviors get released. Lorenz used the lock-key analogy to explain his idea. In this context he also discussed the phenomenon of imprinting. In this case pluripotential unspecific reference patterns are specified by key experiences during a short sensitive period. And once this specification has taken place, the fixation of a response upon an object proves to be quite therapy-resistent to the extent of practical irreversibility. Lorenz furthermore described "instinctive actions or motor patterns" (*Instinkthandlungen*), and he discussed the comparative approach based upon the morphological criterion of homology which he applied in a study to reconstruct the evolu-

---

[3] Form constancy being one criteria of these movements which means that the phase distance of the muscle actions that participate in the movements remaining constant. Amplitude of movement and speech may change with intensity. The pattern remains recognizable the same.

[4] See Eibl-Eibesfeldt 1995 and 1987 for a detailed presentation of the ethological concepts and their empirical backing.

tion of the courtship display patterns in the family of Anatidae (Lorenz 1941a). He also realized that the innate behaviors of animals seemed to be driven by internal motivating mechanisms, causing an animal in "appetitive behavior" (a term accepted from W. Craig 1918) to actively search for stimulus situations which then trigger a run of behaviors culminating in a consummatory act[5].

Even though Lorenz recognized the spontaneity of behavior, he used in his early papers the term chain reflexes for the often more complex "*Instinkthandlungen*." In 1937, he met Erich von Holst, and learned about his investigations into central nervous automatisms underlying the locomotory activity of fishes. Hence, Lorenz hypothesized that not only the motor patterns, but *Instinkthandlungen* in general, were based upon automatic cell groups active within the central nervous system which were responsible amongst other factors for the observed spontaneity.

About the same time a close friendship developed between Lorenz and Niko Tinbergen. In 1938, they were experimenting on the egg rolling movement of the greylag goose, which led to the distinction of taxis and fixed action patterns (*Instinktbewegung*). As a reminder for those less familiar with the concept: If an egg is placed outside the nest of a brooding greylag goose, the goose will reach out with its bill over and beyond the egg, and pull it in with the underside of the bill, balancing it carefully back into the nest. This behavior can be broken down into two components. If one removes the egg after the rolling movement has been started, then the movement continues. The bird behaves as if the egg is still there. However, the lateral balancing movements cease and the neck is pulled back in a strait line to the nest. This movement, which once released will continue in the absence of the triggering stimulus, is the fixed action pattern. The lateral balancing movements are the orientation movements or taxis components, which are also inborn, but responsive to the presence of ongoing stimulation.

The translation of the term *Instinktbewegung* or *Erbkoordination* into "fixed action pattern" led to many misunderstandings. There is variation in the movement by intensity and speed. What is fixed is the score or script of the muscle actions. The movement may just occur as an intention movement or full blown, and it may run off fast or slow. What remains constant is the relative phase distance of the muscle actions involved. Form constancy, furthermore, is characteristic of any stereotyped movement and therefore not only a characteristic for a fixed action pattern. Innateness is the decisive characteristic.

In 1940 Lorenz was called to the University of Königsberg as head of the Institute for Psychology. He thus became the last to occupy Immanuel Kant's chair. Here he laid the foundation of his biological epistemology, a short version of which he published in 1941 in the *Blätter für Deutsche Philosophie*. In agreement with Karl Popper—who was by the way a classmate of Konrad in Vienna (they continued their friendship after the war)—Lorenz accepted critical (hypothetical) realism. All organs of an organism including our brain, sensory organs and the resulting behaviors, according to this view, represent adaptations to an existing world. They depict facets of an extra-subjective reality which are relevant for survival. The fins and swimming movements, which a fish develops within an egg, mirror in this effect fitness-relevant features of the environment, in which the fish will move after hatching. Or in the phrasing of Popper: They constitute hypotheses about certain characteristics of the external world. I want to add: Hypothesis tested by natural selection. This in particular holds true for our perception whose function after all

---

[5] The Lorenz-Craig model: Appetitive behavior – releasing stimulus situation – consummatory act was later modified by N. Tinbergens hierarchy model of appetite behaviors (for details see I. Eibl-Eibesfeldt 1975, 1987).

is to mirror fitness-relevant facets of an outer world. And to those who express their doubts concerning the adequateness of such depictions, Gaylord Simpson used to say: "The monkey which had no correct representation of the reality of the branch on which he was about to jump, does not belong to our ancestry." What is more surprising is that this representation of reality in humans surpasses what would be needed for survival on our planet, since they correspond to reality so closely, that we can send spacecraft to distant planets, which then take and send pictures of distant events, such as the recent vulcano eruptions on the Jupiter moon Jo.

On the other hand there are also assumptions built into our perception, which are, for example, the basis of our visual illusions. Thus we perceive the moon moving against the clouds, since we hypothesize that objects move in a static environment, which fits to our immediate earthly environment. Here it is important to spot a moving object, be it a predator or prey, in minimum time. That we fall prey to an illusion, when we look up at the sky, does not really matter. In addition we know that we categorize and thus impose an order onto and into what we perceive. Colour perception is a good example.

In one way or another, every adaptation reflects some extra-subjective reality, toward which the organism is molded by adaptation during phylogeny and ontogeny. The representation of the outer world becomes in general more differentiated with the evolution of what we term higher organisms. During phylogenetic evolution an organism by mutation, recombination and selection informs itself, so to speak, in a process analogous to learning by trial and error, about its environment. Thus it adapts to and thereby mirrors its surroundings. "*Das Leben ist nicht Gleichnis von irgend etwas, es ist selbst die wissende Wirklichkeit*," said Lorenz many years later in his book *The other side of the mirror*.

His 1943 paper *Die angeborenen Formen möglicher Erfahrung* came out in the *Zeitschrift für Tierpsychologie* (now *Ethology*). Here he demonstrated in which ways the concept of the innate template, key stimulus, releaser and fixed action pattern could contribute to understanding phenomena of human perception and action. Amongst others he presented the *Kindchenschema* well known to all of us. He discussed the functioning of signals of submission, which inhibit intraspecific aggression and how by the invention of armor this inhibition fails to function any more. Particular attention was given to the phenomenon of domestication, a prerequisite for man's evolution as cultural being, but also a source of concern for Lorenz, who as a trained medical doctor, became also aware of degenerative effects.

In 1944, Lorenz was wounded near Witebsk, and was taken by the Russians as a prisoner of war. His weight was down to 55 kg at that time. In the unheated barracks he tried to tame young rats, who wanted to warm themselves on his body, but the mother rat whistled them back. Lorenz did not speak much about this time. He emphasized however that there was no intention to starve the prisoners. The medical doctors and the leaders of the camp did their best, but there was general famine. Nonetheless the civilians, even though on extreme short rations, on occasions passed food over the fence.

2.1. Lorenz's Return

My interest in animal behavior in general and of Konrad Lorenz's approach started in early Spring 1946 when, as a student, I attended a workshop on animal behavior which was held by the ornithologist and ethologist Otto Koenig. He had occupied a few barracks in the Viennese forest opposite Wilhelminenberg Castle and established a biological field station, the *Biologische Station Wilhelminenberg*. Since I was taken with what I heard and since animal behavior had fascinated me already in childhood, I accepted the invitation to

Figure 1. The first lecture at the Wilhelminenberg on April 10, 1948. Lorenz explains how you can start geese to fly by running and suddenly raising the arms. Next to him Otto and Lilly Koenig followed by Kurt Gratzl, Friedrich Haiderer, Eberhard Trumler, my fiancee and future wife Lorle Siegel. (Photograph I. Eibl-Eibesfeldt).

join the team on the Wilhelminenberg. More than half a century has passed since. The field station still exists and is now named the *Konrad-Lorenz-Institut*.

On February 20, 1948, I listened to the radio broadcast sending the list of the names of the 10th home-comer transport from Russia, when I heard Konrad Lorenz announced. I immediately 'phoned Otto Koenig, and we went to the railway station. But alas, the transport had been directly routed to Lorenz's home town of Altenberg. So it took a couple of more days 'till we met. We had sent welcome notes and he visited us at the Wilheminenberg. He was slender but full of vigor. On April 10th he gave his first lecture on the vitalism-mechanism dispute to our small group (fig. 1). From Russia he had brought with him a self-made cage with a tame starling and a manuscript of approximately 750 pages which served as a draft for his lectures. The manuscript was written on paper cut out of cement bags. For writing he used quills home made from bird feathers. The working conditions were extreme. The barracks were poorly heated and lighted, and hunger preoccupied the inmates (the more one is puzzled by the brilliance of Lorenz's formulations!) For a while the manuscript was lost. But in 1991 it was found in a drawer of his writing desk under a pile of proofsheets. Agnes von Cranach (born Lorenz) edited the manuscript which was published in 1992.

Some of you may wonder how it was possible to bring a manuscript of 750 pages from Stalin's Soviet Russia to Austria. It's a remarkable story. When Lorenz learned that he was about to be repatriated, he announced that he had this manuscript and that he wanted to take it back. He was immediately transferred to another camp, prompting him to think that he had spoiled his chances for a return. But to his surprise he was given a typewriter and ordered to copy the manuscript word for word. This done he was sent to report in Moscow, accompanied by a guardian of course. There, a high ranking officer adressed him in perfect German: "Professor, you are not any more a prisoner of war and soon you will be back in Austria. As a man of honour I ask you: Is the copy you made identical with the handwritten manuscript?" Lorenz assured him this was the case. The next question was aimed at learning whether he would change the manuscript after his return. Lorenz responded: "Yes, indeed, I will bring it up-to-date by discussing new publications." Upon which the officer laughed and said: "I did not mean that. Are you going to publish about your time here?" Lorenz stated that this was not his intention. Both exchanged a handshake and Lorenz received written permission. The copy of the manuscript is nowadays in the Moscow Academy of Sciences.

An incident of interest occurred on the way from the camp to Moscow. His guardian fell ill and when the train had a longer stop in Baku, he asked Lorenz to do some shopping. And so it came that Lorenz was strolling through the town, shopping in his shabby looking German uniform without being stopped. Since he had bought soap, he stopped at a fountain to wash himself, when from some distance a one-legged Russian veteran in uniform with many decorations approached shouting, and waving a razor. But it soon became clear that there was no hostile intent, since the veteran said: "You have soap! I have a razor, you give me some soap, I shall shave you!" And indeed in the middle of Soviet Russia, two victims of the war in friendly cooperation began to shave each other. Mankind is perhaps not yet totally lost!

Our relationship was soon a very amiable one. I enthusiastically told Lorenz about my observations on the migration and breeding behavior of toads, on my wild house mice that were running free in my barrack and their interesting closed communities. My nearly grown up pet at that time was a badger which I had raised the year before and from whom I had already learned that nonhuman mammals are able to detach agonistic emotions from play behavior. This allows them to create a field without tension thus favouring free playful experimentation, an ability that is a precursor of the first manifestation of our ability to act freely, in emotional sobriety, a prerequisite for reflective consideration (Eibl-Eibesfeldt 1950 a, b, c).

With Lorenz there, visitors came. Julian Huxley was one of the first. The world started to open up again. Many ties which had been severed during the war became reestablished. Otto Koehler, Erwin Stresemann, Niko Tinbergen, Erich von Holst, Bill Thorpe, all expressed their welcome. The Austrian Academy of Sciences supported Lorenz, but the financial situation was grave at those times, and when Erich von Holst wrote (p. 5) that the Max-Planck-Society was prepared to sponsor an Institute for Behavioral Physiology for both of them, Lorenz accepted the invitation and I was allowed to follow him. From 1951 to 1957 we were located temporarily in an old mill and a servants annex of the water castle Buldern in Westfalia. In 1951 we hosted the first International Ethological Conference in Buldern, and for this occasion the Baron Gisbert von Romberg provided us with a splendid room for lectures within his castle (figs. 2). In 1957, we moved to the newly built *Max-Planck-Institut für Verhaltensphysiologie* in Seewiesen (Bavaria). The years to follow were certainly the high time of the Institute with Erich von Holst, Konrad Lorenz and Jürgen Aschoff in turn as directors. It was a time full of brilliant and sparkling discussions in which visitors from all over the world participated. This creative time certainly was of enormous importance for my own development.

In 1973 Konrad Lorenz, Niko Tinbergen and Karl von Frisch were awarded the Nobel Prize. Shortly afterwards, Lorenz moved back to his home in Altenberg as emeritus[6].

---

[6] Wolfgang Wickler who had joined the Institute as student in 1954 became director of his department.

Figure 2. The first international ethological congress in Buldern 1951, Lorenz presenting his geese. (Photograph I. Eibl-Eibesfeldt).

In Grünau, a field station was established for his geese studies, sponsored by the Austrian Academy of Sciences. Until his late days he was there during summer amidst of his geese (fig. 3). He spent autumn and winter in Altenberg, where philosophical seminars were held. It was to Rupert Riedl's merit to have arranged these regular sessions, and this created an intellectual climate which kept Lorenz in high spirits. In 1973 Lorenz wrote *Die Rückseite des Spiegels (Behind the mirror)* which is one of his best books. Until his very end he was full of enthusiasm. Nothing else could please him more than to hear about a new interesting observation and he was always supportive and encouraging to others. Tinbergen remained his close friend.

3. SOME FOCAL POINTS AND ACHIEVEMENTS OF THE POSTWAR PERIOD

3.1. The Nature-Nurture Controversy

Lorenz continued his research with studies of ducks, geese and cichlid fishes, focussing on social behaviors as well as on imprinting phenomena. My interest pursued two lines: ontogeny and communication, the former being experimental, the latter being more comparative. In both cases, I started out with the ethogram of the species involved, raising hamsters, squirrels, rats, polecats and the like.

In 1953, Daniel Lehrman published in the *Quarterly Review of Biology* his "Critique of Konrad Lorenz' Theory of Instinctive Behavior." The attack was aimed at the concept of the innate, the main argument being that it was only negatively defined as that which is

Figure 3. Nico Tinbergen and Konrad Lorenz in Altenberg 1978 (Photograph I. Eibl-Eibesfeldt).

not learned. Furthermore, Lehrman contended that the deprivation experiment, by which we thought we could prove innateness, was of no value at all, since it was never possible to deprive an animal of all potential external sources of information. After all an animal is always in an environment that exerts some influence, even upon the embryo in an egg or in the uterus. Kuo's observations on chick embryos and the studies of Riess on the nestbuilding behavior of rats were cited as evidence that the pecking response and nestbuilding behaviors commonly considered innate are in fact built up from conditioned precursors. Lorenz's response was to ask, if that were true, why do fresh hatchlings show very different behaviors—chicken which spontaneously peck at kernels, and chaffinches which gape at their parents?

Concerning the value of the deprivation experiment, we pointed to the fact that organisms in their adaptations, be they morphological or behavioral, are molded toward some fitness-relevant features of the environment. And that fit tells us that at some time in the organism's individual or phylogenetic history interaction between the features which an adaptation copies or is molded to must have taken place (see p. 6–7). And one can of course deprive an organism of information relevant to a specific adaptation. If one were to raise a bird isolated in a soundproof chamber from the egg stage on and if it were uttering the species-specific territorial songs or other utterances at the onset of sexual maturity, without ever hearing a conspecific, then it is valid to conclude that patterned information concerning the characteristics of this song need not be fed into the organism during ontogeny, but that the adaptation took place during phylogeny, the pattern thus being a phylogenetic adaptation. Whether "precursors" such as beak movements or any other abilities which constitute necessary preconditions for singing need to be entrained during embryonic development of the bird is irrelevant in this context, since our question aims at a specific level of adaptation. Concerning this level, in the fictive experiment described above, we can conclude that the information concerning the specific patterning must be encoded in the genome of the species. The neuronal networks in their specific wiring with sensory and motor organs grow in such a case in a process of self differentiation under genetic control to functional maturity. I emphasize this point, since even nowadays one reads fairly often the lip service type of interactionism: "Yes, genes contribute a lot, but one can never separate the contribution of genes versus 'experiental' influences." Such statements it seems often intend to blur the issue. Not so seldom, however, they simply reflect a basic ignorance of the issues involved.

Even where birds are known to learn in order to sing according to the species norm, we know by now that this does not usually occur on a blank slate, but that a variety of innate learning dispositions channel learning in specific ways. Concerning this subject, the studies of Peter Marler and his students provided a wealth of information.

Lorenz replied to his critics in his 1961 paper *Phylogenetische Anpassung und adaptive Modifikation des Verhaltens*. It was translated for Chicago University Press as *Evolution and Modification of Behavior* (1965). He presented a definition of the innate as the phylogenetically adapted, pointing out that adaptations mirror facets of the objective world impinging on the organism's fitness.

When Lorenz discussed innate abilities, he postulated that the central nervous structures underlying them must grow in a process of self-differentiation to functional maturity according to the genetically encoded instructions. This process of self-differentiation has in the meantime become well understood down to the molecular level. It started with the pioneering experiments of Roger Sperry (1963, 1971) who demonstrated that the outgrowing nerves "sense" their endorgans apparently chemically and find them even if translocated. If a piece of skin from the back of a frog-embryo is transplanted to the abdomen, the nerves, which would have innervated it at the original spot, now grow to the transferred site. If one tickles the frog on the now ventrally located piece of skin, the frog will scratch its back. Sperry's chemoaffinity hypothesis has been confirmed by studies such as those of Goodman and Bastiani (1984) who demonstrated that from the outgrowing nerve cones long filopodes extend and contract within minutes and thus explore their environment. They demonstrate selective attachment. If the attachment is strong, contraction pulls the nerve cone in one direction, thus guiding it to its target.

3.2. *On Aggression*

In 1963 Lorenz wrote his book *On Aggression* which gave rise to numerous discussions. Perhaps the most common criticism was that by pointing out our innate dispositions for aggression Lorenz was fostering a fatalistic attitude. Thus, in a discussion on research into aggression, Erich Fromm (1974) wrote: "What could be more welcome to people . . who are afraid and feel themselves incapable of changing the course of things leading to destruction than Konrad Lorenz's theory that violence springs from our animal nature and derives from an untamable drive to aggression." But Lorenz never spoke of an "untamable" drive. To those of us who had read the book, this seemed at best a gross misrepresentation. While Lorenz explicitly emphasized that in modern contexts aggression is one of the most dangerous dispositions possessed by human beings, he was far from fatalistic. I quote:

> "With humanity in its present cultural and technological situation, we have good reason to consider intra-specific aggression the greatest of all dangers. We shall not improve our chances of counteracting it if we accept it as something metaphysical and inevitable, but on the other hand, we shall perhaps succeed in finding remedies if we investigate the chain of its

natural causation. Wherever man has achieved the power of voluntarily guiding a natural phenomenon in a certain direction, he has owed it to his understanding of the chain of causes which formed it. Physiology, the science concerned with the normal life processes and how they fulfill their species-preserving function, forms the essential foundation for pathology, the science investigating their disturbances" (pp. 29–30 of the English edition).

The accusation that the findings of ethology serve to justify militarism, racism, and social Darwinism has been perpetuated with rubberstamp-like monotony which unmasks those critics as sloppy or highly biased readers (C. E. G. Tobach et al. 1974). Presently, the main target of attack is the sociobiologists. Since the factual evidence in most cases is difficult to dispute, insinuations of mal-intent are employed. This creates barriers to communication within the scientific community and that is the least productive since we need to talk to each other more than ever. And it has often enough been emphasized that "ought" does not follow logically from "is" (Salter 1995 a). We can indeed learn much from nature, but—as Wolfgang Wickler (1981) emphasized—"also how not to do it." Nature does not necessarily serve as a positive model. I want to emphasize this point since we experience right now a swing from a social or mixed market economy back towards ruthless *laissez-faire* capitalism, natural selection serving as the model. The sterility of argument by accusation was brilliantly formulated by William Charlesworth (1981):

> Speaking of rhetoric, there should be an editorial rule that sentences associated with sociobiology, with effort to justify slavery, imperialism, racism, genocide, and to oppose equal rights should always appear next to sentences associating environmentalist/learning theory with effort to justify propaganda, psychological terror, false advertisement, public indoctrination of hatred of foreigners, class enemies, minority groups, and so on and so on. Juxtaposing sociobiology and learning theory in this manner ought to show how unproductive it is to claim through innuendo or otherwise that science will lead to pseudoscience, will lead to man's inhumanity to man: ergo no science. Actually, one could argue that since man is such a cultural/learning animal we should have greater fear of learning theory since learning has far more power over man's behavior than genes. More specifically, if humans were not such learning animals, they would not learn all that Galton trash: ergo stop learning research so that bad guys will not use the data to teach the trash more effectively (p. 22).

Regarding the motivation of aggression, Lorenz postulated an aggressive drive, building up and inducing an "appetite" for aggression. Indeed studies such as those done by Kruijt (1964) confirmed an appetence for aggression. Jungle fowl cocks reared in social isolation attacked their shadows as substitute objects and tried to fight their own tails, waltzing around in a frenzy. Advances in the study of the brain chemistry provide in the meantime some insight into the processes and driving forces behind such "appetites."

In his introduction to the new Routledge edition of Konrad Lorenz' *On Aggression* (1996) Erich Salzen points to the dopaminergic nigro-striatal-accumbens neurotransmitter system that is involved in locomotory and orientating appetitive behaviors. It receives inputs from lateral hypothalamic systems responsible for the instigation of various consummatory behaviors such as feeding, sex and others. This dopaminergic appetitive motorsystem has been considered as an "incentive" or "go" system and it has been suggested that its general function is to actuate animals to search for rewards. Salzen concludes that the neurochemical and neurological evidence suggests "specific drive systems for consummatory behaviors and that there is also a general appetitive motor activity system which is at the service of the specific drive systems." (p. XVII)

The brain chemistry of aggression seems fairly complex and is in need of further exploration. Dopamin functions as a general energizer in particular motivating behaviors which serve the function to overcome ("conquer") obstacles blocking goal-directed activities. It thus fuels any sort of dominance behavior, including aggression against other human beings the instrumental use of aggression to fight (conquer) problems. In the concrete situation of fighting, epinephrine and norepinephrine prepare us physiologically by activating our metabolistic resources, while endorphines reduce our feelings of pain. In combination with the energizing catecholamines they might be responsible for the phenomenon of aggressive trance during which, besides being insensitive to pain, all inhibitions against aggression are lost. Massacres can thus occur in a fit of blooddrunkeness (delirium). Achievements of dominance finally are rewarded by a hormonal reflex: If a male wins a contest such as a tennis match his blood testosteron level increases significantly; it drops if he loses. The same holds true for other achievements. If male medical students pass their final exams we can observe an increase in the level of male hormones; failure results in falling levels (Mazur and Lamb 1980). Thus success is hormonally rewarded, since those who win as a consequence feel subjectively good. The problem with this positive feedback mechanism is that there is no safeguard against escalation. Power-striving seems insatiable.

4. THE ADVENT OF NEW ETHOLOGICAL SUBDISCIPLINES

4.1. Neuroethology

The choice of the name *Verhaltensphysiologie* for the new Max-Planck-Institute founded in 1957, where Konrad Lorenz and Erich von Holst formed the initial team, emphasizes already the importance of the physiological approach for any "Biology of Behavior." I have already pointed how studies of behavioral embryogenesis confirmed the Lorenzian concept of the innate and how analyses of brain chemistry support his observations on the spontaneity of behavior. In 1951 Niko Tinbergen emphasized the need for neuroethology. The field is a flourishing enterprise initially built upon von Holst's studies of the central nervous automatism (von Holst 1935, 1936 and 1969). Investigations of the "central generator systems" have been conducted down to the cellular level by such researchers as Willows (1971), Stent et al. (1978), and Huber et al. (1989) (for reviews and further references see Kandel 1985 and Eibl-Eibesfeldt 1987). So too the Lorenzian "templates" have been investigated down to the neuronal level by researchers such as Ewert (1974, 1975) and Huber (for further reviews see Huber et al. (1989) and Kandel 1985).

Concerning the phenomenon of imprinting Wallhäuser and Scheich (1987) discovered what happened at the neuronal level when newly hatched chickens were imprinted to acoustic stimuli. Before imprinting ooccurs, the dendrites of the neurons, in processing the acoustic stimuli, show many spines. Following imprinting to a pure sound, most of the spines have disappeared, the neurons being tuned solely to the specific stimulus. If the experimenter uses the normal call of the hen, more spines remain.

4.2. Sociobiology

One of the great achievements in the postwar development of ethology was the foundation of Sociobiology by E. O. Wilson (1975). This new branch of ethology combines theories and approaches from population genetics and ecology. The sociobiological argument starts from the observed fact that differential reproductive success is a fundamental requisite for natural selection. Some individuals are more successful than others in passing on their genes to the next generation and genetic survival is what we measure as fitness. The strategies employed to achieve this vary. It is not necessary for an individual to produce its own offspring to enhance its genetic survival since this can be achieved by assisting close relatives to survive and propagate. This basic insight is based on the fact that copies of an individual's genes are also found in a statistically predictable percentage in his relatives. This allows us to develop cost benefit models which predict under which conditions various strategies work or fail, models which of course have to be tested by experiment and by those ethological methods of field observation which allow for a statistical analysis.

Amongst others the existence of "evolutionary stable strategies" have been observed. That is to say that in many cases not all individuals of a population behave according to the statistical norm, but that a certain percentage deviate and succeed in doing so. Amongst ritualistically fighting animals, for example, we regularly find in some species a minority which fights in a damaging way. These deviants have a definite reproductive advantage and so grow in numbers, but only to a certain proportion of the population. Cost benefit calculations of that sort have been highly inspiring for further research. But the effort to seek in every observed peculiarity an adaptation enhancing the genetic survival of the deviant phenotype sometimes led to forced constructions. The literature becomes increasingly dominated by speculative papers of model constructors (see p. 24). Humans are often considered as fitness maximisers in this context, a view which Buss (1995) labeled as the sociobiological fallacy. Indeed we know that structures of organisms including behavioral characteristics are often far from being maximally adapted. Some are dragged along as historical or phylogenetic burdens having lost their original adaptive value under changed environmental conditions. Others are the result of phylogenetic reorganisation and are thus far from technically perfect. All land vertebrates have fish as their ancestors. Their system of blood circulation was fairly well adapted to an aquatic life. When the ancestral forms exchanged their aquatic habitat for a terrestrial one, the circulation system needed to be transformed and, as we all know, a technician could have found a better solution. But as it were, evolution uses what' is available. Many examples like this are known; just consider the many imperfections in our adaptation to an upright gait. There is furthermore the well known phenomenon of conflict of function to be considered, which leads to many compromises. Last but not least there are detrimental mutations that are weeded out by selection, but occur again and again in a certain percentage of the population, in particular if the mutations involve recessive genes. Sure, these are "experiments of nature" but certainly not fitness-enhancing ones. Mutations and recombination could be considered a strategy of life but if the phenotypic result is a pathological trait it is a failed experiment.

4.3. Human Ethology

"Like rivers, histories of scientific disciplines have many tributaries." This statement of William Charlesworth superbly describes the situation of Human Ethology. I want to add that the tributaries in this case come from different disciplines. This may surprise at first, but evolutionary theory after all is the basic theory of all manifestation of life, and basic therefore for any understanding of human behavior, including those facets of human behavior which are the subject of the various humanities. And indeed, art historians, legal philosophers, political scientists, cultural anthropologists, linguists as well as representatives of the traditional fields studying human behavior such as sociologists and psychologists in increasing numbers begin to ask how behavior and customs evolved, how they function, and thus contribute to fitness. The articles by William Charlesworth (1991), Peter K. Smith (1992) and Daniel G. Freedman (1991) in the *Human Ethology Newsletter* provide interesting insights into how their converging interests led to the establishment of human ethology.

With Charles Darwin, biologists were certainly the first to pioneer this field. And when Konrad Lorenz laid the foundations of ethology, he emphasized it as one of the most important tasks of the new discipline, to test whether the theories deduced from the study of animal behavior had any bearing for the understanding of human conduct. In the late fifties and early sixties psychologists, psychiatrists, and psychoanalysts such as John Bowlby, Corinne Hutt, Bill McGrew, Daniel Freedman and Eckhard Hess, to mention just a few, started to get interested in the ethological approach. In England child ethological studies began.

Konrad Lorenz stimulated my interest in human ethology in our early discussions on the *Wilhelminenberg* and later at Seewiesen. The nature-nurture exchange with Lehrman further stimulated this interest. In 1967 I published the first comprehensive text on ethology, and already at that stage I included a chapter on humans (Eibl-Eibesfeldt 1967[1], English: *Ethology, the Biology of Behavior* 1970 and 1975). Having thus during nearly two decades of animal behavior studies acquired a sound theoretical and practical knowledge of ethology, I thought myself prepared to venture into human ethology. Ever since the Xarifa expeditions in 1953/54 and 1957/58 I had in Hans Hass a close friend and an inspiring discussant. I familiarized him with ethological concepts and in particular with the works of Konrad Lorenz. In the early sixties Hass told me that he would like to make a series of television films on man, and he asked me whether I would like to act as ethological consultant. For this film he wanted documents of unstaged everyday behaviors, which were not as yet available in the large film archives. We had to collect them ourselves. Hans developed for this purpose an unobtrusive method of filming, using a mirror lense which afforded a view at an angle of 90° to axis of the camera. I accepted his offer in the capacity of ethological consultant and on two trips in 1964 to East Africa, and in 1965 around the world, we gathered the documents for the film series, experimenting with slow and fast motion techniques (for details see Hans Hass 1968 and Eibl-Eibesfeldt 1984, 1992). Based upon this experience, I began in 1970 to build up my own enterprise within the Max-Planck-Society. I proposed to the society the establishment of an Institute for Human Ethology, the focal point of my interest being the cross-cultural documentation of unstaged social interactions and rituals. I had already started with the documentation of the behavior of deaf and blind children.

After two further pilot studies in 1967 in New Guinea and 1969 on the upper Orinoko (Yanomami), the Society agreed in 1970 to let me lead a small working unit which in 1975 became an independent research institution of the Max-Planck-Society. For the cross-cultural documentation I had chosen traditional cultures in different continents with different subsistence strategies:

1. The !Ko and G/wi San (Bushmen) of the Central Kalahari. Samples were also taken from the !Kung. These groups represent a model of the hunter-gatherer stage of cultural evolution and thus in many ways an archaic mode of life;
2. The Yanomami from the Upper Orinoko area who represent a transient stage from the hunting and gathering subsistence to horticulture;
3. The Eipo of Western New Guinea who, at the time our documentation began, represented an intact neolithic culture;
4. The Himba, traditional pastoralists in Northern Namibia;

5. The Trobriand Islanders, horticulturists and fishermen;
6. The Balinese as a model for traditional nonwestern peasants.

In these cultures longitudinal studies were carried out. In addition, we took samples in other cultures. In Europe we installed a *Kindergarten* project to study basic strategies of social interaction and the processes of self organization of children groups. For details concerning the documentation program, our interdisciplinary team and the results, I have to refer to my text *Human Ethology* (Eibl-Eibesfeldt 1989). Having pursued this work now for more than 25 years—and the documentation program still continues—we have now collected in our archive about 275 km of 16mm film from which over 200 films have been published[7] by the *Encyclopaedia cinematographica* in Göttingen. The results of our studies are summarized in my *Human Ethology* and in numerous other publications (e.g. Eibl-Eibesfeldt, Schiefenhövel and Heeschen 1989). Our search for universals was successful. Some of the expressive motor patterns upon comparison on a statistical basis prove indeed to be characterized by form constancy, their range of variability being the same in different cultures. Such is for example the case of the "eyebrow flash" which also serves the same range of functions (Eibl-Eibesfeldt 1968, Grammer et al. 1988). There are furthermore many similarities in principle to be observed. On the surface these behaviors appear in a variety of cultural expressions, but they follow the same basic rules, so to speak the same universal etiquette. Thus in greeting rituals agonistic displays are often combined with affiliative appeals, in particular between individuals who want friendly relations with each other but are at the same time not so familiar with each other as to trust each other completely.

There are many basic strategies of social interactions which obey this principle. They follow the same basic rules, so to speak, a universal grammar of social conduct. But there are often a great number of options of how to act within this system of rules. A person can display by ways of a war dance, as the Yanomami males do, when entering the village of an allied group on occasion of a feast. In Europe a visitor of state will be greeted with military honors, in former times even by cannon salute. At the same time the visitor will be greeted by a young lady or a small girl passing a bouquet of flowers—and with the aggressively displaying Yanomami a child will dance waving green palm fronds. It is in principle the same. Behavior patterns of different origin, innate motor patterns such as facial expressions of threat or appeasement, or culturally developed forms of behavior can substitute for each other as functional equivalents which allows for a great variety of expression. People can even translate their actions into words, using "verbal clichés" to trigger responses or fight with words and finally solve disputes verbally. But in doing so they follow the same rules observed when acting nonverbally. The many ways in which human behavior was found to be preprogrammed by phylogenetic adaptations can not even in a most scetchy way be presented here. I have to refer again to my *Human Ethology*.

---

[7] The publications are made from duplicates. The original is never cut, but preserved intact. For the different projects I had the help of different co-workers: For the San project Hans-Joachim Heinz and Polly Wiessner; for the Yanomami Kenneth Good, Harald Herzog, Gabriele Herzog-Schröder and Marie Claude Mattéi-Muller; for the Eipo Volker Heeschen and Wulf Schiefenhövel; for the Trobrianders Ingrid Bell-Krannhals, Gunther Senft and Wulf Schiefenhövel; for the Himba Kuno Budack. The *Kindergarten* project was set up by Barbara Hold and continued by Karl Grammer. All have contributed significantly to the success of our enterprise. I want to thank them fullheartedly.

5. THE PRESENT AND THE PROSPECTS FOR THE FUTURE

Ethology certainly found its way. The basic concepts stood up against challenge. As is usual in a fast developing discipline, focal points of interests shift. The evolutionary approach certainly continues to attract the attention of the other disciplines of human behavior. I just received Hagen Hof's book on the Ethology of Law (*Rechtsethologie*, H. Hof 1996). A pioneering work indeed which explores the rules by which human behavior is regulated without the help of law, in order to understand law in the context of behavior. Another important contribution comes from Frank Salter (1995 b) who studied institutional dominance in naturalistic settings and shows how "social technologies" can substitute for human behaviour in releasing submissive obedience. Polly Wiessners (in press) monograph on the use of rituals by big men to alter norms and values and thereby direct the course of change and Bell-Krannhals' (1990) monograph on gift exchange among the Trobrianders are other fine examples showing how ethological concepts inspired ethnological research. In the United States similar efforts to understand the biological background of political behavior, law, ethnic phenomena and the like are bearing fruit in the published work of Margret Gruter (1983), Roger Masters (1989), Pierre van den Berghe (1981) and others. It is simply impossible to give credit here to all the important contributions which exist.

Human Ethology also profits from the numerous sociobiological contributions. Two international journals – *Ethology and Sociobiology* and *Human Nature* – publish regularly contributions, and the *Human Ethology Newsletter* is developing an important discussion format. There are however also matters of concern as regards methodology.

Konrad Lorenz emphasized again and again the importance of observation and description, but over the years we observe a trend away from original observational data. William Charlesworth (1995) documented this in a recently published investigation in which he analyzed the content of the journal *Ethology and Sociobiology* since its appearance in the late 1970s until 1995. There is a clear trend away from empirical papers toward theoretical ones. Furthermore, within the empirical papers, those based on behavioral observation declined from 39% of the studies for the first half of the period covered to 18% in the second half. Studies based on interviews and questionnaires increased in the same periods from 14 to 37%, and a similar increase was found for predominantly "archival" studies based upon existing data, often collected by others. Charlesworth finds this trend disturbing and so do I.

Another point of concern is a "nothing-but-an-animal" presentation of the human being. When Desmond Morris started to characterize the human being as a "naked ape," it was meant to put the fact of our phylogenetic history into the focus of attention. It was meant to shock as a didactic measure, and as such was appropriate at that time. But if one were to conclude that such is the complete way biologists see the human being, one would be basically wrong. Konrad Lorenz emphasized repeatedly that no one is able to appreciate the uniqueness of the specific human attributes as clearly as those who can perceive them against the background of the much more primitive actions and reactions, which we also share with the higher animals. Reason and the morality based upon it, the capacity to learn a language and to talk about present, past and future events, to pass on knowledge and thus to build up culture, and finally, our ability to set goals for ourselves on the basis of desires and empathy are indeed unique. And this provides us with new potentialities for survival as will be discussed at the end of this contribution.

In spite of this emphasis on the uniqueness of human nature, many biologists perpetuate this nothing-but-an-animal presentation, often by means of appalling illustrations depicting grimacing monkey and human faces, the message being that they are "all the same." This is not just a matter of taste, but harmful to our field since we need to communicate and cooperate more than ever with the humanities. Without understanding culture, we cannot understand the human being who is "naturally cultured." Humans are certainly outfitted with motivations, emotions, biases in perception and with innate motor patterns given as biological heritage. And part of this heritage we share with animals. However, as an old Chinese proverb states, all that is animal is in humans, but not all that is human is in animals.

What are the prospects of human ethology? The investigation of our phylogenetic heritage programming our action and biasing our perception and cognition is still a tremendous task ahead of us. In particular the universal grammar of human social conduct, verbal and nonverbal alike, needs further exploration. Man is a cultural being by nature. He is genetically endowed with the capacity to acquire a language and speak, which allows him to tell others what to do and when, without the need to act as a model. Cumulative culture as a result has made human history a story of success. We have amongst others, however, created ourselves an environment which deviates from that which exerted its shaping influence through selection for most of our ancestral history and this confronts us with some problems. Our ancestors lived in small face-to-face communities with a simple technology, foraging as hunters and gatherers. Modern man lives in anonymous million societies, in urban environments outfilled with the means of technical civilisation. All in all, the development has to be considered as progress, since without societies of millions there were no universities, no large libraries, no concerts or opera houses, and no technical civilization with all the new options including the conquest of distance and space. Within our century, we proceeded from the first clumsy automobiles to space travel, from the mechanical age to the electronic age. We can hardly imagine what a species achieving this in such a short span of time could achieve in another ten thousand or more years, if it were to solve its social and ecological problems.

Overpopulation and environmental destruction threaten the very basis of subsistence for future generations. We certainly need a new generation encompassing a survival ethos, carried by a feeling for moral responsibility for future generations. There are however in our phylogenetic heritage dispositions which we have to face, which in our present situation hamper the very development of such an ethos. One is our being programmed for the sprint in the present. Ever since the first creatures of our planet competed for scarce resources, what alone counted was to win right now. This selected for opportunistic exploitation to the maximum of any opportunity. This was also of advantage to our ancestors until fairly recently. And we are well equipped for this type of competiveness, since we use our strong aggressive dominance striving instrumentally in many ways, not just in the social context, but also to overcome any obstacles. We sink our teeth into problems, we attack them and subdue nature. This together with the fixation on the present seriously hampers the development of an ethos which takes into consideration the fate of future generations.

This trap of short-term thinking has to be avoided. In order to achieve self control, we need to learn about those traits which in certain situations of modern life prove maladaptive, and others which we can tap in order to adapt anew.

We need to learn about the range of modifiability for each of our behavioral characteristics. We must know about innate learning dispositions such as our indoctrinability (Eibl-Eibesfeldt and Salter 1997, in print), be they helpful or maladaptive, and if so in what situations. History and the study of the cultural manifestations of man in historical and prehistorical time provide a wealth of experiments to be studied. As a natural next step, it is my intention to focus on problems such as these within the scope of <u>cultural ethology</u>.

5.1. Cultural Ethology

My team approaches these questions with two main projects which are interconnected and open to wider interdisciplinary cooperation. One of these projects focusses on the ethology of art, the other on urban ethology.

5.1.1. Ethology of Art. In Keesing's excellent anthropology text I read: "It is the anthropologist's special insight, that these internal models we use to create a world of perceived things and events are largely cultural. What we see is what we, through cultural experience, have learnt to see" (Keesing 1981, p. 82).

By now we know that our perception is initially biased by phylogenetic adaptations at different levels. These occur on the basic level of sensory physiology and on the level of neuronal reference patterns (see p. 4/5). Aesthetic perception on the basic level of sensory physiology include those of *Gestalt* perception which Marius Escher plays with in many of his paintings. From these we can distinguish reference pattern in which aesthetic norms are encoded.

We constantly value what we perceive as beautiful, ugly, repulsive, frightening, sinister, cute, and the like. Much of this appreciation is indeed acquired, templates being fixated in imprint-like fashion. There are however strong indications that many templates come preformed in our innate outfit. Such is the case with some of those concerning the aesthetic appeal of human facial and other bodily characteristics. Others concern environmental features. Thus human beings exhibit an aesthetic preference for plants (Phytophilia) which reflect an "archetypical" imprint on features of the environment.

Since many of the aesthetic appeals, be they visual or auditory, arouse specific emotions, they are often employed to bind attention, to appeal to people in specific ways. The state of arousal thus induced is used to communicate political messages. Art in painting, architecture, music and poetry serves communicative functions. Among these, the ideological reinforcement of shared norms and values is of prime importance as well as ideological indoctrination with new values, such as those reinforcing state authority. Even those however seem to tap into existing phylogenetic adaptations, as most of us will be able to experience by introspection. At political rallies, when hymns are sung and when in special rituals people commemorate historical events of importance to the nation or state, and are confronted with the "sacred" symbols which serve identification, many experience the shudder of being touched. This feeling is caused by the contraction of the tiny muscles, which raise our body hairs—our rudimentary fur. It indicates slight arousal of tendencies of collective aggression. Art, as a vehicle of value imprinting and value transportation and the phenomenon of symbol-identification certainly deserves more attention.

This communicative function, to be sure, is not the sole function of art. Aesthetic creation is one of the prime characteristics of man which finds its expressions in nearly every aspect of our daily life. Amongst others we create for ourselves artificial environments with aesthetic appeal. Presently, man experiments in an endeavour to create an urban environment in accordance with aesthetic needs.

The modern urban environment is in many ways experienced as stressful. Traffic limits our freedom of movement, streets are crowded with people we do not know, and many feel irritated and at the same time alone in the crowd. In this context places play an increasing role in society. If constructed with the right aesthetic appeal, they invite people to rest and allow for the development of social contacts. This sociointegrative function is of great importance, in particular in the vicinity of social housing projects. Whether they fulfill a social integrative function depends much of their aesthetic appeal.

5.1.2. Urban Ethology. In a joint enterprise with the Ludwig-Boltzmann-Institute for Urban Ethology[8] in Vienna, Klaus Atzwanger, Kirsten Kruck, Katrin Schäfer and Christa Sütterlin are engaged in the study of the design of public places in Vienna and Munich with reference to the behavior of the people using these places. The design clearly determines whether people are willing to stay and whether they tend to communicate with other users. The possibility to communicate with other users is one of the main factors enhancing bonding and thus to set up stable social networks. Other projects of the Ludwig-Boltzmann-Institute deal with housing, city-specific risk behaviors, and with aggression. People often complain of feeling lonely in the crowd. They want to be embedded in a small community of people which they know. Particularly people of lower income classes need social networks in their neighbourhood. These provide basic trust, a precondition to enjoying the positive aspects of a larger anonymous society and the urban environment. Exploring ways to humanize the urban environment is one of the priorities of urban ethology.

6. CONCLUDING REMARKS

Feelings are prerational, they are there whether we can consciously reflect about them to rationalize our emotional state or not. In fact, emotion can prove an effective adversary of intellect (Hassenstein 1986). In panic or anger, people tend to respond "blindly." In fear, people seek protection and demagogues of all eras have exploited this originally infantile trait to bind people to them as their protector (see bonding via fear in Eibl-Eibesfeldt 1970). Fear blocks intelligence but so do also do positive prosocial emotions as for example love, and prosocial engagements for an idea, a community or any "sacred" thing. Symbols have often led to torture and bloodshed. Strong feelings accompany the attachment to symbols of identification and the ideologies accompanying them.

Little is known about the physiology of enthusiasm, but the shudder of being touched (*der Schauer der Ergriffenheit*, p. 28), which results from the contraction of the erector muscles of our body hair, in particular on our back shoulders and the outer side of our arms, indicates that archaic responses of group defensive behavior are activated.

We defend emotionally everything we love—even our hypotheses—a fact that needs to be borne in mind if we are not to lose the openness needed to solve problems through discussion with people of other theoretical and ideological convictions. We refer often to freedom of speech, a freedom and right which is granted to us by our society, at least under democratic rule. The freedom I mean, however, is the intellectual one, which needs to be constantly reinforced by self awareness and self control. We are phylogenetically prepared for this intellectual freedom. Higher mammals are able to detach their emotions from behavior when they engage in play. This detachment allows them to interact with their environment and with conspecifics in an exploratory fashion. They can decouple be-

---

[8] The *Ludwig-Boltzmann-Institut für Stadtethologie*, Vienna, was established 1991. Directors: Karl Grammer and I. Eibl-Eibesfeldt

haviors of fighting, fleeing, hunting and shift freely from one activity to the other. In man, tool using with lateralisation of the hemispheres enhanced this ability. We talk of objectivity, when with "dexterity" we intellectually investigate a problem (Eibl-Eibesfeldt 1950, 1996). But this ability, even though we are prepared for it, does need individual effort and training. Lorenz emphasized it as the merit of a scientist to be able to refute a beloved hypothesis on the ground of evidence. The whole nature-nurture discussion up to the present is burdened by ideological fixations to the doctrine of "man's unlimited malleability." We need to avoid this "trap of passion" if we want to avoid having harsh lessons taught to us by selection.

As emphasized before our phylogenetic heritage is responsible for traits which as predispositions are in certain contexts maladaptive in our modern world. This holds true for our striving for power as well as our exploitative short time thinking (p. 26). Right now, we are experiencing in Europe a trend away from the social market economy toward ruthless competition. Natural selection thereby serves as a model. True, in times of crisis corrections concerning the costs of labor are necessary. But we must not endanger the social progress achieved. With the slogan of "global development demands global opening to the markets," we allow for ecological and social dumping by importing goods produced by disadvantaged labor. This creates unemployment and misery in countries which produce with ecological and social responsibility, and which as a consequence experience high costs of production. Competition is certainly the driving force of phylogenetic and cultural evolution. Nature in this context however knows no morals, but man does and should act accordingly. If we want to achieve internal and external peace we should continue in our efforts to civilize competition.

Earlier I acknowledged the naturalistic fallacy, that from an is an ought does not necessarily follow, but we always must take the is into consideration. In 1991 I referred to survival as a value which should guide us in order to find acceptable ways to ensure the survival of humankind. For this, I was accused of committing the naturalistic fallacy. In the ensuing discussion Salter (1995a) rightly pointed out that a description of an organism's interest in survival does not break any rules of moral philosophy.

Nature certainly has no interest in any organism. All organisms have, however, been selected to act in such a way as to pursue survival as an individual interest. Phylogeny programmed them as well as human being to such an extent. Humans can reflect on their own interests and state them in abstract terms. This has to be taken into account. In addition, there seems to be a world-wide consensus that the survival of mankind in its ethnic diversity in peace and prosperity should be what we eventually aim for. Again, we need to take biological knowledge into consideration. In face of the presently overwhelming ecological, social and demographical problems it would be absurd not to consult it. It is not enough wanting to be "good." Many of the self imposed gurus of the people, including some of the representatives of critical theory, are of the opinion that it would be enough to obey a moral imperative, ignoring reality. Such attitudes may be permitted when discussing salvation or other religions concepts. But the priorities of politics are the problems of this earthly realm. Whoever pursues the interest of common welfare and happiness must also explore the practicable limits of these goals. Whoever proceeds from conviction without considering knowledge acts irresponsibly and easily becomes what we call "*Überzeugungstäter*," that is to say: the road to ruin is often paved with good intentions. On failing they usually sneak away refraining from responsibility by asserting "this we did not want."

Presently we observe the ideologisation of the discussion of practically all ecological and social problems, be it population control, migration, multicultural coexistence, equal rights, gender roles or general problems of social justice. The noble concepts of freedom and equality are often debased into slogans freely used by totalitarian demagogues. But even more dangerous might be the unreflective use by naive but aggressive moralists. They present us at the turn of this millenium with "politically correct speech" which basically serves as a "newspeak" (Orwell) to veil certain fact considered unpalatable. Who does not obey these self imposed guardians of virtue runs the risk of defamation. Thus Judith Stacey (1993) attacked David Popenoe (1993) for interceding in favor of the family and familial values in the "Journal of Marriage and the Family," arguing from the needs of the children. In the course of her polemic Stacey accused everyone pleading for familial values of racism, sexism and homophobia.

Biology as the science of life certainly occupies a central position in the sciences of man. And Ethology defined as the biology of behavior[9] contributes to this position in ways decisive to an understanding of our conduct. It fulfills this responsible position only if ethologists remain open to the contribution of the other disciplines focussing on the study of humans, including the humanities. They should furthermore avoid sloppy terminology and a presentation of the human being as "nothing but" another ape, otherwise they expose themselves to the accusation of being reductionistic and rightfully so.

REFERENCES

Bell-Krannhals, I. (1990): Haben um zu geben. Haben und Besitz auf den Trobriand Inseln. Basler Beiträge zur Ethnologie 31.
Berghe, P.L. van den (1981): The ethnic phenomenon. New York: Elsevier.
Buss, David M. (1995): Evolutionary Psychology: A New Paradigm for Psychological Science. In: Psychological Inquiry vol. 6, no. 1, 1–30.
Charlesworth, W. R. (1981): Comments on S. L. Washburn's review of Kenneth Bock's "Human Nature History": A response to sociobiology. In: *Human Ethology Newsletter*, Sept. 1981, 22–23.
– (1991): Memoir. Like Thoughts on the Origins and Nature of Human Ethology. In: *Human Ethology Newsletter*, vol. 6, issue 2, June 1991, 1–4.
– (1995): Human Ethology – Still a Good Idea for the Behavioral Sciences and Society. In: *Hum. Ethol. Bull.* vol. 11, issue 1, March 1996, 6–12.
Craig, W. (1918): Appetites and Aversions as Constituents of Instincts. *Biol. Bull Woods Hole*, 34: 91–107.
Dewsbury, D.A. (1985): Leaders in the Study of Animal Behaviour. Autobiographical Perspectives. Lewisbury, P.: Bucknell Univ. Press.
Eibl-Eibesfeldt, I. (1950a): Beiträge zur Biologie der Haus- und der Ährenmaus nebst einigen Beobachtungen an anderen Nagern. Z. tierpsychol., 7, 558–587.
– (1950b): Über die Jugendentwicklung des Verhaltens eines männlichen Dachses (*Meles meles* L.) unter besonderer Berücksichtigung des Spieles. Z. Tierpsychol., 7, 327–355.
– (1950c): Ein Beitrag zur Paarungsbiologie der Erdkröte (*Bufo bufo* L.). Behaviour, 2, 217–236.
– (1967): Grundriß der vergleichenden Verhaltensforschung. 7th ed. 1987 München: Piper.
– (1968): Zur Ethologie des menschlichen Grußverhaltens. I. Beobachtungen an Balinesen, Papuas und Samoanern nebst vergleichenden Bemerkungen. Z. Tierpsychol., 27, 727–744.
– (1970): Liebe und Haß. Zur Naturgeschichte elementarer Verhaltensweisen. 16th ed. 1993, München: Piper.
– (1970): Ethology – The Biology of Behavior. 2nd. ed. 1975, New York: Holt, Rinehart & Winston.
– (1972): Love and Hate. The Natural History of Behavior Patterns. New York: Holt, Rinehart & Winston; 1996, New York: Aldine de Gruyter.

---

[9] The term sociobiology is often employed as synonym for ethology to engulf the field under a new heading. This is an interesting development illustrating that the motivation for dominance finds its outlet even in interdisciplinary exchange. Fortunately the rules of priority still hold and we should stick to them in order to civilize competition.

– (1979): Ritual and ritualisation from an ethological perspective. In: M. v. Cranach, W. Lepenies & D. Ploog (eds.): Human Ethology. Claims and limits of a new discipline. Cambridge, London, New York: Cambridge University Press (1979), 3–55.
– (1984): Die Biologie des menschlichen Verhaltens. 3rd ed. 1995, München: Piper.
– (1987): Grundriß der vergleichenden Verhaltensforschung. 7th ed. München: Piper.
– (1989): Human Ethology. New York: Aldine de Gruyter..
– (1992): Und grün des Lebens goldner Baum – Erfahrungen eines Naturforschers. Köln: Kiepenheuer und Witsch.
– (1994): Wider die Mißtrauensgesellschaft. Streitschrift für eine bessere Zukunft. 3rd ed. 1997, München: Piper.
– (1996): Spiel, Werkzeuggebrauch und Objektivität – Vom instrumentalen Ursprung freien Denkens. In: Matreier Gespräche. M. Liedtke im Auftrag des Matreier Kreises (eds.): Kulturethologische Aspekte der Technikentwicklung. Graz: austria medien service (1996), 60–72.
Eibl-Eibesfeldt, I. & F. Salter (1997 in press): Indoctrinability, Ideology, and Warfare. Evolutionary Perspectives. Oxford: Berghahn.
Eibl-Eibesfeldt, I. Schiefenhövel, W. & V. Heeschen (1989): Kommunikation bei den Eipo. Eine humanethologische Bestandsaufnahme im zentralen Bergland von Irian Jaya (West- Neuguinea), Indonesien. Berlin: Reimer (Schriftenreihe Mensch, Kultur und Umwelt im zentralen Bergland von West-Neuguinea; Beitrag 19).
Ewert, J. P. (1974): The neural basis of visually guided behavior. *Scientific American*, 230, 34–42.
– (1975): Probleme und Lösungswege bei der Untersuchung von Mustererkennungsprozessen. Der Mathemat. u. Naturwiss. Unterricht, 28. Jahrg. (2), 88–95.
Freedman, D. G. (1991): Memoir. My Path as a Human Ethologist. In: *Human Ethology Newsletter*, vol. 6, issue 1, 5–8.
Fromm, E. (1974): Lieber fliehen als kämpfen. Bild der Wissenschaft, 10, 52–58.
Goodman, C. S. & Bastiani, M: J: (1984): How embryonic nerve cells recognize one another. Sci. Amer. 251 (6), 50–58.
Grammer, K., Schiefenhövel, W., Schleidt, M., Lorenz, B., and Eibl- Eibesfeldt, I. (1988): Patterns on the face: The eyebrow flash in cross cultural comparison. *Ethology* 77, 270–299.
Gruter, M. and Bohannan, P. (1983): Law, Biology and Culture. Santa Barbara: Ross-Erikson.
Hass, H. (1968): Wir Menschen. Wien: Molden.
Hassenstein, B. (1986): Widersacher der Vernunft und der Humanität in der menschlichen Natur – zum Menschenbild der biologischen Anthropologie. Heidelberg, Jahrbuch der Heidelberger Akademie der Wissenschaften für 1985, 73–89.
Hof, H. (1996): Rechtsethologie. Recht im Kontext von Verhalten und außerrechtliche Verhaltensregelung. Heidelberg: Riv. Decker.
Holst, E.v. (1935): Über den Prozeß der zentralen Koordination. *Pflügers Arch.* 236, 149–158.
– (1936): Versuche zur Theorie der relativen Koordination. *Arch. Ges. Physiol.* , 237: 93–121.
– (1969): Zur Verhaltensphysiologie bei Tieren und Menschen. Gesammelte Abhandlungen, Bd. 1. München: Piper.
Huber, F., Huber, TH. E. & Loher, W. (1989): Cricket Behavior and Neurobiology. Ithaca – New York: Cornell University Press.
James, W. (1890): Principles of Psychology. New York: Holt, Rinehart & Winston.
Kandel, E. R. & Schwartz, J. H. eds. (1985): Principles of Neural Science. 2nd. ed. New York – Amsterdam: Elsevier.
Keesing, R. M. (1981): Cultural Anthropology. New York: Holt, Rinehart & Winston.
Kruijt, J. (1964): Ontogeny of Social Behaviour in Burmese Red Jungle Fowl (*Gallus gallus spadiceus*). Behaviour Suppl., 12.
– (1971): Early Experience and the Development of Social Behaviour in Jungle Fowl. Psychiatr. Neurol. Neurochir., 74, 7–20.
Kuo, Z.Y (1932): Ontogeny of embryonic behavior. *J. Exp. Biol.*, 61, 395–430, 453–489.
– (1967): The Dynamics of Behavior Development. New York: Random House.
Lehrman, D. S. (1953): A critique of Konrad Lorenz's theory of instinctive behavior. Quart. Rev. Biol. 28, 337–363.
Lewontin, R. C. (1977): Sociobiology – A caricature of Darwinism. In: Suppe, F. and Asquath, P. (eds.), PSA 1976, Vol. 2 PSA Lansing, MI.
Lorenz, A. (1952): Wenn der Vater mit dem Sohne. Wien: Deutike.
Lorenz, K. (1927): Beobachtungen an Dohlen. Journal für Ornithologie 75, 511–519.
– (1931): Beiträge zur Ethologie sozialer Corviden. Journal für Ornithologie 79, 67–127.
– (1932): Betrachtungen über das Erkennen der arteigenen Triebhandlungen bei Vögeln. Journal für Ornithologie 80, 50–98.

– (1935): Der Kumpan in der Umwelt des Vogels. Journal für Ornithologie 83, 137–215 und 289–413.
– (1941a): Vergleichende Bewegungsstudien an Anatinen. J. Ornith., 89, 194–294.
– (1941b): Kants Lehre vom Apriorischen im Lichte gegenwärtiger Biologie. Blätter für Deutsche Philosophie 15, 94–125.
– (1943): Die angeborenen Formen möglicher Erfahrung. Zeitschrift für Tierpsychologie 5, 235–409.
– (1961): Phylogenetische Anpassung und adaptive Modifikation des Verhaltens. Z. Tierpsychologie 18, 139–187.
– (1963): Das sogenannte Böse. Wien: Borotha Schoeler.
– (1973): Die Rückseite des Spiegels. Versuch einer Naturgeschichte menschlichen Erkennens. München: Piper.
– (1992): Die Naturwissenschaft vom Menschen. Eine Einführung in die vergleichende Verhaltensforschung. Das "Russische Manuskript," München: Piper.
Lorenz, K. and N. Tinbergen (1938): Taxis und Instinkthandlung in der Eirollbewegung der Graugans. Z. Tierpsychologie 2, 1–29.
Marler, P. (1976): Sensory templates in species-specific behavior. *In*: Fentress, J. (ed.), Simpler Networks and Behavior. Sunderland, MA: Sinauer Assoc..
– (1978): Perception and Innate Knowledge. *Proc. 13th Nobel Conf. "The Nature of Life."* pp. 111–139.
Masters, R. D. (1989): The Nature of Politics. New Haven – London: Yale Univ. Press.
Mazur, A. & Lamb, Th. (1980): Testosterone, Status and Mood in Human Males. Hormones and Behavior 14, 236–246.
Morgan, C. Lloyd (1894): Introduction to Comparative Psychology. London.
– (1900): Animal Behavior. London.
Popper, K.R. (1973): Objektive Erkenntnis. Ein evolutionärer Entwurf. Hamburg: Hoffmann und Campe.
– (1984): Auf der Suche nach einer besseren Welt. München: Piper.
Salter, F.K. (1995[a]): Comments on the naturalistic fallacy, biology and politics. Social Science Information 34, 333–45.
– (1995[b]): Emotions in Command. A Naturalistic Study of Institutional Dominance. Oxford: Oxford Univ. Press.
Salzen, E. (1996): Introduction to the Routledge edition of Konrad Lorenz: On Aggression. London – New York: Routledge, p. IX-XXV.
Smith, P. K. (1992): Memoir. Human Ethology – Origins in the U.K. *Human Ethology Newsletter*, vol. 7, issue 4, Dec. 1992, 1–2.
Sperry, R. W. (1963): Chemoaffinity in the Orderly Growth of Nerve Fiber Patterns and Connections. Proc. Nat. Acad, Sci. U. S., 50, 703–710.
– (1971): How a brain gets wired for adaptive function. *In:* Tobach, E., Aronson, L. R., and Shaw, E. (eds.), The Biopsychology of Development. London: Academic Press, pp. 27–44.
Stent, G. S., Kristan, W. B., Friesen, W. O., Ort, C. A., Poon, M. & Calabrese, R. L. (1978): Neuronal Generation of the Leech Swimming Movement. An Oscillatory Network of Neurons Driving a Locomotory Rhythm has been identified. Science, 200, 1348-1357.
Tinbergen, N. (1951): The Study of Instinct. New York and London: Oxford Univ. Press.
Tobach, E., Gianutsos, J. Topeff, H. R, and Gross, C. G. (1974): The Four Horsemen: Racism, Militarism and Social Darwinism. New York: Behavioral Publications.
Wallhäuser, E. and Scheich, H. (1987): Auditory imprinting leads to differential 2-deoxy-glucose uptake and dendritic spine loss in the chick rostral forebrain. *Develop. Brain Res.* 31, 29–44.
Watson, J. B. (1919): Psychology from the Standpoint of a Behaviorist. Philadelphia: Lippincott.
– (1930): Der Behaviorismus. Stuttgart.
Wickler, W. (1981): Die Biologie der zehn Gebote. Warum die Natur für uns kein Vorbild ist. München: Piper.
Wiessner, P.& A. Tumu (in press): Historical Vines: The Formation of Regional Networks of Exchange, Ritual, and Warfare among Enga of Papua Guinea. Washington, DC: Smithsonian Institution Press..
Wilson, E. O. (1975): Sociobiology: The New Synthesis. Cambridge, MA: Belknap Press – Harvard University Press.
Willows, A. O. D. (1971): Giant Brain Cells in Mollusks. Scient. Americ., 224, 69–75.

2

RESEARCH ON EMOTIONS AND FUTURE DEVELOPMENTS IN HUMAN ETHOLOGY

Glenn Weisfeld

Department of Psychology
Wayne State University
Detroit, Michigan 48202

ABSTRACT

This chapter presents a model of the basic human emotions which, it is argued, can be taken as comprising the main elements of the human ethogram and hence of human behavior. The chapter begins with a critical review of criteria that have commonly been used in constructing such lists, and argues for heavy reliance on the criteria of adaptive value and affect. Since affect is the most characteristic feature of emotions and seems to induce their particular behavioral tendencies, expressions, and visceral adjustments, affect is perhaps the single best defining feature of each emotional modality. Because of its ethological nature, this sort of list of emotions may be more comprehensive than alternative lists lacking a comparative basis, and it avoids superficial dichotomies between types of emotions. Some general properties of affects are then discussed, such as their valence, intensity, and timing. Some possible applications of this model for studying a wide range of human behaviors are then suggested, including sex and developmental differences, and individual, cultural, and pathological variation. It is argued that ethological methods and a phylogenetic perspective are essential for arriving at a complete description of the human emotions and their facets, while at the same time avoiding an overly cybernetic modularization model.

1. THE HUMAN ETHOGRAM: THE BASIC EMOTIONS

1.1. Plan of This Chapter

I would like to suggest that research and theory about human behavior can best be organized with reference to a list of the basic emotions. I will begin by arguing for this position, proceed to discuss the criteria for composing a list of basic emotions, and then propose a working model of the basic human emotions. Lastly, I will indicate some possible directions for future research on emotion using an ethological perspective.

New Aspects of Human Ethology, edited by Schmitt *et al.*
Plenum Press, New York, 1997

1.2. Status of the Concept of Emotion

Emotions have been neglected for a long time by American psychology. But this was not always the case. Early in this century we had the predominance of William McDougall and his lists of human "instincts." McDougall (1923) drew comparisons with other species to identify 14 universal human instincts, or motives, for which he proposed adaptive explanations. As is well known, instinct theory was eclipsed by behaviorism, whose champions included John B. Watson and B. F. Skinner. The behaviorists downplayed the importance and number of emotions. They sought to show that emotions such as curiosity, parent-infant bonding, and pride/shame were merely the result of secondary reinforcement. Now we are in the throes of a more complex form of learning theory, cognitive psychology in its various forms. There is renewed interest in emotion and motivation, but in many cases emotion is regarded as derivative, as developing from emerging cognitive capacities. Thus, learning and cognition remain ascendant, and emotion recessive.

I think McDougall was right and we should return to comparisons with naturalistic animal behavior, and to the emotions, in our search for the roots of human behavior. That is, we should ground our research in ethological theory and methods. These whole-body movement patterns are the basic units of behavior for our species, the human ethogram.

1.3. Advantages of a Model of Human Behavior Based on the Emotions

Psychology is often criticized for not having provided a basic, general description of human behavior early in its history as a discipline. This deficit is still being felt, in that the field lacks a framework for recognizing the elements of behavior. But ethology can supply this missing perspective by focusing on the emotions.

Why should the emotions provide the elements for that framework? Many other psychological phenomena are species-wide and need to be studied too—sensation, perception, learning, reflexes, and cognition. However, these phenomena are, in general, functionally subordinate to emotion. They only guide the elicitation and execution of emotions, or motives. Our sensory, perceptual and cognitive capacities serve to identify and classify stimuli so that we can react appropriately to them. These same general capacities serve us in modifying, to the extent possible, our behavioral options. The very phenomenon of learning—repeating a response that has been "reinforced"—entails the pleasure of a positive affect; recent research indicates that the amygdala plays a role in these acquired responses (LeDoux, 1993). Also, visceral responses to emotionally salient stimuli, as Pavlov demonstrated, can be classically conditioned so they can be elicited promptly. Likewise, our memories are biased in favor of recording and recalling events of emotional significance (Buck, 1988). Then too, many verbal utterances have clear socioemotional content (Dunbar, 1993), and are in fact supplements to nonverbal expression, e.g., threats, apologies, greetings, expressions of approval, complaints, and attention. In a word, we are almost always feeling something when awake—but we do not always think!

The primacy of emotion is further illustrated by organisms with little in the way of learning and cognition. For example, insects survive and reproduce perfectly well without much flexibility of behavior or awareness of who they are or what they are doing (Izard, 1972). Likewise, people without a functioning neocortex, such as newborns, anencephalics, and unconscious patients, remain alive even without much cognitive capability. On the other hand, great cognitive powers do not ensure fitness unless they enhance actual behavioral outcomes. In short, in the beginning there were tropisms and then fixed action patterns that raised fitness (see Eibl-Eibesfeldt, 1975). Only later did these become elaborated by learning and cognition. Higher learning and cognition are evolutionary upstarts, and are the servants of our ancient emotions. The neocortex is new, whereas the limbic system derives from archicortex and paleocortex (cf. MacLean, 1993). Thus, emerging cognitive capacities are unlikely to be the root cause of any emotion (see Weisfeld, in press, on the emergence of pride and shame).

An ethological view, then, keeps us focused on behavior and its adaptive value. It leads to the realization that human motives, or emotions, raise fitness by serving tissue needs in quite direct ways, more directly than do learning and cognition. This may seem obvious to ethologists, but this functional perspective is virtually absent from most mainstream approaches to emotion and motivation. The proximate causation of emotions and their visceral and expressional correlates is described in detail, but with little consideration of the adaptive value of the behavior or its correlates.

There may be an additional benefit to an evolutionary view and ethological methods: superficial distinctions among types of emotions can be avoided. Mainstream psychologists often distinguish between motives and emotions, or between "biological" drives and emotions, or between primary and social emotions. The "biological" drives are said to be species-wide, affected by endogenous physiological factors, culturally invariant, and of adaptive value. The emotions, or "social motives," are said to be culturally variable, triggered by exogenous elicitors, learned, and so forth. A moment's reflection usually reveals the arbitrariness of these dichotomies. Consider the sex drive—biological or social? Is pain not triggered by exogenous factors? Is hunger not affected by cultural norms and customs? Where does curiosity fit, or fear? They may not reflect conditions of direct tissue deprivation, but surely they ultimately function to maintain tissue integrity. We can dismiss these superficial distinctions by recognizing that all basic, universal emotions are biological; all are adaptive; all are influenced by both exogenous and endogenous factors; all have specific CNS mediators; and all can be modified by learning. A comparative perspective protects one from drawing such dubious distinctions. Human emotions are similar to those of other primates, especially the chimpanzee (Goodall, 1987; van Hooff, 1973), for which such dichotomies are seldom invoked. Why invoke them for humans?

2. CRITERIA FOR IDENTIFYING THE BASIC HUMAN EMOTIONS

2.1. Specific Adaptive Value

Pursuing this functional perspective leads us to the question of lists of the basic emotions and how they are derived. McDougall's (1923) list is a good starting point (Table 1). This list surpasses modern ones that are not informed by functional considerations. Other biologically based lists are similar to McDougall's, thus increasing our confidence in their validity, such as Scott's (1958; Table 2). Scott combined some of McDougall's categories into compounds, e.g., Combat and Escape become Agonistic, but both lists are quite comprehensive, or ethogrammatic. Cattell's (1957) list resembles Scott's but includes Sleepiness and Sensuous Comfort.

Most non-evolutionary lists of emotions, however, would ill prepare the human organism for survival and reproduction. Many lists include happiness or sadness or similar terms (joy, pleasure, distress, tension, elation, satisfaction: see Plutchik, 1994, p. 58).

Table 1. McDougall's (1923) list of human instincts

Parental or protective	Primitive passive sympathy
Combat	Self-assertion and submission
Curiosity	Mating
Food-seeking	Acquisitive
Repulsion	Constructive
Escape	Appeal
Gregarious	Laughter

Table 2. Scott's (1958) list of adaptive behaviors

Ingestive	Care-soliciting
Shelter-seeking	Eliminative
Agonistic	Allelomimetic
Sexual	Investigative
Care-giving	

Table 3. Plutchik's (1980) list of emotions

Subjective language	Behavioral language
Fear, terror	Withdrawing, escaping
Anger, rage	Attacking, biting
Joy, ecstasy	Mating, possessing
Sadness, grief	Crying for help
Acceptance, trust	Pair bonding, grooming
Disgust, loathing	Vomiting, defecating
Expectancy, anticipation	Examining, mapping
Surprise, astonishment	Stopping, freezing

These terms obviously are too vague to correspond to any particular motivated behaviors. What behavior would a sad organism engage in to relieve its distress?[1]

Even lists that do include an appreciation for function, such as Plutchik's (1980), often neglect essential motives such as hunger (Table 3). I am suggesting, then, that emotions be identified partly in terms of their **specific adaptive value**.

2.2. Whole-Organism Behavior

If all species-wide behaviors are adaptive, however, this criterion is not specific for emotions. A stretch reflex is surely adaptive but is not an emotion. We can rule out reflexes as emotions because they do not involve the behavior of the whole organism; they are local. So another criterion of an emotion might be: **involving the whole organism**. In fact, many reflexes can be regarded as part of an emotion complex. The release of saliva in response to

---

[1] Pugh (1977) has suggested, however, that at extreme intensities, various pleasures or displeasures meld into indistinct happiness or sadness. But before these intensities have been reached, discrete affects have been experienced that provide the subject with specific behavioral guidance. Among many others, Fridlund (1994, p. 279) also makes the point that happiness comes in forms that vary with the behavioral context, a difficulty that is resolved by identifying particular emotional modalities.

meat powder in the mouth can be seen as a visceral adjustment that is part of the hunger motive. The same can be said of movements that are subunits of a basic motive, such as the elements of courtship, play, or attack in various species (e.g., Tinbergen, 1951)

2.3. Specific Facial Expression

One common criterion for an emotion is **specific facial expression** (Ekman, 1973). This test does have some utility, as for example in identifying anger, disgust, and fear. There may also be some universal facial expressions in addition to the six identified by Ekman and Friesen (1971). See for example Darwin (1872) on blushing, Davis (1934) on facial expressions of pain and sex, Hess (1965) on pupillary dilation during interest, Provine (1989) on yawning with boredom or drowsiness, and Eibl-Eibesfeldt (1973) on greeting, play invitation, laughter, crying, and flirtation expressions.[2]

But this criterion results in neglect of behaviors that have no particular facial expression but possess other properties commonly attributed to emotions. For example, feeding has adaptive value, is species-wide, has a distinct affect, is associated with particular visceral and hormonal adjustments, and involves the whole body. It even has a distinct expression in infants, the hunger cry, but no distinctive adult facial expression. This example suggests another objection to using facial expression as the sine qua non of an emotion: Some emotional expressions are not facial but rather are vocal, olfactory, tactile, or postural. Why should these expressions not be accorded the same weight as facial ones, since some of them show evidence of universality (e.g., Van Bezooijen et al., 1983)? Ekman (1994, p. 18) now contends that "A universal signal should not *ipso facto* be considered evidence of an emotion. Nor should the lack of a universal signal be used to say a phenomenon is not an emotion." Widespread adoption of this broader position would move us closer to an ethological view of emotion.

2.4. Visceral and Hormonal Patterns

Other common criteria for an emotion are its visceral and hormonal changes. There may be some specificity in **visceral and hormonal patterns** for various emotions, such as fear, anger, and disgust (Ekman, 1994, p. 17). At this point, however, there is only limited evidence for the specificity of these adjustments. One problem is individuals exhibit various patterns of adjustments to the same emotional state.

2.5. Particular Neural Mechanisms

Each emotion is likely to prove to be mediated by **particular neural mechanisms** (Ekman, 1994, p. 18). For example, feeding is (partly) controlled by the lateral hypothalamic area and ventromedial nucleus of the hypthalamus. This criterion for an emotion has been useful for distinguishing among the various types of aggression. Different neural

---

[2] Fridlund (1994) and Russell (1994) offer detailed and important critiques of the notion of universality of facial expressions. However, the inadequacy of experimental methods does not gainsay the likelihood that human expressions evolved from those of our primate ancestors, given their early and stereotypic appearance, apparent mediation by midbrain limbic nuclei (Buck, 1988, pp. 93–98), documentation by observational research in various cultures, and resemblance to those of other species. Certainly, however, Fridlund is correct that these expressions vary markedly with culture and context, to an extent perhaps not covered by Ekman's notion of display rules.

pathways seem to be involved in predatory, defensive, irritable, dominance, and parental aggression (Moyer, 1976). Research on neurotransmitters and hormones may also help in distinguishing the basic emotions. For example, testosterone seems to increase dominance aggression (and play fighting) but not maternal aggression (Ellis, 1986). Serotonin also seems to be involved in dominance (Masters & McGuire, 1994).

2.6. Elicited by Particular Situations

Ekman (1994) suggested that emotions are **elicited by particular situations**. For example, fear is evoked by falling and other dangerous prepotent stimuli. However, the elicitors of a given emotion may not be very specific, as in the case of the emotions of curiosity and pride/shame.

2.7. Specific Behavioral Tendency

On the other hand, our response possibilities are relatively limited. Even though, as primates, we are capable of an enormous range of movements, we only need to fulfill a finite set of biological functions. Our overt behavioral tendencies—our motives—are limited in number. Panksepp (1994) argued that each such a tendency is prompted by a particular affect, so that we have only one response tendency per emotion. Thus, for example, anger prompts us to attack the offending target. We may execute the attack in a variety of ways, but each constitutes an attack. Thus, another useful criterion for an emotion is **specific behavioral tendency**. Wallbott and Scherer (1986) found evidence for cross-cultural consistency in actions specific to emotions, and these actions are recognizable in emotions terms by others from a different culture (Sogon & Masutani, 1989). However, some emotions are characterized by no very distinct action tendency, e.g., interest and the other esthetic emotions. In these cases the behavior consists merely of attending to the stimulus. Also, the behavioral actions associated with some emotions can be highly variable, such as the many forms of food gathering, food preparation, and ingestion. And the same behavior can fulfill various emotional needs, such as suckling behavior.

2.8. Presence in Other Primates

Each emotion is very complex and hence evolutionarily stable. It entails visceral adjustments, motoric responses, an affect, and extensive representation in the ancient subcortex. It is bound to be slow to evolve, hence Ekman's claim (1994, p. 16) that humans are unlikely to have evolved new emotions not found in other primates. Therefore, another useful criterion is **presence in other primates**. However, since every species has some unique characteristics, it is possible that some human emotions are absent from our primate relatives. For example, defecation may be purely reflexive in the arboreal primates but partly voluntary and hence emotional in our own species. Also, pair bonding is seen in humans but not chimpanzees.

2.9. Distinct Affect

The criterion that many theorists seem loath to accept is that of a **distinct affect** (e.g., Lewis & Michalson, 1983, pp. 30f). Yet this may be the most characteristic facet of emotion. When we experience an affect but do not exhibit any emotional expression, visceral changes or overt action, are we not still experiencing an emotion? Furthermore, affects can be rated reliably in other primates, thus permitting cross-species comparison (Goodall, 1987). Admit-

```
Internal Elicitors ──┐         ┌──▶ Overt Behavior
                     ├─▶ Affect ├──▶ Expressions
External Elicitors ──┘         └──▶ Visceral Changes
```

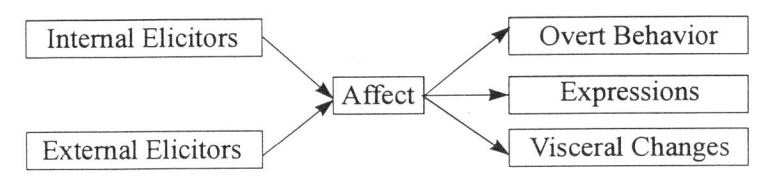

Figure 1. Simple model of the elicitation of emotion.

tedly, self-report data are questionable, but we rely on them nonetheless in studies of sensation and perception. Perhaps another reason that the concept of affect is avoided is that it seems metaphysical. It may be so fundamental that it cannot be defined in terms of other concepts. In this respect it resembles mass, electric charge, time, and other such irreducible notions. But this is no reason not to study it by the best methods available. To the contrary, this suggests that we should accord it a fundamental position in our models of behavior.

One major reason for focusing on affects is that they are the guideposts for adaptive behavior. Organisms do not strive to enhance their fitness directly (unless, perhaps, they have read Darwin), but instead try to satisfy their affective needs in order of urgency.

Another reason for focusing primarily on affects is their central role in relation to the other facets of an emotion. Consider the Cannon-Bard sort of model of the sequence of emotional experience (Figure 1). This is far superior to the James-Lange one (see Buck, 1988) and its modern-day supplement, the facial feedback hypothesis (see Fridlund, 1994, pp. 173–182). This simple figure shows the crucial position of affect. In neuroanatomical terms, the hypothalamus and other limbic structures mediate affect; they contain the brain's pleasure and pain centers. The hypothalamus then orchestrates appropriate **visceral, expressional, hormonal,** and **voluntary behavioral** events through the medulla oblongata, (limbic) central grey of the midbrain, pituitary, and motor cortex, respectively. See Figures 2–4 for details of the first three systems. The mediation of voluntary behavior is more complicated, but current knowledge suggests that the main pathway in primates may proceed from hypothalamus and limbic system to the basal ganglia, next to the thalamus and then to the motor cortex (Apicella et al., 1991).

It is often said that affects are extremely difficult to talk about, that our language for describing them is impoverished (e.g., Plutchik, 1980). But for at least many emotions terms, there is little ambiguity about the affective state referred to. Furthermore, terms referring to affects sometimes help in drawing distinctions among emotions. As Sogon and Masutani (1989) showed, these terms show considerable cross-cultural validity.

Consider Moyer's (1976) distinctions among different types of aggression. Predatory, defensive, irritable, dominance, and maternal aggression have different emotional displays and overt behaviors. These distinctions are corroborated by their having different affects—hunger, fear, anger, dominance, and maternal feelings. For example, food deprivation (hunger) leads to predatory aggression rather than other forms. Likewise, a distinction between angry aggression and play fighting, established by observational research on children (Blurton Jones, 1972; Boulton & Smith, 1992), corresponds with that between anger and playfulness.

2.10. Summary

In summary I would suggest that the basic human emotions be identified by all of these criteria in proportion to their utility, but especially by their affects and functions. How, then, should emotion be defined? I would define a human emotion in affective

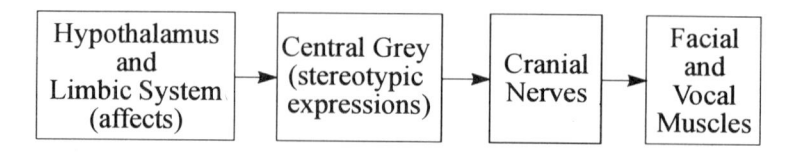

```
┌─────────────┐   ┌─────────────┐   ┌─────────┐   ┌─────────┐
│ Hypothalamus│   │Central Grey │   │ Cranial │   │ Facial  │
│    and      │──▶│(stereotypic │──▶│ Nerves  │──▶│  and    │
│Limbic System│   │expressions) │   │         │   │ Vocal   │
│  (affects)  │   │             │   │         │   │ Muscles │
└─────────────┘   └─────────────┘   └─────────┘   └─────────┘
```

Figure 2. Simple model of the elicitation of spontaneous emotional expressions.

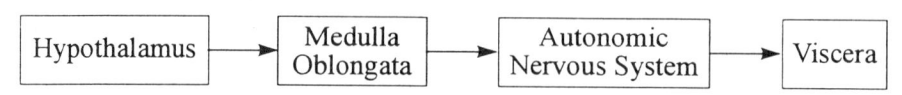

```
┌─────────────┐   ┌─────────┐   ┌──────────┐   ┌────────┐
│ Hypothalamus│──▶│ Medulla │──▶│Autonomic │──▶│Viscera │
│             │   │Oblongata│   │Nervous   │   │        │
│             │   │         │   │System    │   │        │
└─────────────┘   └─────────┘   └──────────┘   └────────┘
```

Figure 3. Simple model of the elicitation of visceral changes in emotion.

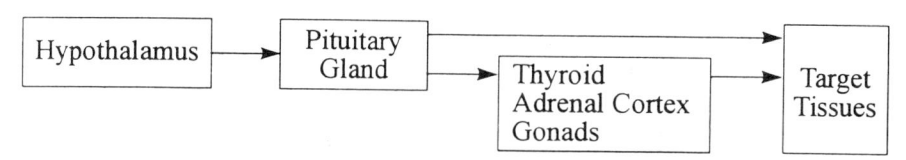

```
┌─────────────┐   ┌──────────┐   ┌──────────────┐   ┌─────────┐
│ Hypothalamus│──▶│Pituitary │──▶│  Thyroid     │──▶│ Target  │
│             │   │  Gland   │   │Adrenal Cortex│   │ Tissues │
│             │   │          │   │  Gonads      │   │         │
└─────────────┘   └──────────┘   └──────────────┘   └─────────┘
```

Figure 4. Simple model of the elicitation of endocrine changes in emotion.

terms, as a **basic, universal and hence adaptive state of substantial pleasure or displeasure.** "Basic" is meant to exclude blends of emotions. "Substantial" eliminates minor irritants or pleasures such as pressure on a limb impeding venous return, or slight preferences for familiar stimuli.

There is obviously more to an emotion than this simple definition conveys. Other common—but not omnipresent—characteristics are: a specific behavioral tendency, a specific emotional expression, and a specific pattern of visceral adjustments. But because an emotion may lack one or more of these facets, they cannot be regarded as part of the strict definition of emotion, but only as descriptive of the *emotion complex*.

Having attacked the thinking of others on emotion, I am obligated to suggest my own list of basic emotions. Since I prefer to think primarily in terms of affects, I like to use affective terms for the emotions. Each emotional modality in such a list should be species-wide and hence adaptive, and include a basic, distinct affect and adaptive (if nonspecific) overt behavioral tendency. The list is also based on evidence for specific neural mediation, presence in other primates, specific situational elicitors, specific emotional expressions, and specific visceral adjustments, but space does not permit systematic review of these data. Taken together, the list is intended to comprise the inventory of whole-body behaviors exhibited by normal adults throughout the world. For each emotional modality, I will mention some of its characteristic features.

3. AN EMOTIONS MODEL OF THE HUMAN ETHOGRAM

3.1. Tactile Pain and Tactile Pleasure

This refers to any of the sensations we localize in our skin and describe as being either pleasant or unpleasant: pain, itching, tickling, hot, cold, softness, smoothness, etc.

The sensory receptors and spinal tracts that convey these sensations are multiple, and the sensations often reach consciousness, in the parietal lobe, as blends of the primary components. Because of the number of receptors and hence of distinct tactile feelings and blends, they are best lumped together.

The adaptive value of perceiving pain is obvious; individuals without pain receptors usually die of injuries. More vulnerable anatomical structures are endowed with more pain fibers and hence are more sensitive. Itching seems to protect us from harmful stimuli such as sharp sticks, debris, and insects. The pleasure of softness, warmth, and smoothness encourages the direct contact necessary for infant care. Both parent and infant are attracted to the other, as seems necessary for any social bond. Tactile stimulation has been shown to increase growth in premature infants (Schanberg & Field, 1987), so affects may be serviceable not only in eliciting adaptive behaviors but also in setting the stage for normal development. The same can be said for the behavioral developmental effects of "contact comfort" on Harlow's rhesus infants (reviewed by Kraemer, 1992).

3.2. Hunger

This is a good model of an emotion because much of the physiology has been worked out. As a result of this research, we know that an emotion can be very complex, can be regulated on a great number of levels. Literally dozens of exogenous and endogenous factors can affect appetite. And yet body weight is extremely stable, seldom varying by more than a few pounds per year for most individuals. The fact that body weight is usually stable in obesity supports the notion that this condition is not pathological but rather represents an adaptive elevation of the set point.[3]

3.3. Thirst

This is a simpler, better understood emotional system. As with probably every other basic emotion, the hypothalamus is involved in thirst. So too is the endocrine system, which is also common for emotions. See Rosenzweig et al. (1996) on thirst, hunger, and temperature regulation.

---

[3] Obesity is found especially in (a) domesticated animals; (b) animals and people inhabiting cold, barren regions; (c) women; and (d) infants and the aged. This distribution suggests the explanation that obesity occurs in sedentary individuals or those under severe starvation pressure. Domesticated animals are usually provisioned and protected, and hence need not flee from predators, hunt, or defend themselves. Hence they can become obese as a hedge against starvation. Arctic animals and peoples tend to be stout and to have ample subcutaneous fat. Obesity is also common in peoples who evolved in cold climates, such as Eastern Europeans whose ancestors came from Siberia. Women, infants and the aged were less dependent on running speed and agility in prehistory than were men who hunted and fought. This explanation of obesity as an evolved response to starvation is consistent with physiological and behavioral data. Obese individuals who diet, i.e, undergo bouts of simulated starvation, tend to lower their metabolic rates with each diet, as if compensating for a shortage of calories by reducing their expenditure of energy. This occurs in starved rats also. Similarly, obese people tend to be opportunistic feeders—to binge when abundant rich, nutritious food is present. This would be adaptive under starvation conditions. By contrast, normal-weight people tend to eat on schedule, and eat moderately. Despite this tendency to overeat on occasion, most obese individuals eat little more than others. Evidently their bodies are efficient in conserving calories, as by lowered the metabolic rate and storing fat.

3.4. Tasting and Smelling

Esthetic emotions such as tasting and smelling are characterized by what Pugh (1977) has termed "concurrent values." These feelings perform a steering function while an overt behavior is being executed. For example, tasting and smelling steer us toward certain flavors and away from others in a way that addresses specific nutritional needs. "Preceding values" are usually unpleasant and prompt us to take steps to satisfy some need, such as thirst. "Trailing values," usually pleasant, provide the payoff for executing a behavior successfully. Pugh's approach is a cybernetic one; he considers the design characteristics necessary to simulate human behavior. But he has been guided by ethological research and evolutionary theory.

3.5. Disgust, or Nausea

This emotion protects us from toxins (Rozin et al., 1993), and may be the main emotional explanation for hygienic practices. It is present from birth (Steiner, 1979). The overt behaviors range from avoidance of the stimulus to ejecting it from the mouth, throat or stomach. Associations between a stimulus that elicits disgust are easy to acquire and hard to extinguish (Rozin et al., 1993). Nature seems to protect us indefinitely from any stimulus that was potentially harmful in the past.

Pugh (1977) has argued that the expression of disgust is similar in form to that of contempt, the distinction between the terms resting on the stimulus: inanimate or human. Perhaps we simulate the expression of disgust when we wish to display our displeasure at someone. On the other hand, Ekman and Friesen (1986) maintained that the expressions of disgust and contempt differ, the former involving the nose and the latter the upper lip; but see Russell (1991). Such issues may perhaps be clarified by asking whether two distinct affects are represented. Disgust seems to represent nausea, at least in its basic form. Contempt may reflect pride (McDougall, 1923, p. 424), anger, lack of empathy (see Oatley & Jenkins, 1996, p. 311), or an anhedonic attitude of disrespect, but seems unlikely to entail a unique affect despite its possibly distinct expression.

3.6. Fatigue

The adaptive value of rest sometimes escapes today's exercise enthusiasts. Animals evolved to be conservative in utilizing their metabolic energy; they obey the law of economy (Stanley, 1898). For example, as they become more practiced in executing a given movement, they perform it more smoothly and accurately, and expend less energy. Cathartic and hydraulic theories of aggression violate the law of economy. Occasionally one still sees an evolutionary argument that violates it. Barber (1991) proposed that animals play partly in order to burn up calories and hence not become obese. To the contrary, animals evolved to conserve calories; they would not incur the costs of foraging only to burn off the resultant calories.

3.7. Drowsiness

Emotions are sometimes said to be physiologically arousing, but sleep is an exception. It is a true motivated behavior because drowsy animals will seek a suitable place to sleep rather than simply collapsing on the spot. Drowsiness can be distinguished from fatigue by its overt behavior. Drowsiness prompts sleep; fatigue prompts rest. But the function of sleep seems to be rest also.

The duration and circadian pattern of sleep have been analyzed functionally, leading to additional insights. Animals sleep when their sensory capacities are less efficient. Thus, diurnal animals are at a disadvantage at night and so they sleep, which usually means they hide. Also, more vulnerable animals, such as prey species that do not hide in burrows, sleep less. Small animals sleep less because they lose heat quickly when relaxed. Animals that must be continuously conscious, such as the porpoise, alternate the sleeping of their cerebral hemispheres. Dreaming (considered below as a vicarious state) seems to aid brain growth and efficiency in various ways. See Hudson (1995); Rosenzweig et al. (1996).

3.8. Sexual Feelings

This is perhaps one of the most neglected emotions, being absent from most psychologists' lists. No other omission better illustrates the folly of ignoring evolutionary considerations in drawing up lists of emotions.

This emotion does not develop fully until puberty (Money & Ehrhardt, 1972). Consequently, it illustrates a flaw in the criterion that a basic emotion must appear at an early age (cited by Mealey, 1995, p. 539).

3.9. Loneliness and Affection-Receiving

Here we have a case of impoverished terminology. One might use the term "love," but that would exclude the pleasure of friendship and of general companionship. It may be best to consider this emotion as comprising several subtypes, such as friendship, filial love, parental love, love for other kin, romantic love (see Jankowiak & Fischer, 1992 on its universality), and desire for companionship not limited to other humans (cf. the extreme distress of prolonged social isolation). However, the similarities among these subtypes may outweigh the differences. All other dyadic social bonds may have evolved from the primordial mother-infant bond.

The intensity of this emotion attests to the adaptive importance of particular social relationships, especially pair bonds and parent-offspring attachment. The expressions that play a role in bond formation, such as flirtation, attention, gentle touching, smiling, as well as behaviors such as performing favors, have been studied by social psychologists but have seldom been brought into an ethological framework (but see Eibl-Eibesfeldt, 1989).

3.10. Interest and Boredom

The more behavioral options an animal potentially has, the more it must learn which responses are appropriate under which conditions. Primates, being generally arboreal and hence acrobatic, have highly flexible behavior. So primates tend to have a lengthy period of immaturity during which to play, explore and learn. They even undergo a growth plateau as juveniles so they can spend extensive time and energy at play (Weisfeld & Billings, 1988).

The things that interest us tend to have relevance for fitness—large size, color (e.g., fruit), violence, sex, infants, faces, projectiles. Dangerous and even horrible things are fascinating, not just pleasant scenes and deeds.

This affect guides learning. It leads us to investigate appropriate stimuli; for example, stimuli that are moderately novel and complex for the subject are the most interesting. In this way, we are drawn to stimuli that are not so simple and familiar that they can teach us little, or so bizarre as to be hopelessly confusing or even frightening. As we try to understand a stimulus or problem, we feel good as we make cognitive progress, and bad if we are thwarted. If we succeed in gaining a sudden insight, we are rewarded by a thrill of discovery. This emotion shows the limitations of dichotomizing behavior into emotion and cognition, or rational and emotional, and of cybernetic models of the mind.

Some lists include surprise, but this cognitive state may entail no particular affect other than interest or, sometimes, fear. See Charlesworth (1969) for a detailed analysis of surprise and the relation of this state to startle, the orienting reaction, and stimulus novelty and expectancy.

3.11. Beauty-Appreciation

Most lists of emotions exclude the esthetic emotions, Pugh (1977) being an exception. Yet how can one otherwise account for the universality of art and music (see also Dissayanake, 1988; Eibl-Eibesfeldt, 1989)? Various animals seem to have a preference for certain geographical features; such a tendency would be adaptive in guiding nomadic species to appropriate habitats, just as humans everywhere seem to perfer savannah-like terrain for their parks. The appeal of ideal features of babies or mates might also be included under this emotion, but perhaps placed elsewhere.

By including this emotion and the next one, the model covers the five senses. Many sensations entail affect; they are not always neutral. These sensory-emotional modalities often convey concurrent values; they steer us toward or from stimuli of adaptive salience.

The esthetic emotions, like interest, follow the cognitive consistency principle: moderate complexity for the subject is most appealing. For example, children tend to appreciate art that is simpler than that favored by adults (Pugh, 1977, p. 332). This suggests that we learn through the arts.

3.12. Music-Appreciation and Noise-Annoyance

Noise can strongly arouse the sympathetic division and hence can be physiologically wearing. As such, it is a significant detractor from health and happiness.

3.13. Humor-Appreciation

I have dealt with this emotion in detail elsewhere (Weisfeld, 1993). It is only some of the evolutionists who have recognized the universality of this affect and its expression, and have proposed adaptive explanations for it (e.g., Eibl-Eibesfeldt, 1989; McDougall, 1923; Pugh, 1977; Alexander, 1986).

3.14. Pride and Shame

This emotion is one of the most neglected; and yet who would deny that it constitutes a distinct affect, or that people everywhere do not seek the approval of others? Darwin (1872) described the characteristic expressions of this emotion—the erect, expansive posture of pride (or dominance), and the antithetical demeanor, and blushing, of shame (or submission). The main objection by mainstream psychologists to accepting pride/shame into the pantheon of basic emotions seems to be the fact that different values prevail around the world, so that what merits approval in one place may warrant disapproval elsewhere. But the same could be said of other emotions such as hunger and interest, which are satisfied in quite different ways around the world too. There may even be some prepotent values for pride/shame, such as pulchritude and bodily grace; see Dong et al. (1996) for evidence of the cross-cultural prevalence of values, and Weisfeld (1980, in press) for an ethological analysis of this emotion.

Recognizing this emotion, like recognizing interest, guards us against believing that rational behavior is not also emotional. Much of our rational planning is aimed at avoiding embarrassment or gaining others' approval. For example, we may stay awake during a boring lecture out of courtesy for the speaker. We do so not because reason is suppressing passion, but because we care more about maintaining others' approval than we do about sleep. Thus, the notion of managing or controlling one's emotions is oxymoronic, and smacks of free will. All of our intentional actions are prompted by a desire to ameliorate our current or future affective state.

3.15. Anger

This emotion seems to be elicited most consistently, in animals and humans, by perceiving a violation of a social norm or expectation (Weisfeld, 1972). For example, the violation of an animal's territory or mating privileges or possession of food often prompts vigorous retaliation. This explanation seems superior to the frustration-aggression hypothesis (Pastore, 1952). Within the context of a human dyad, small group or dominance hierarchy, anger is often triggered by inequities such as failure to reciprocate a favor (Weisfeld, 1980). An individual whose prerogatives do not measure up to his merit may retaliate against the usurper who exploits him (cf. Trivers, 1971). Anger then tends to be dissipated by any course of action that restores equity, e.g., retaliation by the victim against the violator (Hokanson, 1970),[4] retaliation by a bystander against the violator (Baker & Schaie, 1969; Bramel et al., 1968), or the violator furnishing a plausible excuse (Mallick & McCandless, 1966). On the other hand, if the victim attacks an innocent bystander or an inanimate object, or observes an unrelated act of violence, equity is not restored and anger is not reduced. These last examples show that anger need not be "discharged" through taking violent action or by imagining it, but rather that restoring equity is crucial. Violations of social norms evoke anger, and restitution reduces it.

3.16. Fear

As with perhaps all other emotions, fears can be learned or prepotent. The latter include fears of falling, darkness, animals, and strangers. Phobias often develop to these prepotent elicitors rather than to, say, inanimate objects. Because of their adaptive importance, fears are easy to acquire, as by contagion, and slow to be extinguished. Fear can lead to various overt behaviors, such as calling for help, freezing, flight, and defensive aggression (see Kalin, 1993).

---

[4] Hokanson found that women tended to make a friendly, submissive response to mistreatment. They may have felt angry but were afraid to retaliate; even men who retaliated against a high-status tormenter failed to experience a fall in blood pressure, perhaps because they feared the consequences of their defiance. In further support of the notion that both sexes experience both anger and fear in these nasty situations, Hokanson conditioned each sex to show the reaction typical of the other. Women were rewarded for retaliating, and men for friendliness, resulting in reversal of the typical patterns.

4. DIMENSIONS OF THE EMOTIONS MODEL

A few additional properties of this model of emotions will be described at this point.

4.1. Vicarious Emotion

The word "fear" also includes dread of any sort of unpleasant outcome. Thus, we may fear being caught in traffic, or fear a boring social occasion. In these cases we are experiencing the dreaded event vicariously, in our imagination. Likewise, we can anticipate future pleasures, regret past unpleasantness, and reminisce fondly about past joys. These mental rehearsals probably serve to help us to plan our future actions (cf. Damasio, 1994). We play back these scenarios mentally, and thus anticipate the affective consequences of our contemplated action. By extension, we can empathize with others and determine what help they need. We also experience emotions vicariously at the theatre, while dreaming, and while under the influence of hypnosis or psychoactive drugs. When experiencing a vicarious emotion, subjects often flex the corresponding muscles for the imagined movement subtly; also, the visceral and expressional features of the affect may occur (e.g., Fridlund, 1994, p. 157).

Vicarious emotion seems to play a major role in parental behavior in primates, so that it may be inappropriate to postulate the existence of a specific parental affect other than loneliness. In primates and other highly parental mammals, parental behavior seems to entail a generalized capacity for empathy with the offspring. This is indicated by the rather undifferentiated emotional expressions of primate infants, who do little more than signal distress or satisfaction (but see Wolff, 1969 on the distinctiveness of infants' cries). They leave it to the parent to interpret the situation appropriately. By contrast, other mammals such as puppies exhibit a well developed repertoire of discrete distress signals, e.g., cold, hungry, in pain, wet. The mother responds with corresponding stereotypic caring behaviors.

4.2. Valence

Each emotion has a range of valences. These may be exclusively negative, as in anger; exclusively positive, as in humor-appreciation (although some would argue that bad jokes are painful); or both positive and negative, as in interest/boredom. The range of intensities of a given emotion, which includes its threshold and hence its priority, should make adaptive sense.

4.3. Satiety and Deprivation Effects

There are probably satiety and deprivation effects for all emotional modalities, so that the individual neither neglects nor obsesses over any one of them. Pugh (1977) explained that there should also be a tendency for each emotion to drift back toward the middle range of the scale, so that the individual does not become stuck at either extreme and hence become either complacent or despondent. Lewin et al. (1944) described a similar idea: level of aspiration and level of attainment. The function of this mechanism seems to be to maintain the subject's effort by appropriate goal setting.

4.4. Interactions between Emotional Modalities

There also are lawlike interactions between particular emotions, such as the inhibition of interest by fear, or the enhancement of sexual arousal by love. And of course we may experience blends of affects as we contemplate or execute a course of action (see McDougall, 1923 on blends). It is well known to ethologists that blends of emotional expressions can occur in humans and animals (e.g., Lorenz, 1953, cited in Eibl-Eibesfeldt, 1989).

4.5. Ontogenesis

This model pertains to the adult emotions. Major developmental changes occur in the types and thresholds for various emotions. For example, infants seem to possess a sucking drive and desire for rhythmic vestibular stimulation such as occurs when being carried (Pugh, 1977). Infants' pleasure from human contact is very pronounced. Pride/shame emerges between 2 and 3 years of age (reviewed by Weisfeld, in press). The emotional life of adolescents features maturation of the sex and pair bonding emotions, and by intensification of pride and shame, reflecting the emergence of competitiveness for mates (Weisfeld & Billings, 1988). Senescence brings adaptive adjustments in the thresholds for various emotions.

4.6. Sex Differences

Then too, sex differences doubtlessly exist for many emotions, such as the esthetic emotions (Hutt, 1972; Velle, 1992), responsiveness to infants (Goldberg et al., 1982), and dominance aggression (Ellis, 1986). These may be more substantial and more consequential than sex differences in cognition.

4.7. Individual and Population Differences

Similarly, individual and population differences may exist for the threshold for various emotions; temperamental traits tend to have appreciable heritabilities (Plomin, 1990). These personality differences, again, are probably more significant than cognitive differences. One can conceive of different individuals being specialists at particular emotions, such as fear, anger, curiosity, and the esthetic emotions, and thus playing various roles in a society. The specialist would provide his or her emotional sensitivity on behalf of other group members, who would then reciprocate.

5. POTENTIAL UTILITY OF THE MODEL

Many of the distinctions in this list of emotions are debatable. Should air hunger and elimination feelings be included? They certainly reflect tissue needs and are not merely reflexes. Is parental aggression prompted by anger or by vicarious experience of fear? The important thing is not these distinctions, but rather the ideal of inclusiveness of the human ethogram. Once we can agree on the set of observable universal behaviors, we will have a working model of the main units of human behavior, a framework for studying developmental change, sex differences, individual differences, and pathological variation.

Such an inventory that stresses emotional well-being might be useful for evaluating the well-being of individuals or populations; it might be superior to monolithic measures such as income, gross national product, happiness, health, longevity, or number of offspring.[5] Lewis and Michalson (1983) are particularly critical of cognitive measures such as IQ as predictors of success and happiness. They point out that most measures of emotional well-being are very general. However, it is specific sources of unhappiness such as loneliness, low social power, anxiety, and anger that are stressful. Each of these can increase adrenal gland activity and hence depress the immune system and lower resistance to disease. Personality inventories that assessed each of the emotions directly might offer greater precision than global measures for predicting morbidity and suggesting preventive and remedial measures. For example, depression might be broken down into loneliness and shame types. The latter might be better treated by serotonin reuptake inhibitors.

I would suggest that this emotions model or a similar one with an ethological basis could be used not only as a framework for organizing research on human behavior, but also as a heuristic device for identifying potential areas for future research. To illustrate the breadth of the emotions model, I will try to fit some examples of recent sociobiological research into it.

6. THE EMOTIONS MODEL AND SOME EXAMPLES OF RECENT RESEARCH

6.1. Jealousy

The topic of sexual jealousy has received attention from sociobiologists. Jealousy may be a blend of different basic emotions, probably anger, shame, fear, and loneliness. These emotionals may be evoked by a single prepotent stimulus, the perception that one is losing one's mate. Presumably, with the evolution of pair bonding, selection strongly favored taking various defensive actions (hence the multiplicity of emotions) when in danger of losing one's mate, and so this connection between that perception and these pre-existing affects arose.

6.2. Mate Choice

Other research by sociobiologists fits into the emotions model. The criteria for mate choice change during speciation, but sexual motivation probably remains rather stable. Some evolutionary psychologists are studing the criteria, or releasers, of sexual attraction. Recognition of them as releasers, or prepotent values, may offer protection against suggesting the existence of improbable adaptations such as wealth detectors. It is more likely that, insofar as husbands are chosen for wealth, this may occur on a cultural basis or because the dominance displays of wealthy men attract women (Weisfeld et al., 1992).

---

[5] Reproductive success measured where there is effective contraception is, of course, of limited value in assessing the evolutionary utility of a given human behavior. Under these modern conditions it may be better to measure mate value, i.e., sexual attractiveness, as a proxy for reproductive success (e.g., Dong et al., 1996; Weisfeld et al., 1992). Sexual attractiveness, of course, refers to an emotion and as such is a proximate cause of ultimate biological success.

6.3. Reciprocal Altruism

Trivers' description of human reciprocal altruism certainly gives prominence to emotions. This behavioral system may have evolved from dominance hierarchization, an ethological concept (Weisfeld, 1980). There seems to be an opportunity to analyze other social psychological phenomena in adaptive terms; usually they are explained only in terms of very narrow, specific mini-theories. Roes (1993) has offered an evolutionary explanation of reactance theory, for example.

6.4. Modules of Behavior

Some of the sociobiologists have written about adaptive "modules" of behavior. These modules, or algorithms, are unlikely to have evolved from scratch as free-standing stimulus-response homunculi or neural networks. Rather, they probably represent newly evolved perceptual, or cognitive, sensitivities that prompt pre-existing emotional reactions.[6]

Just as new anatomical adaptations usually evolve from pre-existing structures whose functions have become obsolete, new behavioral adaptations probably represent re-tooling or elaboration of old ones. The number of emotions and their corresponding behavioral tendencies is limited, as argued above, but these can be triggered by new releasers. By considering a limited number of ancient emotions instead of a very large number of modules, we can construct a more parsimonious model of human behavior that also possesses phylogenetic, proximate, and ontogenetic dimensions.

Some examples from mammalian parental behavior may be useful here. Although human parental behavior seems to rely heavy on empathy and our general capacity for flexible behavior, parental behavior in nonprimate mammals seems to consist mainly of an array of specific responses to particular releasers. For example, licking the birth membranes and perineal region of newborns seems to depend on salt deprivation in the mother, so evolution has recruited her need for salt in order to motivate her to perform these parental behaviors. Similarly, lionesses tear out the fur around their nipples before giving birth, presumably because it becomes itchy and not because they anticipate their cubs' need for access. Rat mothers build nests more vigorously in cold weather, mouse mothers defend young they have suckled, and so forth. Nature seems to be very pragmatic in adopting existing motivational tendencies in the service of new goals.

7. THE EMOTIONS MODEL AND POSSIBLE EXAMPLES OF FUTURE RESEARCH

7.1. Infantile Releasers

Research based on the ethological tradition also can be placed within the emotions model framework, suggesting new avenues of research. There has been a great deal of research on infantile releasers of parental behavior—the infant schema, infant vocalizations,

---

[6] The term "cognitive" is potentially misleading here if it means "not affective," or "not emotional." In fact, these reactions are eminently emotional: appreciation of the beauty of face or figure, or of some chemical moiety that proves to be nutritious; or aversion to the sound of a distressed baby's cry or the smell of human waste.

and so forth. This research might be extended developmentally to see what particular caregiving behaviors are elicited by what filial characteristics, in that caregiving changes qualitatively as well as quantitatively as the child ages. What are the effects of prolonged separation on parental and filial behavior? How do facial features connoting, for example, infantilism or dominance alter our perceptions of children and adults? For example, feminine features cause girls to be perceived as feminine in behavior (Jackson, 1992, p. 66); is this stereotype veridical?

7.2. Reconciliation

In conflict resolution, the role of conciliatory gestures is now being studied as de Waal (1989) has done for chimpanzees. Do these work by restoring the perception of equity? Sackin and Thelen (1984) have demonstrated that reconciliation between children was usually followed by resumption of peaceful interaction.

7.3. Analyzing Particular Emotions

Other practical issues can be addressed by focusing on particular emotions. How can fatigue be reduced while driving? How much does ambient noise contribute to muscular tension and hence to fatigue? What are the emotional advantages and disadvantages of crowds? Are certain types of clothing perceived as dominance signals and hence more unsettling to observers? Which emotions inhibit which others, and which are facilitative? Sequential analysis of naturalistic behavior is called for to address this question. In addition, many important findings emerging in the neural sciences and endocrinology on various clinical conditions, such as pain, panic, and obsessive-compulsive disorder, need to be incorporated into our understanding of normal emotions. At the same time, the naturalistic perspective can be useful in learning about the context and possible evolved basis of such conditions.

7.4. Studying Emotional Expressions

How do emotional expressions change immediate behavior and not just judgments on questionnaires? Do some people exhibit a consistent background expression, such as anxiety or anger? If so, does this alter how they are treated? There may be quite a bit of research on such topics, but seldom are the results interpreted in evolutionary terms; see, for example, the *Journal of Nonverbal Behavior*, and *Motivation and Emotion*. Expressions are, after all, good examples of human fixed (or modal) action patterns. There is much work to be done on the evolution of particular expressional forms, including vocalizations and bodily postures and movements. Why do we raise our eyebrows in fear? Might this be antithetical, sensu Darwin, to the lowered brows of anger? Alternatively, is this movement a vestige of retraction of the ears in alertness (Fridlund, 1994)? Do the tears, vocalization, and grimace of crying derive from the appearance of a baby at birth, which had to evoke maternal care aboriginally, as Roes (1989) has suggested? These issues might be addressed in part by developing and applying versions of Ekman and Friesen's Facial Action Coding System to other primates, as suggested by Fridlund (1994). In addition, the effect of emotional expressions on the observer's behavior has seldom been studied, either naturalistically (e.g., Zivin, 1977) or semi-naturalstically (e.g., Kahn & Weisfeld, 1993). This sort of observational research reduces the ambiguities introduced by language and various experimental biases. Another technique is to use moving (video) images rather than still photos of posed expressions. Videotaped images sometimes yield higher recognition scores than photos (Shimoda, Argyle & Ricci Bitti, 1978).

7.5. Social Bonds

How are social bonds formed? Grammer and Eibl-Eibesfeldt (in press) have made some interesting discoveries on the role of humor in pair bond formation. Since bonds must be mutual to be effective, how does one partner enhance the likelihood of reciprocity? Does the pair bond weaken with the advent of children (i.e., monotropy) or just with the passage of time (cf. Fisher, 1992)? Russell and Wells' (unpublished) data on British couples suggest that both factors operate. How is mother-infant bonding affected by oxytocin, which is released during parturition into the mother's circulation and also transmitted through the milk during nursing (Becker el al., 1992)? What is the function of the release of oxytocin during sex? Also, observational research is very important for studying social behaviors such as marrital interaction, but has been neglected (but see Gottman, 1991; Schell & C. C. Weisfeld, 1994).

8. CONCLUSION

Now that mainstream social science is beginning to recognize "neo-Darwinism" as at least an acceptable fad, perhaps we can gain more respect for ultimate explanations of human behaviors. But we should remember Tinbergen's four questions and not neglect developmental and proximate factors. Observational research using ethologically valid concepts is greatly needed to address questions on all four levels. At the same time, research in neuroscience, behavioral genetics, and developmental psychobiology is offering many clues for understanding emotions. Methods combining physiological and self-report measures with observational research are very promising in the biomedical fields. Combinations of these methods are being employed by human ethologists, as evidenced by some of the presentations at the congress of the International Society for Human Ethology in Vienna in 1996.

In conclusion, the emotions model, with its neural, visceral, affective, situational, behavioral, and expressional facets, can offer a functional, comparative framework for viewing behavior as a whole and integrating many disciplinary approaches. If we can succeed in constructing a detailed, generally accepted model of the human emotions, we will perhaps also gain a greater acceptance of our human nature. A famous scientist made this same point. He first explained that approach and avoidance tendencies evolved like tropisms to draw us to food and other resources and to repel us from dangers:

> "...only those individuals could continue to exist who detested certain highly pernicious influence[s] with the full intensity of their nerve power and tried to prevent them, and who also sought out, with the same nerve power, other influences necessary for the preservation of themselves or of the species. Thus we understand how the whole intensity and power of our emotional life evolved—joy and pain, hate and love, desire and fear, delight and despair."

He concluded on a hortatory note:

> "Just as with our bodily diseases, we cannot get rid of the entire spectrum of our passions, but we can learn to understand and bear them."

That scientist was Ludwig Boltzmann (*Principles of Mechanics*, 1900).

REFERENCES

Alexander, R. D. (1986). Ostracism and indirect reciprocity: the reproductive significance of humor. *Ethology and Sociobiology*, 7, 253–270.
Apicella, P., Ljungberg, T. Scarnati, E., & Schultz, W. (1991). Responses to Reward in Monkey Doral and Ventral Striatum. *Experimental Brain Research*, 85, 491–500.
Baker, J. W., & Schaie, K. W. (1969). Effects of aggression "alone" or with "another" on physiological and psychological arousal. *Journal of Personality and Social Psychology*, 12, 80–96.
Barber, N. (1991). Play and energy regulation in mammals. *Quarterly Review of Biology*, 66, 129–147.
Becker, J.B., Breedlove, S. M., & Crews, D. (1992). *Behavioral Endocrinology*. Cambridge, MA: MIT Press.
Blurton Jones, N. (1972). Categories of child-child interaction. In N. Blurton Jones (Ed.), *Ethological Studies of Child Behaviour* (pp. 97–127). New York: Cambridge University Press.
Boulton, M. J., & Smith, P. K. (1992). The social nature of playfighting and play chasing: mechanisms and strategies underlying co-operation and compromise. In J. H. Barkow, L. Cosmides & J. Tooby (Eds.), *The adapted mind*. New York: Oxford University Press.
Bramel, D., Taub, B., & Blum, B. (1968). An observer's reaction to the suffering of his enemy. *Journal of Personality and Social Psychology*, 8, 384–392.
Buck, R. (1988). *Human Motivation and Emotion*, second edition. New York: Wiley.
Cattell, R. B. (1957). *Personality and Motivation: Structure and measurement*. New York: Harcourt Brace Jovanovich.
Charlesworth, W. R. (1969). The role of surprise in cognitive development. In D. Elkind & J. Flavell (Eds.), *Studies in Cognitive Development: Essays in honor of Jean Piaget* (pp. 257–314). New York: Oxford University Press.
Damasio, A. R. (1994). *Descartes' Error: Emotion, reason, and the human brain*. New York: Grosset/Putnam.
Darwin, C. (1872). *The Expression of the Emotions in Man and Animals*. Reprinted—New York: Philosophical Library, 1955; Chicago: University of Chicago Press, 1965.
Davis, R. C. (1934). The specificity of facial expressions. *Journal of Genetic Psychology*, 10, 42–58.
de Waal, F. B. M. (1989). *Peacemaking among Primates*. Cambridge, MA: Harvard University Press.
Dissanayake, E. (1988). *What Is Art For?* Seattle: University of Washington Press.
Dong, Q., Weisfeld, G., Boardway, R., & Shen, J. (1996). Correlates of social status among Chinese adolescents. *Journal of Cross-Cultural Psychology*, 27, 476–493.
Dunbar, R. I. M. (1993). Coevolution of neocortical size, group size, and language in humans. *Behavioral and Brain Sciences*, 16, 681–735.
Eibl-Eibesfeldt, I. (1973). The expressive behavior of the deaf-and-blind-born. In M. von Cranach, & I. Vine (Eds.), *Social Communication and Movement* (pp. 163–194). New York: Academic Press.
Eibl-Eibesfeldt, I. (1975). *Ethology: The biology of behavior*, 2nd ed. New York: Holt, Rinehart & Winston.
Eibl-Eibesfeldt, I. (1989). *Human Ethology*. New York: Aldine de Gruyter.
Ekman, P. (Ed.) (1973). *Darwin and Facial Expression*. New York: Academic Press.
Ekman, P., & Friesen, W. V. (1986). A new pan-cultural expression of emotion. *Motivation and Emotion*, 10, 159–168.
Ekman, P., & Friesen, W. V. (1971). Constants across culture in the face and emotion. *Journal of Personality and Social Psychology*, 17, 124–129.
Ellis, L. (1986). Evidence of neuroandrogenic etiology of sex roles from a combined analysis of human, nonhuman primate and nonprimate mammalian studies. *Personality and Individual Differences*, 7, 519–552.
Fisher, H. (1992). *Anatomy of Love: The mysteries of Mating, marriage, and why we stray*. New York: Norton.
Fridlund, A. J. (1994). *Human Facial Expreession: An evolutionary view*. New York: Academic Press.
Goldberg, S., Blumberg, S. L., & Kriger, A. (1982). Menarche and interest in infants: Biological and social influences. *Child Development*, 53, 1554–1560.
Goodall, J. (1986). *The Chimpanzees of Gombe: Behavior patterns*. Cambridge, MA: Harvard University Press.
Grammer, K., & Eibl-Eibesfeldt, I. (in press). The ritualization of laughter. In *Naturlichkeit der Sprache und der Kultur. Bochumer Beitrage zur Semiotik*. W. A. Koch (Ed.), Bochum: Brockmeyer, Vol. 18.
Goodall, J. (1986). *The Chimpanzees of Gombe: Patterns of behavior*. Cambridge, MA: Harvard University Press.
Gottman, J. M. (1991). Predicting the longitudinal course of marriages. *Journal of marital and Family Therapy*, 17, 3–7.
Hess, E. H. (1965). Attitude and pupil size. *Scientific American*, 212, 46–54.

Hokanson, J. E. (1970). Psychophysiological evaluation of the catharsis hypothesis. In E. I. Megargee & J. E. Hokanson (Eds.), *The Dynamics of Aggression* (pp. 74–86). New York: Harper & Row.
Hudson, J. A. (1995). *Sleep*. New York: Freeman.
Hutt, C. (1972). *Males and females*. London: Penguin.
Izard, C. E. (1972). *Patterns of Emotions: A new analysis of anxiety and depression*. New York: Academic Press.
Jackson, L. A. (1992). *Physical Appearance and Gender: Sociobiological and Sociocultural Perspectives*. Albany: State University of New York Press.
Jankowiak, W. R., & Fischer, E. F. (1992). A cross-cultural perspective on romantic love. *Ethos, 31*, 149–155.
Kahn, E. S., & Weisfeld, G. (1993, August). Facial expressions that influence subjects' postural erectness. Paper presented at the Human Behavior and Evolution Society convention, Binghamton, New York.
Kalin, N. H., (1993). The neurobiology of fear. *Scientific American, 268*, 54–60.
Kraemer, G. W. (1992). A psychobiological theory of attachment. *Behavioral and Brain Sciences, 15*, 493–541.
LeDoux, J. E. (1993). Emotional networks in the brain. In M. Lewis & J. M. Haviland (Eds.), *Handbook of Emotions* (pp. 109–118). New York: Guilford Press.
Lewin, K., Dembo, T., Festinger, L., & Sears, P. S. (1944). Level of aspiration. In J. McV. Hunt (Ed.), *Personality and the Behavior Disorders, Vol. 1*. New York: Ronald.
Lewis, M., & Michalson, L. (1983). *Children's emotions and moods*. New York: Plenum Press.
Lorenz, K. (1953). Die Entwicklung der vergleichenden Verhaltensforschung in den letzten 12 Jahren. *Stud. Generale, 9*, 36–58.
MacLean, P. D. (1993). Cerebral evolution of emotion. In M. Lewis & J. M. Haviland (Eds.), *Handbook of Emotions* (pp. 67–83). New York: Guilford.
Mallick, S. K., & McCandless, B. R. (1966). A study of catharsis of aggression. *Journal of Personality and Social Psychology, 4*, 591–596.
Masters, R. D., & McGuire, M. T. (1994). *The Neurotransmitter Revolution*. Carbondale: Southern Illinois University Press.
McDougall, W. (1923). *Outline of Psychology*. New York: Charles Scribner's Sons.
Mealey, L. (1995). The sociobiology of sociopathy: An integrated evolutionary model. *Behavioral and Brain Sciences, 18*, 523–599.
Money, J., & Ehrhardt, A. A. (1972). *Man and woman, boy and girl: The differentiation and dimorphism of gender identity from conception to maturity*. Baltimore: Johns Hopkins University Press.
Moyer, K. E. (1976). *The Psychobiology of Aggression*. Ne York: Harper and Row.
Oatley, K., & Jenkins, J. M. (1996). *Understanding Emotions*. Oxford: Blackwell.
Panksepp, J. (1994). Evolution constructed the potential for subjective experience within the neurodynamics of the neomammalian brain. In P. Ekman & R. J. Davidson (Eds.), *The Nature of Emotion: Fundmental questions* (pp. 396–399). Oxford: Oxford University Press.
Pastore, N. (1952). The role of arbitrariness in the frustration-aggression hypothesis. *Journal of Abnormal and Social Psychology, 47*, 727–731.
Plomin, R. (1990). *Nature and Nurture: An introduction to human behavioral genetics*. Pacific Grove, CA: Brooks/Cole.
Plutchik, R. (1980). *Emotion: A Psychoevolutionary Synthesis*. New York: Harper and Row.
Plutchik, R. (1994). *The Psychology and Biology of Emotion*. New York: Harper Collins.
Provine, R. R. (1989). Contagious yawning and infant imitation. *Bulletin of the Psychonomic Society, 27*, 125–127.
Pugh, G. E. (1977). *The Biological Origin of Human Values*. New York: Basic Books.
Roes, F. L. (1989). On the origin of crying and tears. *Human Ethology Newsletter, 5*, 5–6.
Roes, F. L. (1993). Reactance theory and Darwinism: an example of theoretical reduction. *Human Ethology Newsletter, 8*, 6–7.
Rosenzweig, M. R., Leiman, A. L., & Breedlove, S. M. (1996). *Biological Psychology*. Sundeland, MA: Sinauer.
Rozin, P., & Fallon, A. E. (1987). A perspective on disgust. *Psychological Review, 94*, 23–41.
Russell, J. A. (1991). Confusions about context in the judgment of facial expression: A reply to "The contempt expression and the relativity thesis." *Motivation and Emotion, 15*, 177–184.
Russell, J. A. (1994). Is there universal recognition of emotion from facial expression? *Psychological Bulletin, 115*, 102–141.
Schanberg, S. M., & Field, T. M. (1987). Sensory deprivation, stress and supplemental stimulation in the rat pup and preterm human neonate. *Child Development, 58*, 1431–1447.
Schell, N. J., & Weisfeld, C. C. (1994, August). Marital power dynamics: A multimeasure approach. Paper presented at the conference of the International Society for Human Ethology, Toronto.
Scott, J. P. (1958). *Animal Behavior*. Chicago: University of Chicago Press.

Shimoda, K., Argyle, M., & Ricci Bitti, P. (1978). The intercultural recognition of emotional expressions by three national racial groups: English, Italian and Japanese. *European Journal of Social Psychology, 8*, 169–179.

Sogon, S., & Masutani, M. (1989). Identificaion of emotion from body movement. *Psychological Reports, 65*, 35–46.

Stanley, H. M. (1898). Remarks on ticklng and laughing. *American Journal of Sociology, 9*, 235–240.

Steiner, J. E. (1979). Human facial expressions in response to taste and smell stimulation. In H. Reese & L. P. Lipsitt (Eds.), *Advances in Child Development and Behavior* (pp. 257–293). New York: Academic.

Tinbergen, N. (1951). *The Study of Instinct*. New York: Oxford University Press.

Van Bezooijen, R., Van Otto, S. A., & Heenan, T. A. (1983). Recognition of vocal dimensions of emotion: A three-nation study to identify universal characteristics. *Journal of Cross-Cultural Psychology, 14*, 387–406.

Van Hooff (1973). A structural analysis of the social behaivour of a semi-captive group of chimpanzees. In M. von Cranach & I. Vine (Eds.), *Social Communication and Movement*. New York: Academic.

Velle, W. (1992). Sex differences in sensory functions. In J. M. G. van der Dennen (Ed.), *The nature of the sexes: The sociobiology of sex differences and the 'battle of the sexes'* (pp. 29–54). Groningen: Origin Press.

Wallbott, H. G., & Scherer, K. (1986). How universal and specific is emotional experience? Evidence from 27 countries in five continents. *Social Science Information, 25*, 763–795.

Weisfeld, G. E. (1972). Violations of social norms as inducers of aggression. *International Journal of Group Tensions, 2*, 53–70.

Weisfeld, G. E. (1980). Social dominance and human motivation. In D. R. Omark, F. F. Strayer & D. G. Freedman (Eds.), *Dominance Relations: An ethological view of human conflict and social interaction* (pp. 273–286). New York: Garland.

Weisfeld, G. E. (1993). The adaptive value of humor and laughter. *Ethology and Sociobiology, 14*, 141–169.

Weisfeld, G. E. (in press). Discrete emotions theory with specific reference to pride and shame. In N. L. Segal, G. E. Weisfeld & C. C. Weisfeld (Eds.), *Genetic, Ethological and Evolutionary Perspectives on Human Development: Essays in honor of Dr. Daniel G. Freedman*. Washington, DC: American Psychological Association.

Weisfeld, G.E., & Billings, R. L. (1988). Observations on adolescence. In K. MacDonald (Ed.), *Sociobiological Perspectives in Human Development*, pp. 207–233. New York: Springer-Verlag.

Weisfeld, G. E., Russell, R. J. H., Weisfeld, C. C., & Wells, P. A. (1992). Correlates of satisfaction in British marriage. *Ethology and Sociobiology, 13*, 125–145.

Wolff, P. H. (1969). The natural history of crying and other vocalizations in early infancy. In B. M. Foss (Ed.), *Determinants of Infant Behavior* (Vol. 4). London: Methuen.

Zivin, G. (1977). Facial gestures predict preschoolers' encounter outcomes. *Social Science Information, 16*, 715–730.

PLAY FIGHTING AND REAL FIGHTING

Perspectives on Their Relationship

Peter K. Smith

Department of Psychology
Goldsmiths College
University of London
New Cross, London SE14 6NW, England

ABSTRACT

Rough-and-tumble play (R&T) is a distinctive form of behaviour, prominent in children. It has been studied by a variety of methods which complement each other in interesting ways.

Although superficially similar to real fighting, play fighting is distinct from it, and there are recognised cues which can be used in telling these two behaviours apart. These have been discerned by observational studies, and children too can verbalise many of these cues.

Play fighting is much more frequent than real fighting, in playgrounds. A proportion of play fighting can, however, turn into real fighting. The most usual reason, at least in middle childhood, appears to relate to 'honest mistakes' or accidental injury. However in some children, and more frequently by early adolescence, R&T may be used more deliberately as a social tool, consistent with a 'cheating' hypothesis.

Teachers make somewhat biased judgments about the relative frequency of play fighting and real fighting, and how often play fighting becomes real; possibly basing their perceptions on the small number of more aggressive children they may come into contact with in a disciplinary way.

Views on the functions of R&T may need modification. Social bonding and social skills may be incidental benefits of R&T, but fail to explain its distinctive features or the sex differences. Practice in fighting/hunting skills is a functional hypothesis consistent with design features and sex differences, but lacks direct support. The age changes and existence of 'cheating' suggest that, at least for some children and many adolescents, R&T can be used as a social tool in establishing or maintaining dominance in peer groups. The technique of interviewing participants in play fighting appears to have promise for further work in this area.

New Aspects of Human Ethology, edited by Schmitt *et al.*
Plenum Press, New York, 1997

1. INTRODUCTION

Rough-and-tumble play provided one of the paradigmatic examples of the application of ethological methods, back in the 1970's. Since then, a modest number of researchers have developed our knowledge of this kind of activity, using a variety of methods, and addressing some quite fundamental questions about age changes, sex differences, nature and function of behaviour. In this chapter I will review work on this topic, mentioning particularly the interest in comparing results from different informants and different methods of investigation.

Briefly, rough-and-tumble play (or R&T for short) refers to a cluster of behaviours whose core is rough but playful wrestling and tumbling on the ground; and whose general characteristic is that the behaviours seem to be agonistic but in a non-serious, playful context. The varieties of R&T, and the detailed differences between rough-and-tumble play and real fighting, will be discussed later.

2. A BRIEF HISTORY OF RESEARCH ON R&T

In his pioneering work on human play, Groos (1901) described many kinds of rough-and-tumble play. However, R&T was virtually an ignored topic from then until the late 1960's. There was, of course, a flowering of observational research on children in the 1920s and 1930s, especially in North America; but this research had a strong practical orientation, and lacked the cross-species perspective and evolutionary orientation present in Groos' work. Mainly based in the new Institutes of Child Study and Child Welfare, these studies (reviewed at the end of this period by Arrington, 1943) were designed to give basic knowledge about the nature and abilities of preschool children at different age levels, for purposes of designing good preschool facilities from the point of health and education, and for alerting teachers to those children with developmental delays. Toy preferences, friendships, and aggression were all covered, but R&T was hardly mentioned (see Smith & Connolly, 1972; Fassnacht, 1982).

The 1940s to 1960s covered what to many human ethologists seems a rather arid period in developmental research on social behaviour, in which artificial experimental paradigms predominated. Human ethology in it's more modern form started up in the late 1960s, guided by then influential work in non-human and primate ethology. Indeed, it was in a volume entitled *Primate Ethology* (1967) that we find the first detailed observational/ethological study of R&T, by Nick Blurton Jones (who had previously done his doctoral thesis on the great tit, under Tinbergen's supervision).

More detailed descriptive work was published by Owen Aldis in his book *Play Fighting* (1975). Since then, the topic has been pursued by several investigators, including Tony Pellegrini, Doug Fry, Kevin MacDonald, Michael Boulton, Sean Neill, Annie Humphreys, Angela Costabile, and myself. In addition, work by play theorists such as Robert Fagen, and animal ethologists such as Marc Bekoff and many others, have fed into our work on human R&T. We have not solved all the issues and problems concerning R&T, but the problem landscape is well delineated.

3. METHODS IN STUDYING R&T

Natural observation has, rightly, been the prime mover in studying R&T. The descriptive work of Blurton Jones and Aldis provided, in the first instance, the knowledge that here was a definite kind of behaviour pattern in children's play, which merited study. Traditional ethological methods were used by Blurton Jones (1967) and later by Smith (1973), Fry (1987) and Boulton (1991a,b) to show that certain behaviours clustered together to form R&T, that there are different kinds or clusters of R&T, and that there are certain characteristics associated with R&T which make it distinct from real fighting.

Direct observation can be usefully supplemented by film or video recordings (as done by Aldis, amongst others), since R&T is typically fast moving, and repeated playback can be very useful for noticing and recording details of interactions accurately. Long distance filming can be combined with radio microphone recordings to get details of concurrent verbal interactions (as done by Debra Pepler in Toronto).

Video recordings of R&T have also been used in another way; to play episodes of R&T directly to children (or adults) and ask them to interpret what is going on. Used first by Smith and Lewis (1985), this method has also been used by Costabile, Boulton and Pellegrini. It is especially useful in finding out how children (or adults) distinguish between R&T and real fighting—what cues they report using, in watching actual recorded occurences.

Yet another technique has been to interview children or adults (such as teachers, Schäfer & Smith, 1996) directly, either using questionnaires (Smith et al., 1992), or interviews (Boulton, 1992a). Clearly, people's answers may not be accurate in terms of what we observe, but this method should inform us about what they think and perceive. This has it's own intrinsic interest, and also interest in terms of discrepancies between beliefs and behaviour. In the case of older children, especially adolescents, we may get uniquely useful insights, while bearing in mind the possible distortions due to selective perception and memory, limited insight to motivation, and social desirability in responses, that bear on verbal report data (Boulton, 1992a).

Mention should also be made of the more ethnographic reports of playground life, such as that of Sluckin (1981). These have used observation, but also placed considerable weight on the conversations of children and, sometimes, short interviews at the time or soon afterwards. While there are several relevant studies in playgrounds, none have really focussed in detail on peer R&T however.

Together with Tony Pellegrini, Rebeccca Smees and Ersilia Menesini, we have tried going one or two steps further. First, we combined observational and questionnaire data with children's responses to video-recorded episodes; and in addition, we distinguished between responses of child participants in the R&T bout, and child nonparticipants, to the episodes on the videofilms. In this way, we tried to examine the unique insights which participants may have into episodes, while keeping the grounding in actual (video-recorded) behaviour which may help to minimise the biases in verbal reports not so firmly based in immediate behavioural experience. I will mention some findings from this study later.

I will also try to illustrate how the combination of methodologies is a strength, provided we take each data source on it's own terms.

4. BASIC DATA ABOUT R&T

The descriptive work on R&T has suggested several overlapping clusters of behaviour. Play fighting, the core concept, involves such behaviours as hitting, kicking, grappling, wrestling to the ground. Somewhat distinct from this is play chasing, a primarily non-contact version of R&T. Besides these very vigorous forms of R&T, we have more gentle forms, such as rolling and tickling, and the kinds of physical play (sometimes gentle, sometimes less so) between parents and children. In effect, we have two gradations here: between contact and non-contact play, and between rough and gentle play.

There is a third gradation—between playful and real fighting. R&T is usually playful, as evidenced by the playface and other cues. But some episodes are ambiguous in the motivation of participants, and some play fighting does end in real fighting. Especially by adolescence, the boundary does get rather blurred.

Through middle childhood, however, R&T is generally a friendly activity. Indeed, it is often done with friends (Smith & Lewis, 1985; Humphreys & Smith, 1987). It is often dyadic, but can occur in larger groups, from which dyads may often split off for short bouts before rejoining in a longer group episode.

Generally, across a variety of cultures (Western and non-Western), boys engage in more R&T than girls do; a difference which probably has a hormonal basis (Collaer & Hines, 1995).

While this is uncontentious, there are a number of areas where informants disagree, and others where as researchers we are still uncertain. I shall discuss five topics:

- the ways in which playful fighting and real fighting can be distinguished;
- the relative frequency of playful and real fighting;
- the extent to which play fights turn into real fights;
- why some play fights turn into real fights;
- what are the developmental and evolutionary functions of R&T?

5. THE WAYS IN WHICH PLAYFUL FIGHTING AND REAL FIGHTING CAN BE DISTINGUISHED

5.1. Playful and Real Fighting Can Appear Similar, and Are Sometimes Confused

Playful fighting superficially resembles real fighting, since many of the component behaviors, such as wrestling, grappling, hitting, kicking, and restraining are the same at least at a level of first-order behavioral description. Play fighting is in fact sometimes confused with real fighting by playground supervisors and teachers:

"Well, if you do see them, and think they're fighting you go and say, 'no fighting'; and they say, 'it's all right, we're only playing,' and I let them get on with it" (interview with school playground supervisor cited in Sluckin, 1981, p.41);

"It's the games boys want to play. For the most part this is fighting, violence, jumping on each other and 'killing' each other. There's so much fighting in the name of play .... The children will tell you they're only playing, but it can move into a real fight" (interview with junior school teacher cited in Blatchford, 1989, p.29).

Schafer and Smith (1996) interviewed a sample of 30 primary school teachers; they said that they had difficulties distinguishing play fighting from real fighting on about one-third of occasions. Similar results were obtained by Smees (1992) in a sample of 17 primary school teachers; 41% said they found it difficult to distinguish between them. In the past, some psychologists too have confounded the two activities; for example Ladd (1983, p.291) defined rough-and-tumble as "unorganised agonistic activity with others (e.g., fights or mock-fights, wrestling, pushing/shoving)."

5.2. Observational Studies of the Distinction between Playful and Real Fighting

A number of observational studies are consistent in showing that researchers can distinguish playful and real fighting, by a number of co-occurring features (Blurton Jones, 1967; Smith & Lewis, 1985; Humphreys & Smith, 1987; Fry, 1987; Pellegrini, 1988; Boulton, 1991b). A range of these is shown in Table 1.

These cues or features have been validated at a detailed, quantitative level in children in the U.K., U.S.A., and the Zapotec people of Mexico; while at a more qualitative level the distinction between playful and real fighting can be found reported in psychological and anthropological studies in a considerable number of human cultures. Indeed, the apparent cross-cultural ubiquity of play fighting suggests that it may be a cultural universal, whose development is canalised to a considerable degree. This would lead to the prediction that play fighting should be readily recognised across different national and cultural groups, using common cues or criteria.

Table 1. Criteria discriminating between R&T and real fighting in children, from observational studies (adapted from Smith, 1989)

Criterion	R&T	Real fighting
Circumstances leading up to an encounter	no conflict over resources	frequent conflict over resources
How an encounter is initiated by one child and responded to by another	one child invites another, who is free to refuse	one child frequently challenges another, who must respond or lose face
Facial and vocal expression of participants during encounter	play face, laugh or smile; or neutral expression	staring, frowning, red face, puckering up and crying
Number of participants involved	often two, but may involve several more in connected short episodes	seldom more than two children involved
Reaction of onlookers	little if any interest from non-participants	attracts attention from non-participants; sometimes a crowd of onlookers draw round
Self-handicapping	a stronger or older child will often self-handicap, not using maximum strength	self-handicapping normally absent
Restraint	participant will often show restraint in the force of a blow	restraint occurs to a lesser extent or is absent
Reversals	participants may take it in turns to be on top or underneath, or to chase or be chased	turn-taking not usually observed
Relationship between participants immdiately after encounter has ended	participants often stay together in another activity	participants often separate

5.3. Children Can Usually Distinguish Playful and Real Fighting, and Explain How

Some studies have obtained the perspectives of children directly. Do children themselves think they can distinguish playful and real fighting? Which cues do they actually use, and how does this vary with age? It is important to know more about this, in the light of evidence that some children may misinterpret cues of hostile or friendly intent (particularly sociometrically rejected children who are aggressive, Dodge & Frame, 1982) and that some rejected children engage in play fighting of a less cooperative nature (Pellegrini, 1988, 1994).

In the first study to report on this, Smith and Lewis (1985) prepared a videotape of 4 episodes of real fighting interspersed with 16 episodes of playful fighting (as judged by the researchers). They played this to 8 preschool children aged 4 years, and asked them whether each episode was playful or really fighting, and how they could tell. Six of the children showed considerable consistency in their judgments; some were unable to give

Table 2. Definitions of criteria used by children to distinguish between playful and real fighting (extended from Costabile et al., 1991)

Length of episode. Reference to the duration of the encounter, e.g. "if it was a fight it would have lasted longer." "it didn't last long enough to be a fight."

Stay together or separate. Reference to what the participants do or try to do at the end of the encounter, e.g. "it was a fight 'cos that boy was trying to get away," "it was only a mess about 'cos that boy stayed with him."

Facial expressions. Reference to the facial expression of the participants, e.g. "it were a mess about 'cos he were smiling," "you could tell it was a real fight because he had a right angry face."

Physical actions. Reference to the presence and/or absence of physical acts, and/or to the qualification of physical acts, e.g. "it was only a play fight 'cos he didn't hit him hard," "that was a real fight because he ran after him and kicked him in leg."

On the ground. Reference to whether or not the participants were upright or on the ground, e.g. "he was really fighting 'cos he got the other boy on the ground."

Crowd. Reference to the presence or absence of spectators, e.g. "it was real 'cos they all came round." "that were a toy fight as the other boys didn't watch them."

Actions of non-focal peers. Reference to the actions of other children, e.g. "it were real because his friend tried to break it up," "that were only a mess about 'cos another boy tried to join in'"

Inference about affect. Reference to the (presumed) affect of the participants, without reference to other criteria, e.g. "it was a real fight because they were both angry," "that was a toy fight 'cos he didn't mind him doing that."

Inference about action and/or intent. Reference to the (presumed) actions and/or intentions of the participants, e.g. "it was a play fight 'cos that probably didn't hurt him," "that were a true (fight) 'cos he tried his best to get him."

Stereotyped knowledge. Reference to what children think usually happens, e.g. "that were a mess about 'cos girls don't fight with boys," "it was really aggressive 'cos boys usually do that when they are mad."

Knowledge about focal child(ren). Reference to how the participants have been seen to behave in the past, e.g. "that were a toy fight 'cos I know them and they're best friends," "real fight because he is always picking on him, he's a bully."

Other. Any other response not covered by the above.

Can not give reason. Child is unable to express why they think an episode is playful or real fighting

---

Rules of game. Explicit mention of nature of game, or rules of fight, e.g. "it's only playing because he's taking him to prison."

---

reasons, but when they could, they usually referred to the physical characteristics of actions, or restraint, e.g. "they were hitting hard," or "he was only doing it gently."

Costabile, Smith, Matheson, Aston, Hunter and Boulton (1991) extended this technique to look at the judgments made by 8 and 11 year olds in two countries, Italy and England. They used videofilms made in both Italy and England, and played them to children in both Italy and England. The videofilm was played to children individually. After each episode, the child was asked whether the incident was playful fighting or real fighting; and how they could tell. The child's responses were content analysed, using the first 13 criteria shown in Table 2.

There was high consensus in describing most videotaped episodes as either playful, or real. Results were similar for Italian and English children, irrespective of whether the videotape of Italian or English children was used. The 11 year old children (with 87% accuracy) performed slightly but significantly better than the 8 year olds (84% accuracy), but both age groups were very good at the task. The most frequently cited cues were the physical characteristics of actions/restraint, and inference about affect or intent. Very similar results were obtained by Boulton (1993a) in another sample of English 8 and 11 year olds. Studies on adults by Boulton (1993b) and on teachers by Schäfer and Smith (1996) suggest that they have comparable accuracy to 11 year old children in judging these video episodes.

We have examined children's views further in an interview study with English and Italian 5, 8 and 11 year olds (Smith, Carvalho, Hunter & Costabile, 1992). When asked if they could tell play fighting and real fighting apart from each other, 70% of 5 year olds, and 90% of 8 and 11 year olds replied that they could. Many 5 year olds could give some ways of doing so, though 40% could not. Almost all the 8 and 11 year olds could give reasons, including those obtained from the earlier videotape study but with a higher frequency of cues such as facial expression, and verbalisation.

6. THE RELATIVE FREQUENCY OF PLAYFUL AND REAL FIGHTING

In school peer groups, R&T takes up some 5 to10% of free playground time (Humphreys & Smith, 1987; Fry, 1987; Boulton, 1992b). It is therefore much more common in school playgrounds than is real fighting. According to recent observational data, time spent in real fighting appears to be less than 1% (Humphreys & Smith, 1987; Pellegrini, 1989, 1994; Boulton 1992b). Thus at most 20% of fighting would be real. (Fry's 1987 data on Zapotec children in Mexico similarly found some 10 to 20% of fighting to be real).

Schäfer and Smith (1996) interviewed a sample of UK primary school teachers. When asked what proportion of all playful and real fighting was real, the mean answer was 37%, with a range of individual replies from 5% to 80%). Similar data from Smees (1992) found a mean of 31% and a range from 10% to 80%. The figure obtained from observational research is however at *maximum* 10–20% (Humphreys & Smith, 1987; Pellegrini, 1988, 1994; Schäfer & Smith, 1996). Perhaps, teachers are basing their judgments on the small proportion of highly aggressive children who are more likely to come to their attention. Generalising from these children, to children generally, would be a form of stereotyping. It seems that many teachers, who in England are seldom on outside playground duty, may rely on stereotyped judgments about playfighting.

7. THE EXTENT TO WHICH PLAY FIGHTS TURN INTO REAL FIGHTS

We do not have extensive estimates from observational work of the proportion of play fighting turning into real fighting; but for most studies of primary school children up to age 11 years, the proportion seems very small: a handful of instances in hundreds of observations, or around 1%. Humphreys and Smith (1987) found that only 3.7% of play fights resulted in injury, and this was often unintentional (the partner staying to help the injured child); only one-third, or 1.2%, were likely to be caused intentionally. The around 1% figure (actually, 0.6%) also applies to popular children in Pellegrini's (1988) study; although this rises to a quarter for rejected children (28% in Pellegrini, 1988, with 5 to 9 year olds; 26% in Pellegrini, 1994, with 13 year old boys). Thus the best guess so far would be a very small percentage, of the order of 1%, perhaps rising to around 25% in children with particular difficulties in peer relationships, or aggressive tendencies.

In their interviews with primary school teachers, Schäfer and Smith (1996) found that these teachers reported that around 29% of play fighting episodes turned into real fighting. They thought this change from play to real to be appreciably more likely in boys than in girls (39% for boys, 19% for girls). As with the data on frequency of real fights, this appears greater than is actually the case, unless one were routinely considering difficult or aggressive children.

Interviews with children suggest that most children recognise that play fights can turn into real fights. Smith et.al. (1992) asked 5, 8 and 11 year olds "can playfighting lead to a serious fight?"; about 80% replied 'yes.' Similarly when Boulton (1992a) asked this question of 13–16 year olds, 80% said "yes." Neither of these studies found a significant sex difference in this response.

7.1. Theoretical Perspectives on Why Play Fighting May Turn into Real Fighting

There are two main theoretical arguments as to why play fighting and real fighting are sometimes confused (Smith & Boulton, 1990). One corresponds to what Fagen (1981, p.338), in his review of animal play fighting, calls 'honest mistakes.' For example, a child may respond to a playful initiation as if it were hostile, because he or she does not interpret a play initiation or play face correctly. This could be an example of lack of appropriate social skills.

An alternative explanation for playfights which 'go wrong' corresponds to what Fagen (1981, p.337) calls 'cheating.' This suggests that some children, probably quite sophisticated in understanding and manipulating playground conventions, deliberately misuse the expectations in a playfight situation to hurt someone, or display social dominance.

There is some evidence for both of these possibilities. Regarding the 'honest mistakes' hypothesis, Pellegrini (1988, 1994) has found that the likelihood of play fighting turning to real fighting is much greater for children who are sociometrically rejected, than for those who are popular. It is a widely held hypothesis that rejected children may lack social skills (Asher & Coie, 1990), and there is some evidence for the specific link to play fighting discrimination. Pellegrini (1989) showed a videotape of 11 episodes to children aged 5 and 9 years; he reported that rejected children used a more restricted range of cues than did popular children, and were less successful at making the discrimination in agreement with adult judges. Boulton and Smith (submitted) similarly found that sociometri-

Table 3. Average number of errors made on videotape test to distinguish playful and real fighting, for children of differing sociometric status at 8 and 11 years of age (from Boulton & Smith, submitted)

Sociometric status	Popular	Average	Other	Rejected	Neglected
8 years	3.0	3.7	3.6	4.8	5.8
11 years	1.6	2.7	2.1	3.1	3.3

(age effect significant, $p<.01$; sociometric status effect significant, $p<.05$)

cally neglected and rejected children made more errors on the videotape test than did popular children (Table 3), in the sense of less consensus with majority verdicts.

There is some evidence for the second, 'cheating' hypothesis, especially for older children. Neill (1976) reported direct observations of play fighting in 12–13 year old boys. In some episodes, it was apparent that one child might be taking advantage of a playful bout to deliberately inflict hurt on a partner, before turning the bout back to 'play' again.

In our interview study (Smith et al., 1992), we asked children why playfighting might lead to a real fight. The most common reply (over 50% of respondents) was 'accidental injury,' or 'misinterpretation of accidental injury,' for example "if someone accidentally hurt the other and the other didn't take it as an accident and hit back." This is consistent with the 'honest mistakes' hypothesis. However there were also reasons given to do with 'hitting too hard,' and 'disgreement over play,' which could be more consistent with the 'cheating' hypothesis; and a minority of respondents (around 15% of 8 and 11 year olds) said they liked playfighting because they could show how strong or dominant they were.

In a similar study with 13–16 year olds, Boulton (1992a) found that accidental injury was a reason given by 24% of respondents, but 'hitting too hard' and similar reasons were given by 37%, suggesting that purely accidental injury may figure less strongly as a reason for play fights turning to real fights, in adolescence.

8. WHAT ABOUT INTERVIEWING THE PARTICIPANTS?

The studies reviewed have used either observations by experienced adult researchers, or interviews with children who are either questionned generally, or asked to comment on videotaped episodes which they did not participate in themselves. What they do not do is to use data based on interviews with the participants. Yet, participants may have insights into what it going on in a bout, which outside observers, whether adult or child, may not have.

Boulton (1992a) carried out detailed interviews with a small number of 15 year olds. They sometimes drew on their own experiences to illustrate a point, as in the following example (p. 322):

Craig: ...we're only playing but you can still tell who's the hardest.
MB: How can you tell?
Craig: Well, you try and get them down, or try and hold them down, or get them round the neck, things like that.
MB: So how can you tell who is the strongest?

Craig: It's the one who can do those things. Like, the other day, Peter came up to me from behind and grabbed me round [my] neck, so I tried to get him off and get him down. We were only messing around, but I still got him down.
MB: So what does that show?
Craig: It shows that I'm harder than him. I could probably win him in a fight.

In research I have done with Rebecca Smees, Tony Pellegrini (University of Georgia), and Ersilia Menesini (University of Florence), we used videotapes of 42 episodes of what may be playful or real fighting, shown to both participants and non-participants in the bouts. When shown to participants, it was shown both on the same day (short-term interview) and one week later (long-term interview). 44 children took part, all boys aged 5 to 8 years.

To categorise the 42 videotaped episodes, we looked for agreement between the researchers (all experienced in this area), and the participants. There was complete agreement between the researchers, and all participants, for two-thirds of the bouts, which is statistically significant. But there was some disagreement on 19 episodes. We looked to see how non-participant children rated these episodes. The result was clear cut; when there was disagreement between researchers and participants, non-participant childrens' ratings differed significantly from those of participants, but not significantly from those of the adult researchers. That is, we found a participant/non-participant difference, not an adult/child difference.

Why is there a participant/non-participant difference? We looked further at the reasons children gave for making their decisions—the cues they used, using our previous category system (Table 4). The participants made more use of cues such as 'quality of physical actions,' presumably having better knowledge of whether one child was really holding back from hitting hard, or not. Also, they made more use of cues to do with 'rules of the game,' for example "it's only playing because he's taking him to prison." Given the short excerpts available on the videotaped episodes, such insights into the pretend nature of the game are much more available to participants, and the contrast with non-participants is large; indeed, we had not separated this category in our previous work with non-participants, because of its low frequency of occurence with them.

Non-participants, by contrast, made more use of 'inference about affect or intent'; for example, "that was a play fight because it probably didn't hurt him." In effect, the non-participants more often fall back on general statements about emotion or intention, in place of giving more objective or specific reasons. We also found that the long-term (one week later) interviews with participants shifted them somewhat to the non-participants perspective, presumably as their memories of the specific episode faded.

The detailed interviews with participants are still being processed, but so far we have found rather little evidence for 'cheating' at this age (5 to 8 years). In cases where there is disagreement or uncertainty about the status of episodes, the participants stated that 'no-one is being hurt' 75% of the time, compared to non-participants who make this

Table 4. Some differences in cues used by participants (short-term and long-term) and non-participants, in distinguishing playful and real fighting

	Participant (short-term)	Participant (long-term)	Non-participant
Physical characteristics of actions	55%	48%	35%
Inference about affect, or intent	9%	18%	14%
Rules of game	18%	10%	7%

judgement only 45% of the time. Possibly, some episodes which appear of mixed motivation to an outsider, are seen as part of a more complex game by participants. We are examining our data in more detail to look at this.

The results emerging from this study suggest that participants may have unique insights into the nature and motivation of playfighting (and real fighting) bouts. This insight is best tapped shortly after the incident (we filmed in the morning, interviewed in the afternoon), but is still available to some extent after a one week period. Non-participants fall back on less useful cues for their judgments. Adult researchers can of course, by repeatedly replaying and comparing videotaped episodes, get more insight into the issue than most outsiders, and their judgment is probably more useful than that of teachers, who often spend little time in the playground, and who have a different set of concerns about play fighting.

9. WHAT ARE THE DEVELOPMENTAL AND EVOLUTIONARY FUNCTIONS OF R&T?

R&T is one form of vigorous physical activity, and as such may share the benefits for muscle strength and metabolic capacity which probably selected for such general forms of play in most mammalian species. However, this does not explain the more specialised forms of vigorous play which constitute R&T, nor its social nature.

Numerous benefits have been proposed for engaging in R&T, but the four main contenders have been:

- that is serves some social bonding or social cohesion function;
- that it serves as practice for certain social skills, such as role taking/social problem solving, or opportunity to encode and decode social signals;
- that it serves as practice for fighting and/or hunting skills;
- that it serves to establish or maintain dominance, or to give information relevant to dominance encounters.

9.1. Social Bonding Hypothesis

One hypothesised function of R&T is social bonding—forming close relationships with playmates. The main evidence in favour of this would be the consistent finding that R&T partners are often (more often than by chance) close friends (Smith & Lewis, 1985; Humphreys & Smith, 1987; Boulton, 1991a). However, there is no particular evidence to show that the R&T causes the friendship, any more than any other kind of association or playing together would tend to do. In fact, Boulton (1991a) found that many 8 and 11 year olds had the same partners in a range of playground activities as in R&T, and provisionally concluded that "choice of partners in r/t (and other activities) is a reflection of existing friendships/affiliations rather than providing a context in which they may be formed" (p.188). There are many ways of making friends. Children who are popular, do not exhibit more R&T than those of other sociometric status. Girls, who engage in less R&T, do not show less cohesive friendships; in fact, girls dyadic relationships may be more intense than those of boys (Lever, 1978).

We may rather suspect the inverse direction of causation, that the nature of R&T is such that it is desirable to play with friends. Firstly, if R&T confers benefits, it is best to give/share such benefits with friends (or kin, such as siblings or relatives). Second, R&T will be safer with friends; we have seen from the questionnaire studies how play with friends is less likely to lead to retaliation in the event of misunderstandings. The likelihood of "cheating," or exploiting R&T for dominance exhibition ends, is low (Axelrod & Hamilton, 1981), since the friendship would be jeopardised by acting in this way. Also, the likelihood of misperceptions is reduced by the familiarity friends have with each other, by virtue of the fact that they meet each other repeatedly, and have intimate knowledge of each other.

9.2. Social Skills Hypothesis

A second hypothesis has related R&T to (non-agonistic) social skills. Here, the argument might be supported by some design characteristics. R&T is characterized by reciprocal role taking; children engage in different roles and in many cases take turns at playing the same role; switching between superior and inferior roles or self-handicapping. These R&T bouts afford opportunities for players to view events from multiple perspectives. The ability to view events from different perspectives and use play strategies to maintain social interaction might be beneficial in other aspects of social interaction, such as perspective taking and social problem solving.

Pellegrini (1988, 1993, 1995a) has reported that the frequency with which boys engage in R&T does relate to their social problem solving ability. However it is not the best predictor; enactment of a varied repertoire of play behaviors is a more robust predictor of social problem solving in his results. Again, many social play activities give rise to different roles and turn taking, and no case is made for the priviliged position of R&T or its particular characteristics; nor for why there are sex differences.

A related hypothesis, which builds on the fact that the play face or similar signals designate R&T, supposes that R&T has a function in practicing encoding and decoding of emotional signals. Research by Parke and colleagues has suggested that the ability to encode and decode play signals originates in socially vigorous play between parents (primarily fathers) and their children (primarily sons), in infancy and throughout childhood (Carson, Burks & Parke, 1993; MacDonald & Parke, 1984; MacDonald, 1987). The amount of time spent in vigorous play bouts was positively related to preschool children's ability to decode emotional expressions, and children's expression of emotional states was related positively to bout length. Involvement in R&T with peers has also been related to primary school children's ability to decode play signals (Pellegrini, 1988).

It may be the case that parent–child play provides the ground work for children's ability to encode and decode emotions, with this ability later being practiced in physical activity play with peers. But again, we do not know cause and effect here. It may be that children less able to encode/decode emotions are less willing to engage in physical play. We do not have adequate longitudinal data to distinguish these two possibilities. It is also worth noting that not all the predicted correlations in these studies were significant, and that the Pellegrini (1988) results related only to the ability to distinguish R&T and real fighting, not a more general ability. Finally, there is no explanation of why girls should engage in less R&T, since emotional encoding/decoding should be just as important for them, and they are certainly no worse at it than boys.

9.3. Practice Fighting Hypothesis

A decade and a half ago, I reviewed functions of play in animals and humans (Smith, 1982). At that time, I argued for the third hypothesis as regards R&T: that it served as safe practice during childhood for what, in the environment of evolutionary adaptedness, would have been the expectation that we would need to fight and hunt as an adult. This hypothesis explains the form of R&T (similar to fighting, and hunting so far as play chasing is concerned), and it explains the characteristic sex difference (given that fighting and hunting are primarily male activities). The sex difference is important, as the first two hypotheses fail to predict it; it is not clear why males need more social cohesion than females, or why they need more non-agonistic social skills.

Since that time, however, more evidence has appeared concerning both mammalian and specifically human R&T, which leads to some revision of this position. The clearest test of the fighting/hunting practice hypothesis would be that individuals who engage in more R&T, will be better fighters or hunters. In fact, this evidence is virtually non-existent in animal studies. We lack longitudinal studies in humans to test this, but the frequency of R&T does not appear to relate strongly to sociometric status, which might be taken as a possible proxy measure for later aggressiveness. In effect, this hypothesis awaits further testing.

Even if this hypothesis is still a contender for explaining the origins of R&T, it appears insufficient to explain some age changes which we have discussed earlier. There appear to be solid grounds for invoking a dominance function for R&T, which may be instead of, or supplementary to, any function for R&T in training fighting/hunting skills (Pellegrini & Smith, submitted).

9.4. Dominance Hypothesis

Symons (1978) was critical of the hypothesis that R&T related to dominance, because of the self-handicapping and restraint which characterises much R&T; if a stronger child self-handicaps, Symons argued, how can this be functional from the point of view of dominance? However while this may be valid for normal R&T, it does not explain the blurring of R&T and real fighting which occurs in adolescence, and the findings that older children especially, to varying extents, maintain the appearance of play while in fact ensuring that they test their opponent's strength and, where appropriate, demonstrate that they are stronger.

The observational and interview findings suggest that R&T may be involved in dominance in various ways. First, it may provide a way of assessing the strength of others, so as to decide one's strategy vis-a-vis dominance competition—a form of 'ritualised aggression' as described in other animal species, which only leads to real fighting in certain circumstances. Second, it may be a way of establishing dominance, by using the play convention to lull an opponent into a false sense of security and then pinning them to the ground. Third, it could be a way of maintaining status in a friendly way, by subtly demonstrating one's superior strength in R&T with others in one's group or large peer network. Observations of play fighting in chimpanzees by Paquette (1994) supports the idea that young chimpanzees can use R&T to learn the strengths and weaknesses of others in planning challenges for dominance; Paquette suggests that R&T is more useful for this can real fights, because of the lesser chance of injury, and also the leser likelihood of intervention by a third party.

Some early evidence for this view came from Neill's (1976) observations of 12- 13-year-old boys. He reported a mix of playful and hostile episodes to be not uncommon, and observed some boys deliberately use R&T conventions to continue an interaction whose overriding intent was to assert dominance. Thus, "the attacker might start vigorously, causing distress, and then become more playful, often after pinning the other boy down" (p. 217).

Such observational findings are not typical of research with younger children (Smith, 1989); but, evidence for the association of dominance functions of R&T with age, came from work by Humphreys and Smith. They observed play partners in R&T at 7, 9 and 11 years, and also obtained peer nominations of dominance, and friendship. They then examined the choice of R&T partners in relation to the peer nomination characteristics. At all ages, children tended to choose friends as R&T partners; but relations to dominance changed with age. At 7 years, and largely at 9 years, R&T partners were random with respect to dominance. But by 11 years, children tended to choose R&T partners who were near themselves in dominance status. Furthermore, where there was an asymmetry, the person initiating the bout tended to choose a slightly less dominant partner (as seen by peers). This is consistent with R&T being used to establish or maintain dominance, at this older age only.

Boulton (1991a) examined in more detail the choice of play partners in a sample of 8 and 11 year old English children. He too found that partners tended to be similar in strength, although relations to dominance were not as clear as in the Humphreys and Smith study; sometimes children chose weaker partners, but sometimes weaker partners themselves initiated bouts. He concluded that his data were consistent with "r/t being used as a "social tool" to achieve certain social goals with specific peers on different occasions" (p.191).

More support for age changes comes from Fry (1987), in his observations of play in Zapotec children. He commented that "the clear distinction between play and aggression apparent among 3 to 8 year old Zapotec children may blur somewhat for Zapotec 12–16 year olds. At times, horseplay among teenagers appeared on the basis of facial expression to become somewhat serious, but then shifted back to obvious play once again" (p. 298).

An observational study by Pellegrini (1994) also supports this conclusion. Pellegrini made observations and sociometric measurements on 305 thirteen-year-old boys. He examined separately chasing forms of R&T, and rough forms of R&T which in his study often involved vigorous but disorganised ball play. Irrespective of sociometric status, he found that the frequency of rough R&T correlated with dominance status. The frequency of chasing R&T did not correlate with dominance, (except for rejected children). Furthermore (with the exception of popular children), children engaging in rough R&T tended to choose less dominant children as partners. This supports the result of Humphreys and Smith, and suggests that in adolescence many children are, with some success, using R&T to maintain dominance (though this may not make them particularly popular).

Pellegrini makes a somewhat stronger link to rejected sociometric status in this connection. He argues that rates of R&T drop when boys become adolescents and enter secondary school, compared to occurrence during primary school. When adolescents, compared to primary school boys, go on to the playground for recess, they typically exhibit less vigorous behavior and interact in mixed gender groups (Pellegrini, 1995a). Thus, in adolescence, most boys become interested in heterosexual relations and they do not frequently exhibit in social physical activity behaviors. Those boys who continue to engage in R&T, use R&T to establish and maintain dominance (Pellegrini, 1993, 1995b), but are rejected by most of their peers (Pellegrini, 1993). Specifically, these boys use R&T and physical aggression as a way in which to bully their peers (Pellegrini, 1995b).

The detailed interviews with children provide some support for a dominance function of R&T, as reviewed earlier. A majority of children in the studies by both Smith et al. (1992) and Boulton (1992a) said that you *could* tell who is stronger in a play fight; and the detailed interviews by Boulton, such as that cited earlier, suggest that this information can be used in dominance challenges. Here is another example from Boulton's work (p. 324):

> MB: If two people were play fighting to see who was the strongest, what would happen if one of them did not give in?
>
> Peter: That happened last week. Craig and Imran were messing about, and it went on for ages. Imran got Craig down but he wouldn't give in, so Imran got Craig round the neck and got it tighter and tighter until Craig couldn't talk. He wouldn't let go. He was right mad.
>
> MB: Is Imran always like that?
>
> Peter: No way, I think he wanted to show Craig that he was [the] hardest.
>
> MB: Was it a real fight?
>
> Peter: Sort of. They know each other and don't fight, but it came close that time. It started off as a toy fight, but Craig wouldn't give in.

The ethnographic work provides some support for this view, as well.

Sluckin's (1981) study of 5- to 9-year-olds falls more into the ethnographic category, since he tried to record the main features of behavior sequences in context, rather than the occurrence of specific categories, and also interviewed the children extensively. Besides the "honest mistakes" in R&T that he observed, described above, it is clear from his account that some children used, or tried to use, R&T play conventions in deceitful, intimidating or manipulative ways. For example:

> Ivan grabs Ashok (9:6) round the neck and says: we're only playing, aren't we Ashok? Ashok refuses to accept this verbal definition of the situation and defensively shouts: No. Ivan goes off" (p. 40).

Sluckin argued that many 7-, 8- and 9-year-olds seemed adept at moving between R&T and actual aggression as a way of trying to control others--for instance, by redefining a situation to their advantage as Tim does in the following extract:

> Nick (9:9) tells Tim who is holding Bob: You'd better leave him alone. Nick grabs Tim who lets Bob go. Tim: as soon as you've let go of me, I'll hurt him some more. Nick grabs Tim again. Tim: It's only a game, it's only a game. (p. 36).

An ethnographic study of 6- and 10-year-olds in a school in West Berlin, by Oswald, Krappmann, Chowduri, and von Salisch (1987), refers to "fooling around" or "horseplay" (p. 212) in the children. They commented that for 6-year-olds:

"there were instances when fooling around degenerated into annoyance or a quarrel, and others where actions designed to indicate fooling around were met with irritation on the part of the intended recipient. In most cases, however, the fooling was clearly high-spirited and, unlike such play among older children, was devoid of elements that could be perceived as encroachment. In contrast, the fooling around we observed in 10 year-olds often incorporated elements that could be regarded as disturbing, annoying, or even trespassing beyond limits. The children were able to present their behaviour in a manner that clearly indicated that they were fooling around, and not serious or hurtful in their intentions. Nonetheless, over a protracted period of fooling around this scheme could go awry; suddenly a thump was painful, or it was no longer sufficiently clear whether or not a gesture was intended to be disparaging. In such cases the nature of the interaction was reversed and the situation ended in verbal abuse or a scuffle. With some social skill, however, it was often possible to restore the interaction to the level of fooling around. In some sequences we gained the impression that children deliberately held the interaction at the boundary between playfulness and seriousness, between fooling around and annoyance." (p. 213).

There seems to be some consensus among the small number of relevant studies that deceptive, intimidative, and manipulative forms of R&T become more frequent with age. Nevertheless, observational/ethological studies tend to suggest that such forms of noncooperative R&T are infrequent up to 10 or 11 years of age, whereas the ethnographic studies tend to suggest that they are more frequent in the middle school period (not all ethnographic studies do so; for example Finnan, 1982, and Thorne, 1986 do not comment on deception, intimidation, or manipulation in chasing games).

10. SUMMARY

Rough-and-tumble play is a distinctive form of behaviour, prominent in children. Although superficially similar to real fighting, there are recognised cues which can be used in telling these two behaviours apart. These have been discerned by observational studies, and children too can verbalise many of these cues.

Play fighting is much more frequent than real fighting, in playgrounds. A proportion of play fighting can, however, turn into real fighting. This is a small proportion for most children, but can be higher for a few. The most usual reason, at least in middle childhood, appears to relate to 'honest mistakes' or accidental injury. However in some sociometrically rejected children, and more frequently by early adolescence, R&T may be used more deliberately as a social tool, consistent with a 'cheating' hypothesis.

These findings—based on observations and interviews with children—suggest that views on the functions of R&T may need modification. Social bonding and social skills may be incidental benefits of R&T, but fail to explain its distinctive features or the sex differences. Practice in fighting/hunting skills is a functional hypothesis consistent with design features and sex differences, but lacks direct support. The age changes and existence of 'cheating' suggest that, at least for some children and many adolescents, R&T can be used as a social tool in establishing or maintaining dominance in peer groups.

Data from teachers, while agreeing with observation and with children's views in some respects, reveals systematic biasses. Specifically, primary school teachers overestimate the relative frequency of real fighting compared to play fighting, and overestimate the frequency with which play fighting turns into real fighting. It is as if they were basing their judgments of play fighting on the few aggressive or rejected children for whom observational studies do reveal higher frequencies of real fighting.

One explanation may be in terms of when episodes of play fighting are drawn to teacher's attention. Often, this will be on those occasions when a play fight has 'gone wrong' in some way—either through accidental injury, or through a deliberate fight emerging out of it—and a supervisor or teacher may be called in to help resolve the conflict, or give first aid. Even if proportionately infrequent, these instances will be the ones often noticed and recalled, and perhaps given undue weight in their general perceptions of play fighting.

These different perceptions are important for policy issues, such as how R&T is treated in school playgrounds, and training of teachers and playground supervisors in this respect. MacDonald (1992) has suggested a positive role for R&T, and the benefits of training ADHD children in managing R&T, and shy and neglected children in engaging in R&T. Generally, a functional view suggests positive benefits of R&T in early and middle childhood. However, the more manipulative functions of R&T in later childhood and adolescence do possibly suggest more need for social control or restraint. More research is still needed, especially on the age changes into and through adolescence. A variety of methodological approaches, now well tested out, are available to complement each other in this endeavour. The technique of interviewing participants in play fighting appears to have promise for further work in this area.

REFERENCES

Aldis, O. (1975). *Play fighting*. New York: Academic Press.
Arrington, R. (1943). Time sampling in studies of social behavior: a critical review of techniques and results with research suggestions. *Psychological Bulletin*, 40, 81–124.
Asher, S.R. & Coie, J.D. (1990). *Peer rejection in childhood*. Cambridge: Cambridge University Press.
Axelrod, R. & Hamilton, W. D. (1981). The evolution of cooperation. *Science*, 211, 1390–1396.
Blatchford, P. (1989). *Playtime in the primary school*. Windsor: NFER-NELSON.
Blurton Jones, N. (1967). An ethological study of some aspects of social behaviour of children in nursery school. In D.Morris (Ed.), *Primate ethology* (pp.347–368). London: Weidenfeld & Nicolson.
Boulton, M.J. (1991a). Partner preferences in middle school children's playful fighting and chasing: A test of some competing functional hypotheses. *Ethology and Sociobiology*, 12, 177–193.
Boulton, M.J. (1991b). A comparison of structural and contextual features of middle school children's playful and aggressive fighting. *Ethology and Sociobiology*, 12, 119–145.
Boulton, M. J. (1992a). Rough physical play in adolescents: Does it serve a dominance function? *Early Education and Development*, 3, 312–333.
Boulton, M.J. (1992b). Participation in playground activities at middle school. *Educational Research*, 34, 167–182.
Boulton, M.J. (1993a). Children's abilities to distinguish between playful and aggressive fighting: A developmental perspective. *British Journal of Developmental Psychology*, 11, 249–263.
Boulton, M.J. (1993b). A comparison of adults' and children's abilities to distinguish between aggressive and playful fighting in middle school pupils: implications for playground supervision and behaviour management. *Educational Studies*, 19, 193–203.
Boulton, M.J. & Smith, P.K. (submitted). Peer social status, rough-and-tumble play and aggression: social skill and behaviour. Submitted for publication.
Carson, J., Burks, V. & Parke, R.D. (1993). Parent-child physical play: Determinants and consequences. In K. MacDonald (Ed.), *Parent-child play* (pp. 197–220). Albany: State University of New York Press.
Collaer, M.L. & Hines, M. (1995). Human behavioral sex differences: A role for gonadal hormones during early development. *Psychological Bulletin*, 118, 55–107.
Costabile, A., Smith, P.K., Matheson, L., Aston, J., Hunter, T. & Boulton, M. (1991). Cross-national comparison of how children distinguish serious and playful fighting. *Developmental Psychology*, 27, 881–887.
Dodge, K. & Frame, C. (1982). Social cognitive biases and deficits in aggressive boys. *Child Development*, 53, 620–635.
Fagen, R. M. (1981). *Animal play behavior*. New York: Oxford University Press
Fassnacht, G. (1982). *Theory and practice of observing behaviour*. London: Academic Press.
Finnan, C. R. (1982). The ethnography of children's spontaneous play. In G. Spindler (Ed.), *Doing the ethnography of schooling* (pp.358–380). New York: Holt, Rinehart & Winston.
Fry, D.P. (1987). Differences between playfighting and serious fights among Zapotec children. *Ethology and Sociobiology*, 8, 285–306.
Groos, K. (1901). *The play of man*. London: Heinemann.
Humphreys, A.P. and Smith, P.K. (1987). Rough and tumble, friendship, and dominance in schoolchildren: Evidence for continuity and change with age. *Child Development*, 58, 201–212.
Ladd, G. (1983). Social networks of popular, average and rejected children in a school setting. *Merrill-Palmer Quarterly*, 29, 283–307.
Lever, J. (1978). Sex differences in the complexity of children's play and games. *American Sociological Review*, 43, 471–483.
MacDonald, K. (1987). Parent-child physical play with rejected, neglected and popular boys. *Developmental Psychology*, 23, 705–711.
MacDonald, K. (1992). A time and place for everything: A discrete systems perspective on the role of children's rough-and-tumble play in educational settings. *Early Education and Development*, 3, 334–355.

MacDonald, K. & Parke, R. D. (1984). Bridging the gap: Parent-child play interactions and peer interactive competence. *Child Development, 55*, 1265–1277.

Neill, S.R.StJ. (1976). Aggressive and non-aggressive fighting in 12 to 13 year-old preadolescent boys. *Journal of Child Psychology and Psychiatry, 17*, 213–220.

Oswald, H., Krappmann, L., Chowduri, F. & von Salisch, M. (1987). Gaps and bridges: Interactions between girls and boys in elementary school. *Sociological Studies of Child Development, 2*, 205–223.

Paquette, D. (1994). Fighting and playfighting in captive adolescent chimpanzees. *Aggressive Behavior, 20*, 49–65.

Pellegrini, A.D. (1987). Rough-and-tumble play: Developmental and educational significance. *Educational Psychologist, 22*, 23–43.

Pellegrini, A.D. (1988). Elementary school children's rough-and-tumble play and social competence. *Developmental Psychology, 24*, 802–806.

Pellegrini, A.D. (1989). What is a category? The case of rough-and-tumble play. *Ethology and Sociobiology, 10*, 331–341.

Pellegrini, A.D. (1994). The rough play of adolescent boys of differing sociometric status. *International Journal of Behavioral Development, 17*, 525–540.

Pellegrini, A.D. & Smith, P.K. (submitted). Physical activity play: The nature and function of a neglected aspect of play. Submitted for publication.

Power, T. G. and Parke, R. D. (1981). Play as a context for early learning. In L. M. Laosa & I. E. Sigel (Eds.), *Families as learning environments for children* (pp.147–178). New York: Plenum.

Rutter, M., Graham, P., Chadwick, O. & Yule, W. (1976). Adolescent turmoil: fact or fiction? *Journal of Child and Psychology and Psychiatry*, 17, 35–56.

Schåfer, M. & Smith, P.K. (1996). Teachers' perceptions of play fighting and real fighting in primary school. *Educational Research, 38*, 173–181.

Sluckin, A.M. (1981). *Growing up in the playground: The social development of children*. London: Routledge & Kegan Paul.

Smith, P.K. (1973). Temporal clusters and individual differences in the behaviour of pre-school children. In R. P. Michael & J. H. Crook (eds), *Comparative Ecology and Behaviour of Primates*, (pp751–798). London & New York: Academic Press.

Smith, P. K. (1982). Does play matter? Functional and evolutionary aspects of animal and human play. *The Behavioral and Brain Sciences, 5*, 139–184.

Smith, P.K. (1989). The role of rough-and-tumble play in the development of social competence: Theoretical perspectives and empirical evidence. In B.H.Schneider, G.Attili, J.Nadel & R.P.Weissberg (Eds.), *Social competence in developmental perspective* (pp.239–255). Dordrecht: Kluwer.

Smith, P.K. and Boulton, M. (1990). Rough-and-tumble play, aggression and dominance: perception and behaviour in children's encounters. *Human Development, 33*, 271–282.

Smith, P.K., Carvalho, A.M.A., Hunter, T., & Costabile, A. (1992). Children's perceptions of playfighting, play-chasing and real fighting: a cross-national interview study. *Social Development, 1*, 211–229.

Smith, P.K. & Connolly, K. (1972). Patterns of play and social interaction in pre-school children. In N.G. Blurton Jones (Ed.), *Ethological studies of child behaviour*. Cambridge: Cambridge University Press.

Smith, P.K. & Lewis, K. (1985). Rough-and-tumble play, fighting, and chasing in nursery school children. *Ethology and Sociobiology, 6*, 175–181.

Symons, D. (1978). *Play and aggression: A study of rhesus monkeys*. New York: Columbia University Press.

Thorne, B. (1986). Girls and boys together but mostly apart: Gender arrangements in elementary schools. In W. W. Hartup & Z. Rubin (Eds.), *Relationships and development* (pp. 167–184). Hillsdale, NJ.: Erlbaum.

4

GAMES EVOLUTION PLAYS

Karl Sigmund

Institut für Mathematik
Universität Wien
Strudlhofgasse 4, A-1090 Vienna, Austria

ABSTRACT

The evolution of cooperation is frequently analysed in terms of the repeated Prisoner's Dilemma game. Computer simulations show that the emergence of cooperation is a robust phenomenon. However, the strategy which eventually gets adopted in the population seems to depend sensitively on fine details of the modelling process, so that it becomes difficult to predict the evolutionary outcome in real populations.

1. EVOLUTIONARY GAME THEORY

Every science has a predilection for singularities, for phenomena which are not buried in the ceaseless stream of day-to-day routine, but stand distinctly out. Physicists addressed the motion of heavenly bodies long before they turned to falling leaves or bubbles of foam or the mechanics of human locomotion, and economists dealt with the impact of monetary devaluation upon national wealth centuries before they discovered the continuous trading of services and goods that pervades the humblest household. One reason is, of course, that a phenomenon which stands out is likely to be fairly isolated from its environment, and therefore easier to analyse. There are far fewer forces acting on a planet than on a knee-cap.

It took economists a long time to appreciate that their science has to be founded on day-to-day individual behaviour. The tendency to act in groups—as firms, or guilds, or nations, and so on—is so widespread that such groups were often viewed as natural units. They are not, of course: they result from the aggregation—spontaneous or enforced—of human individuals, and have to be understood on this level. This point took a surprisingly long time to sink in. A similar delay occured in evolutionary biology: group or species selection arguments held their own, against gene or individual selection arguments, with a considerable tenacity.

Interestingly, it was the same methodological instrument that helped, both in mathematical economics and in theoretical biology, to bolster the approach based on the level of the individual. This is the use of *game theory*, a mathematical discipline founded more

New Aspects of Human Ethology, edited by Schmitt *et al.*
Plenum Press, New York, 1997

than fifty years ago by John von Neumann and Oscar Morgenstern, who were motivated by games like poker and chess to attempt an analysis of all sorts of conflicts of interest. However, note that many of the high hopes originally raised by game theory as a means of solving social, economic and military conflicts have been disappointed. But later, after game theory was amended in some essential aspects by John Nash and John Maynard Smith, it provided an extremely useful framework for discussing the social results of individual moves. Roughly speaking, John Nash introduced the main equilibrium concept to analyse non-cooperative games, which allowed to deal with non-zero sum situations; and John Maynard Smith formulated a population dynamics which permitted to get rid of the rationality assumption. We will encounter these two ideas in the following investigation of *reciprocal aid*, which is a basic concept both in economy and in ethology, and as essential to the social sciences as is chemical binding to chemistry.

But first, let us briefly describe the main aspects of the new, *evolutionary* game theory (see Maynard Smith, 1982, Hofbauer and Sigmund, 1988, Weibull, 1996). In its simplest form, one no longer assumes that the players engaged in such games are endowed with perfect foresight and rationality, able to find their best moves and to out-guess the best moves of their adversaries. Rather, one considers players programmed to one specific type of inborn behaviour. Thus a strategy is no longer viewed as a sequence of well-plotted moves, but as a hard-wired program, a behavioural phenotype. Depending on their strategy, and that of their adversary, the players obtain a higher or lower payoff. This payoff, now, is not money, and it has little to do with an individual utility scale. It is simply an increment in Darwinian fitness, i.e. in average reproductive success. Depending on their payoff, the players have therefore more or less offspring, and the offspring inherit their strategies. Thus the frequencies of the strategies will change in the population. This, of course, can mean that the success of a strategy can change, from generation to generation. A strategy is said to be evolutionarily stable if, whenever it prevails in the population, a small minority playing an alternative strategy cannot invade under the influence of natural selection. But a given game need not admit any evolutionarily stable strategy at all; alternatively, it can admit several such strategies; and it is not always the case that a population close to an evolutionarily stable strategy will converge to it under the combined effects of mutation and selection. It is quite possible that the evolutionary path will not settle down to an equilibrium. Game theorists used to show this by means of a fictitious example called the Rock-Scissors-Paper game—three strategies in a cyclic rank ordering (Rock beats Scissors, Scissors beats Paper, Paper beats Rock). It was well understood that this was a mathematical oddity which ought never to worry real ethologists; but a few months ago, it was found that the side-blotched lizard engages in it (see Sinervo and Lively, 1996). More precisely, there exist three morphs—male mating strategies conveniently associated with different throat colours—which are exactly in such a cyclic ordering. Males with dark blue throats defend small territories. They can be invaded by more agressive males with orange throats. But these more aggressive males can no longer control their larger territories efficiently, and can in their turn be invaded by 'sneaker' males with yellow stripes (looking like females). Once there are many sneakers around, the more modest blue throated males can spread again, etc. Similar dynamic complexities may well be expected whenever there are lots of conceivable strategies—as, for instance, in most issues studied in human ethology.

2. MUTUAL AID AND THE PRISONER'S DILEMMA

Humans need tools to survive. Arguably their most important tools are fellow humans, and an essential function of language consists in properly handling these human tools—witness the myth of the tower of Babylon. Charles Darwin has certainly not been the first to stress the importance of cooperation in human societies when he wrote, in the *Descent of Man*, that 'the small strength and speed of man, his want of natural weapons etc are more than counterbalanced by his ... social qualities, which lead him to give and receive aid from his fellow-men.'

Mutual aid, of course, is not restricted to humans. Ethologists list striking examples of mutual aid in many species, under the headings of joint hunting, helping in fights, predator inspection, warning the flock, teaching the young, feeding and grooming. Some social insects display even higher degrees of cooperation than humans do. Many instances of mutual aid can be easily explained by kin selection—helping a relative is helping a watered down copy of oneself. Since human households are often organised along family lines, this may well account for some instances of human cooperation. But there are also households which seem to work just as well without any family ties—sailors on a fishing vessel, hunters on an expedition, robbers on the run. This kind of mutual aid has to be explained without genetics. It obviously is based on an economic exchange. But such a trading of assistance is much more vulnerable to abuse.

The first to point this out may well have been David Hume, whose *Treatise of Human Nature* ought to rank as one of the great books of human ethology. Some ten years ago, Robert Sugden (1986) wrote a most remarkable sequel, *The economics of rights, cooperation and welfare*, where he couched many of Hume's arguments in simple game theoretical terms and showed how relevant they still are for the study of human societies. Take the following passage, for instance:

'Your corn is ripe to-day, mine will be so tomorrow. 'Tis profitable for us both, that I shou'd labour with you to-day, and that you shou'd aid me tomorrow. I have no kindness for you, and know you have as little for me. I will not, therefore, take any pains upon your account; and shou'd I labour with you upon my own account, in expectation of a return, I know I shou'd be disappointed, and that I shou'd in vain depend upon your gratitude. Here then I leave you to labour alone: You treat me in the same manner. The seasons change; and both of us lose our harvest for want of mutual confidence and security.'

Let us translate this thought experiment into the language of game theory. The two farmers, now, are the players of a game. They both have two options, or strategies: to cooperate (play **C**) or not to cooperate, i.e. to defect (play **D**). Depending on their decisions are their respective payoffs. If both cooperate, they each obtain a reward R which is better than the penalty P obtained if both defect and lose their harvest. But if one helps the other, but receives no return, he obtains the sucker's payoff S which is even lower than P, whereas the other player receives the highest payoff T, the temptation for unilateral defection: his harvest is safe, and he has spared himself the labour of helping his neighbour. In addition to this rank ordering of the payoff values $T>R>P>S$, we also have $2R>T+S$. This last condition means that the assistance yields more than it costs (if the two farmers were obliged to share their harvest, joint cooperation would have been better than unilateral exploitation).

Whenever the payoff values satisfy these two inequalities, we have what game theorists call a *Prisoner's Dilemma game* (see Trivers, 1985). It is encountered very frequently: indeed, whenever support costs less to the donor than it benefits the recipient, mutual cooperation is obviously advantageous for the two players. However, each player could do better still by not returning the other's help; and in case no such help is forthcoming, there is of course all the more reason to defect. The message, as Hume showed, is clear: no mutual aid. The Prisoner's Dilemma game epitomises the clash between what is best from an individual's point of view and from that of a collective, a conflict that threatens countless forms of cooperation, including trade and mutual aid.

We can illustrate the two features added by Nash and by Maynard Smith at this stage already. First, the game is not zero-sum, and not even constant sum: indeed, the sum of the payoffs for the two players is larger if both cooperate than if both defect. Most parlour games are zero-sum, since the gain to one player is the loss to the other. John von Neumann, who was a passionate if somewhat amateurish poker player, elaborated his notion of a minimax solution as a central concept for zero-sum games, but this concept is of little interest for the vastly more frequent real-life interactions which are not zero-sum. Secondly, we need not assume that our players are rational and can analyse the game in advance, like Hume did, by effectively outguessing their antagonist. In the framework of evolutionary game theory, one considers (fictitious) populations consisting of programmed players—mere robots. Each of these robots is firmly wedded to one fixed strategy and will either always cooperate or always defect. They engage in a round-robin tournament of the Prisoner's Dilemma game. For each contestant, the total payoff will depend on which other players he encountered, and therefore on the composition of the population. A defector will, however, always achieve more than a collaborator would earn in its stead. We can assume that the more successful players have more offspring, which inherit their strategy (or alternatively that the less successful players have the tendency to switch their strategy and imitate their more successful rivals). The members of the newly composed population will again engage in a round-robin tournament, etc. In this caricature of an evolving population, where success means becoming more numerous, the outcome is clear: defectors will steadily increase, and eventually swamp the population. This outcome is inevitable, and has nothing to do with rationality or foresight.

3. RECIPROCAL ALTRUISM AND THE REPEATED PRISONER'S DILEMMA

The drab outlook of Hume's thought experiment is obviously contradicted by the many instances of cooperation found in human and animal societies. In fact, we stopped too soon with our quote of Hume. It is not true, he argues, that 'I shou'd in vain depend upon your gratitude'. In fact, I can depend on it because I can depend on your needing my services for the next harvest, or, in Hume's words, 'because I foresee that you will return my service, in expectation of another of the same kind'. But foresight is not actually needed; and neither is the 'teaching of moralists or politicians' which Hume seems to believe essential. All that is needed is the next harvest.

More precisely, as soon as one assumes that the same two players are engaged in a *repeated* Prisoner's Dilemma game, with a large random number of rounds, then the strategy of unconditional defection is no longer the inevitable outcome (see Axelrod, 1984). It can easily be seen that there exists no strategy which is best against all comers. If the opposite player, for instance, decides to always defect (or always to cooperate), it will be best to always defect. But if your adversary decides to cooperate until you defect and from then on never to cooperate again, you will be careful not to spoil the partnership: the temptation to defect in one round is more than offset by the prospect of permanently losing a cooperative partner. There is no best strategy in the repeated Prisoner's Dilemma game (see Axelrod, 1984).

One can, of course, adopt the evolutionary viewpoint again. It reduces to *drowning by numbers*: have lots of players play lots of Prisoner's Dilemma games for lots and lots of generations. More precisely, one can simulate (in a computer) populations where each individual is matched against randomly chosen opponents for a large random number of rounds of the game. Assume again that each player follows certain 'inbuilt' rules, and in this way accumulates benefits and costs, which are measured in reproductive success—the only valid currency in a Darwinian world. At the end of each generation, the players produce offspring in proportion to their overall balance. The offspring inherit their parents behaviour, except for a few mutants testing out new rules. And so on through the next generation and the next and beyond. We may watch the population evolve for as long as we care.

The main snag here is that there exist so many conceivable strategies to play the repeated game. Evolutionary trial and error can never explore all theoretical possibilities. But biological constraints help to reduce the rule-space to a manageable size. We need probably only consider strategies (a) consisting of a finite number of states (which we interpret as internal motivational variables), (b) making decisions on whether to cooperate or not in the next round, and (c) having a transition rule which leads, depending on the outcome of the current round, to the next internal state (see Nowak and Sigmund, 1995)

The famous Tit For Tat rule is an example: it starts cooperatively and then simply imitates the other player's previous move. This retaliatory rule will invade a population of defectors, as soon as its frequency has overcome some very small threshold: cooperation, then, spreads with increasing momentum through the population. A population of stern retaliators, on the other hand, cannot be invaded by defectors. But it is victimised by its own strictness: any unintentional defection (due, for example, to a mistake in implementing a move, or an erroneous perception of the other's move) entails a costly series of reprisals. Stern retaliators therefore give way to more tolerant strategies, prepared occasionally to forget or to extend an olive branch. Such generous behaviour, however, can only thrive on a back-cloth of cooperation. Since it could never spread in a defector's world, it needs Tit For Tat to pave the way. The situation is, in fact, somewhat similar to ecological succession. It leads to a cooperative community whose members are able to retaliate *judiciously*, not too much and not too little. However, such an equilibrium is not necessarily proof against an eventual increase, through random drift, of over-gentle strains of players. Once the community is sufficiently softened up, defectors can cash in and ultimately take over.

No two evolutionary chronicles played out on a computer are alike: contingency rules. But behind the vagaries of historical accident, some common traits show up (see, e.g. Lindgren, 1991, Nowak and Sigmund, 1992, 1993, 1994, Lindgren 1996, Boerlijst et al, 1996). Most conspicuous is the *bang-bang*-principle: Either almost all members cooperate almost all the time, or they almost always defect. The change from one regime to the other takes usually not more than a few hundred generations and occurs only rarely; it can be triggered by complex events—in most cases, polymorphism in the population builds up before a shift occurs—and occasionally, it can even change direction—a sharp plunge in cooperation can be reversed in mid-fall—but the population almost never settles down to a medium level of collaboration: it is all or nothing. This is a game-theoretical version of *punctuated equilibrium*: abrupt transitions between extended periods of stasis. As more and more mutant strategies are tested, the probability of being in a cooperative regime increases; the episodes of cooperation become more frequent and tend to last longer.

4. STABLE STRATEGIES FOR COOPERATION

There are many behavioural rules which can sustain a cooperative society. Generous Tit For Tat—always repay cooperation in kind, but tolerate defection with a certain probability—is but one example (see Nowak and Sigmund, 1992). An even more frequent outcome of the evolution towards cooperation is the strategy which has been christened, somewhat unfortunately, Pavlov (see Nowak and Sigmund, 1993). A Pavlov player starts with a cooperative move and henceforth cooperates if and only if, in the previous round, the other player used the same move than himself. This seems at first sight a strange rule. It is reasonable not to break a string of mutual cooperation, but why should one cooperate after a round where both players have defected? The rule becomes more transparent if one realises that it consists in repeating a move that has been rewarded, and in switching if the move has been punished. This rule knows how to let bygones be bygones. A mistaken defection between two Pavlovians leads to one round of mutual defection (since the successful defector persists and the sucker shifts), but in the next round, both players will make terms and resume cooperation, and continue happily with it. Intuitively, a defection is followed by a domestic quarrel and then cooperation is resumed. The Pavlov strategy embodies the 'win-stay, lose-shift' principle which psychologists view as the very simplest learning rule, and philosophers as the root of hedonistic morals. Natural selection has often been compared with learning by trial and error. Here is an instance where such a generation-wise 'learning process' leads a population of primitive robots to the most basic *individual* learning rule.

In a cooperative world, both Pavlovians and Generous Tit For Tat players do better than Tit For Tat, because they are immune to errors. Pavlov has an additional advantage: it ensures that the population cannot be subverted by unconditional cooperators. Indeed, after a mistaken defection, Pavlov will not resume cooperation, but blithely continue to exploit the sucker. This is important, since otherwise neutral drift will cause unconditional cooperators to spread and thus 'soften up' the population to such an extent that it loses its power to retaliate against exploiters.

In a world of defectors, both Pavlov and Generous Tit For Tat cannot catalyse the first step towards a cooperative society. They are too sanguine, and try too trustingly to resume cooperation. It needs stern retaliators to prepare the ground: the grim law of *An Eye For An Eye* was probably at the origin of every self-supporting community.

Do there exist strategies which are both able to invade, in small clusters, societies of defectors and able to resist, once they have reached predominance, every invasion attempt of a mutant strategy? Yes. One such strategy has been proposed by Robert Sugden. This is *Contrite Tit For Tat* (see Sugden, 1986, Boyd, 1989, and Wu and Axelrod, 1996). In contrast to the strategies considered so far, the 'internal state' of a Contrite Tit For Tat player does not only depend on the last moves in the game, but also on the *standing* of each player, which can be good or bad. A player is in good standing if he has cooperated in the previous round, or if he has defected while provoked (i.e. while he was in good standing and the other player was not). In every other case defection leads to a bad standing. Contrite Tit For Tat begins with a cooperative move, and cooperates except if provoked. There are three possible internal states for such a player, which we can label, in antropomorphic terms, as 'content', 'guilty' and 'provoked'.

If two Contrite Tit For Tat players engage in a repeated Prisoner's Dilemma, and if the first player defects by mistake, he loses his good standing. In the next round, he is 'guilty' and will cooperate, whereas the other player is 'provoked' and will defect without losing his good standing. Then both players will be in good standing again; both are 'content' and resume their mutual cooperation in the following round.

A population of Contrite Tit For Tat players cannot be invaded—it is evolutionarily stable, at least if we assume—as we should certainly do—that there is always a small possibility for making mistakes by mis-implementing a move. (This is the 'trembling hand' approach due to the game theorist Reinhard Selten, see Selten, 1975, and Boyd, 1989.) Moreover, Contrite Tit For Tat is as good as Tit For Tat itself at invading a population of defectors—as soon as a small cluster of such players exists, it will grow with increasing momentum.

To recapitulate: evolutionary simulations of the repeated Prisoner's Dilemma show that a robust cooperative regime is highly probable. What is less clear is whether there exist theoretical or empirical grounds for favouring one specific cooperative strategy. Contrite Tit For Tat is certainly a contender. But there exist other evolutionarily stable strategies which lead to a cooperative population—for instance, Pavlov (for certain values of the payoffs R, S, T and P), and *Remorse* (for the complementary values) (see Boerlijst et al, 1996). *Remorse* is the strategy that cooperates after a round of mutual cooperation or when in a bad standing. Statistical explorations suggest that Contrite Tit For Tat is about twice as likely to occur as an evolutionarily stable outcome than either Pavlov or Remorse. This is essentially due to the fact that Contrite Tit For Tat, which needs no catalyser to invade a society of defectors, has a head-start. (It should be noted that paradoxically, for very *low* values of the temptation payoff T to defect unilaterally, the strategy *Weakling* has a chance of about one in ten to get established in the population. In this case, the average payoff per round is not the reward R for mutual cooperation, but the average of R and P, the punishment for mutual defection. *Weakling* is the strategy that cooperates only after being in bad standing.) Unfortunately, such statistical explorations seem to depend sensitively on the technical details of the numerical exploration, and cannot offer robust predictions, so far.

To give a sample of the kind of technical details, take for instance that most numerical experiments assume that the mistakes in the interactions are mis-implementations of an intended move. A bit of introspection might suggest that *mis-perceptions* of the other player's actions are at least as likely to cause trouble. As soon as we introduce this type of mistakes, Contrite Tit For Tat loses much of its lustre. If, in a match between two Contrite Tit For Tat players, one player mistakenly believes that the other is in bad standing, this leads to a sequence of mutual backbiting, just as with Tit For Tat. In contrast, Pavlov can easily be shown to be immune to errors in perception. Remorse, on the other hand, is not. Furthermore, neither Contrite Tit For Tat nor Remorse are prepared to exploit unconditional cooperators the way Pavlov does. We have seen that if neutral drift (which is usually not included in the computer simulations, but certainly very plausible in natural populations) can allow suckers to spread, the cooperative regime becomes threatened by all-out exploiters.

Another aspect which has certainly to be considered concerns the inherent plausibility of the contending strategies—their suitability for human players. We can, for instance, easily sympathize with states like 'guilty' or 'provoked'. Does that mean that the evolution of such feelings has been an outcome of selection (it need not be selection for the Prisoner's Dilemma, but possibly for another purpose; human cooperation, of course, is so important that it can certainly be expected to shape our 'internal states' to a large degree.) Many other finite state strategies for the repeated Prisoner's Dilemma are also evolutionarily stable, but their description often sounds arid and artificial. However, one could also describe Contrite Tit For Tat in such a way that it would not be recognizable as a natural strategy. Similarly, the rule for Pavlov can be either defined as 'win-stay, lose-shift', in which case it seems eminently sensible, or as 'cooperate after an R or a P', in which case it seems silly to most people (it has, originally, been termed *Simpleton* for this reason). An argument strongly in favour of Pavlov, of course, is that the 'win-stay, lose-shift' rule can be applied in many other situations, independent of the Prisoner's Dilemma game: it could have evolved in the context of foraging strategies, for instance. On the other hand, internal states like 'guilty' or 'provoked' do also play a role in human interactions having nothing to do with mutual help.

5. ALTERNATING ASSISTANCE

Another detail concerns the timing of interactions. So far, we have considered the so-called 'simultaneous Prisoner's Dilemma', where the two players make their choices in the same time instant. The reason for this is mostly conventional. If we take a closer look at Hume's example, we actually see that the two farmers cannot help each other at the same time: they have to take turns. Thus we ought to have modelled this interactions by an 'alternating Prisoner's Dilemma' (see Nowak and Sigmund, 1994, and Frean, 1994). In fact, the partners alternate in their roles of donor and recipient in many if not most instances of reciprocal help. Robert Trivers summarizes that 'reciprocal altruism is expected to evolve when two individuals associate long enough *to exchange roles frequently* as potential altruist and recipient' (Trivers, 1985).

Thus we ought also to study the iterated Prisoner's Dilemma when the players have to take turns. The slight modification in such an *alternating* Prisoner's Dilemma game can affect the interaction to a considerable extent. For instance, if two Tit For Tat players engage in a Prisoner's Dilemma of the usual *simultaneous* kind, and if one of them defects by mistake, both players will subsequently cooperate and defect in turns. On the other hand, if two Tit For Tat players engage in an *alternating* Prisoner's Dilemma, and a unilateral defection occurs inadvertently, then the outcome will be an unbroken sequence of mutual defections.

In the alternating Prisoner's Dilemma, the roles of the two players in each round are asymmetrical. One of the player is the 'leader' (to use a game-theoretic expression) and able to decide what the outcome is going to be. In a single round, the leader obtains a higher payoff if he defects than if he cooperates. But if the leader cooperates, the other player receives a payoff which is higher than what he would receive if the leader defects; and this increment of his payoff is higher than the loss to the leader. If we consider one 'unit' of two consecutive rounds, and assume that each player finds himself during one of the two rounds in the role of leader, we find that their payoff, if both help each other, is larger than their payoff if both defect, but that if one player helps (when a leader) and the other does not (when it is his turn), then (a) the helper gets the lowest payoff—he has been had for a sucker—and (b) the defector obtains a payoff even higher than the *reward* for mutual cooperation, whereas (c) the average payoff for both players is lower than this reward. These are exactly the conditions defining the Prisoner's Dilemma. However, the interactions for the iterated game are now quite different.

In the alternating game, Pavlov is no longer error-correcting. A mistake by one of the players results in a run where each player defects every third round. Accordingly, evolutionary chronicles of the alternating game do not lead to the establishment of Pavlov. Much more successful is a strategy which always cooperates, except if it has been suckerpunched in the previous round. This strategy, which one may call *Firm But Fair*, is more tolerant than Tit For Tat, since it does forgive a defection by the opponent if this was in answer to the own defection. It is error-correcting.

In an interesting series of experiments, Claus Wedekind and Manfred Milinski have tested humans (more specifically, first year students) in both the simultaneous and the alternating Prisoner's Dilemma (see Wedekind and Milinski, 1996). In both cases the subjects became more and more cooperative and successful as the experiment went on. It turned out that their strategies eventually clustered around two types which could be reasonably interpreted as versions of Pavlov respectively Firm But Fair. The players, however, were rather inflexible: they did not distinguish whether they were playing a simultaneous or an alternating game. This missing flexibility, according to Wedekind and Milinski, 'suggests either that people might have preferred niches, which contain either the simultaneous or the alternating situation, or that the game situation [in the experimental setup] did not offer the natural clues by which humans would recognise the mode and trigger their response conditionally... Pavlovian player appeared to gain more in the simultaneous game and [Firm But Fair]-players in the alternating game.' Apparently our propensity for cooperation has not been fine-tuned to a large extent. It should be noted that Milinski and Wedekind concentrated on finding strategies which depended only on what happened in the last round, and not on the players' standing. They did not consider Contrite Tit For Tat, for instance. But computer simulations show that Contrite Tit For Tat does very well (and even better than Firm But Fair) in the alternating Prisoner's Dilemma.

It seems rather difficult to test whether human subjects play Contrite Tit For Tat, by the way. The experimenter has to cause the subject to mis-implement a cooperative move in such a way that he feels guilt (rather than annoyance at the experimenters' interference).

6. THE SNOWDRIFT GAME

Another detail likely to plague ethologists on the look-out for mutual aid is the similarity between the Prisoner's Dilemma game and a close relative which has been termed the *Snowdrift* game by Robert Sugden: 'Suppose you are driving your car on a lonely road in winter, and you get stuck in a snowdrift, along with one other car. You and the other driver have both sensibly brought shovels. It is clear, then, that you should both start digging. Or is it? The other driver cannot dig his own way out of the drift without digging your way out too. If you think he is capable of doing the work on his own, why bother to help him?' (Sugden, 1986).

If we assume that it is better to do half the digging and get out of the snowdrift than to remain stuck, but that it is so important to get out of the snowdrift that each player would rather do all the digging himself than remain stuck, we obtain a rank ordering of the payoff given by $T>R>S>P$. This looks like the Prisoner's Dilemma, except that the payoffs P and S have been interchanged: the outcome for mutual defection is less good than that for unilateral cooperation. Better do the job by yourself than not get it done at all. This may seem, at first glance, a minor variation, but it is an entirely different ball-game, in fact.

As Sugden has pointed out, the Snowdrift game has the same structure as the much studied *Chicken* game, which played an essential role in providing an individual (rather than group) selection argument for the evolution of constraints in innerspecific fights (see Maynard Smith, 1982). In the Chicken game, the players have to decide whether to escalate a potentially dangerous conflict or whether merely to display (and retreat if the other player starts getting mean). If we label escalating by **D**, and displaying by **C**, we obtain the same rank ordering of the outcomes as with the Snowdrift game, in spite of the fact that the options in the two games have a totally different meaning: refusing help is certainly something else than offering an all-out fight.

Since the Chicken game has been thoroughly studied, we can use the corresponding results. It follows that for one round of the Snowdrift game, it is not necessarily the best to play **D**, i.e. to defect. The evolutionarily stable outcome will eventually depend on some asymmetry between the players, a seemingly arbitrary labelling of the players resulting from the more or less contingent history of the evolving population, reminiscent in human terms of a 'norm' or convention like driving on the left hand side on British roads. An asymmetry (in age, or strength, or ownership, etc.) decides which one of the two players opts for **C** and which for **D**. This outcome is entirely different from the seemingly similar Prisoner's Dilemma: we can no longer expect mutual cooperation, or a strategy like Pavlov or Contrite Tit For Tat.

This may shed some light on some recent controversy. A group of behavioural ecologists discovered that among lions, the principle of reciprocity seemed not to hold. 'That', it was claimed, 'throws a monkey wrench into the classic explanation for the evolution of cooperation in a selfish, dog-eat-dog (or lion-eat-gazelle) world' (see Morell, 1995).

Female lions typically live in prides of two to seven members, around which cluster their dependent offspring and a coalition of immigrant males. The males defend the pride against incursion by other males, whereas the females defend the territory against incursions by other females. Territorial incursions can be simulated by the playback of recorded roars, and these routinely elicit cooperative defense. In response, female lions pair off to approach what must seem to them an unknown and potentially dangerous enemy (see Heinson and Packer, 1995).

This is similar to the *predator inspection* whereby pairs of stickleback gingerly approach a pike to test its current mood and motivation. As long as the stickleback approach together, their risk is more than halved, and we can speak of cooperation. If one stickleback consistently lags behind and gains information by waiting in the wings, this is a clear case of defection. A beautiful series of experiments by Milinski has shown that this is an instance of the repeated Prisoner's Dilemma game (of the simultaneous sort, by the way), and that the stickleback use a strategy based on reciprocation (see Milinski, 1987).

The lions, in an apparently very similar situation, seem not to use the principle of reciprocity. Indeed, some of the lions turned out to be consistently laggards, others consistently bold defenders of the home range. The bold lions did not change their behaviour, even after having experienced in many 'rounds' the cowardly behaviour of the laggards. (Some of these laggards, incidentally, seem to use a conditional strategy, rushing forward only at the last minute.) More surprising than the diversity of behaviour among lions is its relative fixity: each lioness has her character—that of an intrepid leader attacking unconditionally, or that of a feet-dragging laggard shying away from every risk of an all too direct confrontation with a territorial intruder. The fact that the laggards are not punished, and that the leaders do not stop their advance, but just glance back, is a strong argument against a strategy based on Tit For Tat or Pavlov.

Does this imply the 'inadequacy of current theory to explain cooperation'? We believe that this is not the case. Indeed, remember that in the Prisoner's Dilemma, one of the essential prerequisites is that a player matched against a defector obtains a higher payoff if he defects than if he cooperates. Can we assume that this condition is met here? What happens to a lioness paired with a defector? If she cooperates by advancing boldly, she certainly runs a risk to be wounded in a fight against the intruder. But if the lioness retreats, this means that the pride is giving up its territory. As the authors of the lion study write, however, 'territory is essential for successful breeding'. It could well be that this is a graver threat to the reproductive success than the risk of squaring off against an intruder.

This is quite different from the scenario with the two stickleback approaching their pike. If a stickleback approaches all by himself, he risks a lot and gains relatively little; it is certainly better, in this case, to give up at an early stage of the approach, rather than learn about the motivational state of the pike at one's own expense. Just swimming quietly away does not mean giving up the territory. In contrast, if both lions defect, their situation becomes critical: 'Without a territory, a female lion has little chance of raising her cubs and so loses the chance to pass her genes on to the next generation—the bottom line of evolutionary success.'(see Heinson and Packer, 1995).

As soon as we admit this, we are faced with the Snowdrift game. Two lions belonging to the same pride and called upon defending their territory will know each other well and be aware of many asymmetries in their strength, age, status etc. It seems almost inescapable that the two lions find themselves in different roles, and hence will adopt different strategies for a Snowdrift game. Thus we ought not to be surprised when one of the lions is chickening out.

We have seen that seemingly slight variations in the game—leading to an alternating Prisoner's Dilemma, for instance, or to a Snowdrift game—can have remarkably different outcomes. There are other factors that are just as likely to affect the outcome in a fundamental way: for instance, if we assume that players can watch how their rivals do against each other, or if we allow that a player can break off a repeated game and start anew with another partner. These variants are likely to play an important role, especially in sophisticated communities—for instance, human tribes. The analysis of the repeated Prisoner's Dilemma game by means of theoretical or experimental mathematics provides at best a first approximation to the issue of mutual aid among humans, something comparable to the role of motion without friction in classical mechanics, which is valid for planetary motion, but hardly for anything more down-to-earth. Nevertheless, it provides an essential jump-off point for experimental research.

7. REFERENCES

Axelrod, R. (1984) *The Evolution of Cooperation*, Basic Books, New York (reprinted 1989 in Penguin, Harmondsworth).

Boerlijst, M., Nowak, M. and Sigmund, K. (1996), The logic of contrition, to appear in *Journal of Theoretical Biology*.

Boyd, R. (1989), Mistakes Allow Evolutionary Stability in the Repeated Prisoner's Dilemma Game, *Journal of Theoretical Biology* **136**, 47–56.

Frean, M.R. (1994), The Prisoner's Dilemma without synchrony, *Proceedings of the Royal Society London B*, **257**, 75–79.

Heinson, R. and Packer, C. (1995), Complex cooperative strategies in group-territorial african lions, *Science* **269** 1260–2.

Hofbauer, J. and Sigmund, K. (1988), *The Theory of Evolution and Dynamical Systems*, Cambridge UP.

Lindgren, K. (1991) Evolutionary phenomena in simple dynamics, in *Artificial life II* (ed. C.G. Langton et al), Santa Fe Institute for Studies in the Sciences of Complexity, X, 295–312.

Lindgren, K. (1996), Evolutionary Dynamics in Game-Theory Models, to appear in *The Economy as an Evolving, Complex System II*, ed. W. Brian Arthur et al.

Lindgren, K. and Nordahl, M.G. (1994), Evolutionary dynamics of spatial games, *Physica D* **75**, 292–309.

Maynard Smith, J. (1982) *Evolution and the theory of games*, Cambridge UP.

Milinski, M. (1987), Tit for tat in sticklebacks and the evolution of cooperation, *Nature* **325**, 434–435.

Morell, V. (1995), Cowardly lions confound cooperation theory, *Science* **269**, 1216–7.

Nowak, M. and Sigmund, K. (1992), Tit for tat in heterogeneous populations, *Nature* **355**, 250–2.

Nowak, M. and Sigmund, K. (1993), Win-stay, lose-shift outperforms tit-for-tat, *Nature*, **364**, 56–8.

Nowak, M. and Sigmund, K. (1994), The alternating Prisoner's Dilemma, *Journal of Theoretical Biology* **168**, 219–26.

Nowak, M., Sigmund, K. and El-Sedy, E. (1995), Automata, repeated games, and noise, *Journal of Mathematical Biology* **33**, 703–32.

Selten, R. (1975), Re-examination of the perfectness concept for equilibrium points in extensive games, *International Journal of Game Theory* **4**, 25–55.

Sinervo, B. and Lively, C.M. (1996), The rock-paper-scissors game and the evolution of alternative male strategies, *Nature* **380**, 240–243.

R. Sugden (1986), *The economics of Rights, Co-operation and Welfare*, Blackwell, New York.

Trivers, R. (1985), *Social Evolution*, Menlo Park CA, Benjamin Cummings.

Wedekind, C. and Milinski, M. (1996), Human cooperation in the simultaneous and the alternating Prisoner's Dilemma: Pavlov versus Generous Tit For Tat. *Proceedings of the National Academy of Science of the USA*, **93**, 2686–2689.

Weibull (1995) *Evolutionary Game Theory*, MIT Press, Cambridge, Mass.

Wu, J. and R. Axelrod (1995), How to cope with noise in the iterated Prisoner's Dilemma, *Journal of Conflict Resolution* **39**, 183–9.

5

GROUPS, GOSSIP, AND THE EVOLUTION OF LANGUAGE

R. I. M. Dunbar

ESRC Research Centre for Economic Learning and Social Evolution
Department of Psychology
University of Liverpool
P.O.Box 147, Liverpool L69 3BX, England

1. INTRODUCTION

No one doubts that language is the single most important evolutionary development in our history as a species. So much flows from it in terms of our culture that it is difficult to imagine what a languageless human society would really be like. However, the very fact that language is so important in our daily lives raises questions about its function and origin. Just what does language do for us? Why did it evolve? And why did language evolve in our lineage but not in any of our sibling species such as chimpanzees?

In this paper, I shall try to answer these questions. To do so, however, I have to go back in time, as it were, to look at our origins among the primates. To understand the evolution of language, we first need to understand the social behaviour of primates and the way in which evolutionary processes arising out of these social considerations influenced the primate (and eventually the human) mind. These will, then, guide us towards an answer for the questions I posed above. In doing so, I hope I shall also be able to demonstrate that there is an underlying unity to psychology as a discipline. For in working through to our conclusion, we shall traverse almost all aspects of the behavioural sciences, from neuroanatomy to ethology, cognition to developmental psychology and social psychology to evolutionary biology.

2. WHY DO PRIMATES HAVE BIG BRAINS?

We have been aware ever since Jerison's (1973) seminal analyses that primates have unusually large brains for their body size. The central question is why this should be so. A number of alternative hypotheses have been proposed over the years, but recent analyses seem to be converging rapidly on the view that, in primates at least, the brain evolved to solve social problems (Brothers 1990, Sawaguchi & Kudo 1992, Dunbar 1992, Barton

1996). In contrast to previous analyses (which have all used total brain size: see Jerison 1973, Clutton-Brock & Harvey 1980), these newer analyses have considered only neocortex volume. There are two reasons for doing so. One is that it is the neocortex that has grown out of all proportion during the course of primate evolution, whereas other sub-cortical and cortical parts of the brain have remained relatively constant. Although there is a general allometric relationship between neocortex volume and total brain volume (see Aiello & Dunbar 1993, Barton 1996), there is a significant degree of variance around this regression line such that some species (e.g. gorilla) have much smaller neocortices than would be expected on the basis of their brain size. The other is that the neocortex does seem to be the seat of conscious brain activity; since it is this that we are mainly interested in rather than those parts of the brain that are responsible for managing the day-to-day machinery of the body's physiology, it makes sense to focus on the neocortex (see also Jerison 1973).

When neocortex volume is plotted against group size, a significant relationship is found. Fig. 1 shows the data for anthropoid primates. This relationship has proved robust for a number of alternative analytical methods (residuals from regressions against non-cortical volume, ratio of neocortical volume to the volume of the rest of the brain, residuals to body size, residuals to medulla volume) and several alternative statistical methods for dealing with phylogenetic inertia (generic means, independent contrasts of values for individual species). In contrast, neocortex volume shows no relationship with ecological variables, including diet type, size of ranging area (when scaled for body size) and style of foraging (extractive vs non-extractive) (Dunbar 1992, 1995, Barton 1996).

These results are interpreted as implying that neocortex evolution has been driven by the information-processing demands imposed by increasing group size. Why group size should have increased in particular lineages or taxa of primates is another matter, though it is almost certainly a consequence of one or more ecological variables (see van Schaik 1983, Dunbar 1988). The point at issue here is, rather, that, given that these ecological

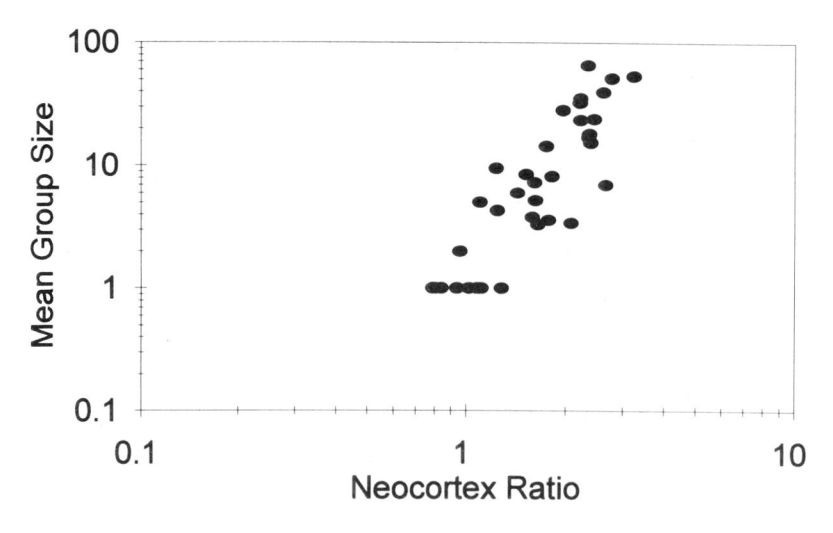

Figure 1. Mean group size for individual primate genera plotted against relative neocortex size. The index of relative neocortex size used here is the ratio of neocortex volume to the volume of the rest of the brain. In most cases, each genus is represented by a single species. Source: Dunbar 1992, with brain part volumes from Stephan et al. 1980.

pressures selected for increasing group size, the demands for increased group size in turn imposed selection on brain size evolution in order to provide the processing capacity necessary to create and maintain groups of the size required.

I had originally assumed that this relationship reflected something about the peculiar nature of primate societies (notably their tight bonding based on intense social knowledge: Byrne & Whiten 1988, Harcourt 1992). However, there is growing evidence to suggest that this relationship may extend beyond the primates to include at least two other mammalian orders (bats: Barton & Dunbar 1996; carnivores: Dunbar & Bever submitted). Thus the relationship may be a very general one concerning the computing power required to handle social relationships in mammalian groups.

2. NEOCORTEX SIZE AND SOCIAL SKILLS

2.1. Social Skills

This does, however, raise a number of interesting questions about the way in which social skills relate to neocortex volume. There are, for example, at least three reasons why we might find a relationship between group size and neocortex size (assuming that the latter actually does measure computing power). One is that the animal has to remember all the other individuals with whom it interacts; a second is that it has to be able to manipulate the information relating to these individuals in a complex cognitive hyperspace; a third is that it has more to do with how the individual relates to a smaller subset of special allies.

We can, I think, dismiss the first of these fairly easily because it seems that this is not a pure memory problem. For one thing, memory for individuals is probably stored elsewhere in the cortex (just as spatial memory is located in the hippocampus: O'Keefe & Nadel 1978, Barton & Purvis 1996). More importantly, perhaps, the whole essence of the social brain hypothesis is that it is the manipulations that individuals do with the information they have that is important rather than just remembering who's who. Finally, all the work on theory of mind in humans has shown quite conclusively that the problem lies in understanding another individual's mental state not in remembering factual information (Happe 1995, Kinderman et al.. in press) (see below, section 2.2).

Testing between these alternative options is not easy, however. Nonetheless, there are three lines of evidence available, all of which point us in the direction of the third hypothesis (that it is the social skills inherent in managing a subset of close relationships that is important rather than knowledge of all the relationships within the group).

First, a careful consideration of the relationship between group size and neocortex volume in the original analysis (Dunbar 1992) revealed that there appear to be distinct grades in the relationship: prosimians appear to have larger groups for a given neocortex ratio than monkeys, while monkeys in turn have larger groups than apes. This suggests that the computing power required to maintain a group of a given size is larger for apes than it is for monkeys, and, in turn, larger for monkeys than for prosimians. These distinctions seem to correlate uncontroversially with the perceived social complexity of these three taxonomic groups.

Second, Kudo et al. (submitted) have shown that neocortex ratio correlates with with grooming clique size. This result suggests that species with larger neocortices can handle proportionately more relationships and keep them simultaneously functional. In this context, grooming cliques can be interpreted as alliances on which members rely in negotiating their relationships within the group. Primates are in many ways unique in the extent to which they use coalitions to buffer themselves against the costs of living in groups (Dunbar 1988, Harcourt 1990). Harcourt (1992) has pointed out that these alliances are unusually complex in both their structure and dynamics. They depend on exploitation of detailed knowledge about prospective members' behaviour and propensities and, unusually, they are established well ahead of the time at which they are actually needed. In other words, their future need is anticipated and steps are taken ahead of time to prepare the necessary relationships (for one example of this in chimpanzees, see Koyama & Dunbar 1996). The cognitive skills involved are clearly considerable.

Finally, Pawlowski & Dunbar (submitted) have shown that, among polygamous primates, neocortex size correlates negatively with the correlation between male rank and mating success. The significance of this finding is that it implies that males from species with large neocortices are able to use more subtle social strategies to undermine the simple power-based monopolisation strategies of high-ranking males. Such strategies include the exploitation of both alliances with other males and female's mating preferences (i.e. female choice).

All three sets of data suggest that species with larger neocortices can handle a larger number of more complex or more sophisticated relationships.

2.2. The Role of Theory of Mind

Perhaps the most important development in psychology in the past decade has been the recognition of a phenomenon now known as "theory of mind" or ToM (Premack & Woodruffe 1978). ToM is the ability to understand another individual's state of mind. It requires second order intensionality (*sensu* Dennett 1984), that is the ability to "understand that someone else understands that something is the case." (Note that, following the original practice, I use *intensionality* with an *s* to denote the mental states associated with mind-reading, as distinct from the everyday sense of *intentionality* with a *t* meaning a specific kind of intensionality.) Developmental psychologists have identified "false belief" (recognition that another individual's beliefs about the world can be different from yours and/or that they can hold beliefs that you know to be false) as the key bench mark for ToM (Leslie 1987). Intensionality is a particularly convenient concept in this context because it gives us a hierarchically structured index of a species' cognitive abilities that is directly relevant to social behaviour.

Unfortunately, we cannot at present determine in any detail the intensional abilities of different species. Nonetheless, we know enough to be certain that humans are able to achieve much higher levels of intensionality than other species, with fourth order intensionality being well within most individuals' everyday capabilities. This is equivalent to my "understanding that you believe that I think that you assume that something is the case." Apes seem able to achieve second order intensionality without too much difficulty, with some circumstantial evidence to suggest that they might be able to achieve third order intensionality. Monkeys, on the other hand, are clearly confined at best to second order intensionality, though there remains some doubt as to whether they might in fact be limited to first order intensionality (Byrne 1995). The distinction hinges critically on whether or not monkeys use theory or mind when engaging in tactical deception (the use of false information to mislead an opponent: see Whiten & Byrne 1988). Because it involves the use of false information, tactical deception was taken to be an analogue benchmark for false belief in animals. This is not to say that monkeys cannot engage in deceptive tactics designed to outwit opponents or inveigle friends (they clearly do), but it may mean that they engineer these strategies by using simpler behavioural rules rather than using theory of mind. In other words, to quote Cheney & Seyfarth (1990), monkeys are good behaviourists (they can read another animal's behaviour correctly) but poor psychologists (they cannot read the mind that underpins the behaviour); as a result, they can operate competently in a complex social world, but occasionally are caught out because they focus on the superficialities of the observed behaviour rather than trying to understand the mental states and intentions of their opponent.

To be truly effective, tactical deception depends on the possession of second order intensionality, since it involves the ability to inveigle someone into behaving in a particular way by feeding them false information. It seems that it may be possible to mimic this with first order intensionality, although the kinds of tactical deception that result are likely to be of poor quality and relatively superficial. Tactical deception (while important in itself) may thus not be a cast-iron benchmark for ToM. There is some uncertainty at present as to whether the kinds of tactical deception seen in monkeys is of the first or second order variety (Byrne 1995). This caveat aside, however, the data on tactical deception in primates provide us with an intriguing light on primate social and cognitive skills. Of particular significance in this present context is the fact that Byrne (1995) has shown that a measure of the relative frequency with which tactical deception has been observed in different primate species correlates with neocortex ratio. Thus, having a large neocortex appears to open the way for greater use of tactical deception (and presumably other comparably sophisticated social devices).

In contrast to the uncertainty with respect to monkeys, the situation in respect of great apes (or at least chimpanzees) is more clear cut. Chimpanzees emerged as the most frequent users of tactical deception in Byrne's (1995) analysis. In addition, evidence from experimental studies by Povinelli et al. (1990) and O'Connell (1996) provide convincing evidence that these great apes at least do possess formal theory of mind. Children are not born with a theory of mind ability, but acquire it at about the age of 4 years (Astington 1994). Some individuals (whom we label autistic) never develop this ability (Leslie 1987, Happe 1994). O'Connell (1996) devised a mechanical analogue of the standard false belief test which she applied to chimpanzees as well as normal children and autistic adults. Her results demonstrate rather clearly that chimps do better than autistic adults and about as well as 4-year-old children on the same test. In other words, chimps perform about as well as children who have just acquired basic theory of mind.

This does not, of course, mean that chimpanzees perform as well as normal adult humans. Normal adult humans are capable of fourth order intensionality (two orders above minimal ToM) as a matter of regularity, and can on occasion aspire to sixth level intensionality. Chimpanzees probably aspire only to second order as a matter of course, with third order intensionality as the absolute upper limit.

The important point in the present context is that these cognitive abilities do not come for free. It is clear that high levels of intensionality are extremely difficult to cope with in computational terms. Kinderman et al. (in press), for example, tested normal adults with a series of tests similar to those used in standard ToM tests but which allowed for up to fifth order intensionality (as opposed to the conventional second order of standard ToM false belief tests). At the same time, subjects were also given tests of environmental causal relationships that required only memory of a sequence of events. Memory tests involved causal relationships of up to sixth orders of embeddness ("A caused B which caused C which .... caused F"). Error rates on memory tasks varied fairly uniformly between 5–15% across the six levels of embeddness with no significant trends; in contrast, error rates on the ToM tasks increased exponentially with order of embeddness (i.e. intensionality).

While the error rates on the ToM questions were of similar magnitude (5–20%) up to fourth order intensionality, those for fifth order tasks soared to nearly 60%. We take this as strong evidence for the cognitive difficulty involved in sorting out mental states in other individuals.

Additional evidence to support this claim comes from the fact that there was a significant correlation between subjects' performance on the ToM tasks and their scores on an attributional scale questionaire. This showed that individuals who tended to make more errors on the ToM tasks were also more likely to misattribute the blame for any unfortunate events that befell them, being prone to blaming others rather than themselves or unavoidable environmental events.

These findings suggest that trying to coordinate and hold together very large groups depends on advanced cognitive skills like theory of mind and that it is to provide for this that the vast computing power of the human brain is required. Without wishing to endorse any form of progressivist views of evolution, it seems to me clear that the history of primate brain and social evolution can be seen as a continuing attempt to enlarge on very special cognitive abilities in order to facilitate the bonding and temporal coherence of increasingly large social groups.

3. WHERE DOES THIS LEAVE HUMANS?

We can now return to the relationship shown in Fig. 1 and ask what implications it has for modern humans. Since the same database on which Fig. 1 is based also provides data on human neocortex volumes, it is easy to plug the values for humans into the regression equation and predict a group size for modern humans (Fig. 2). This turns out to be approximately 148.

It turns out that this value is a surprisingly common size of social grouping in human societies of all kinds (Dunbar 1993). It seems to correspond to that set of people that any one individual knows well. It seems to correspond, for example, to the number of indi-

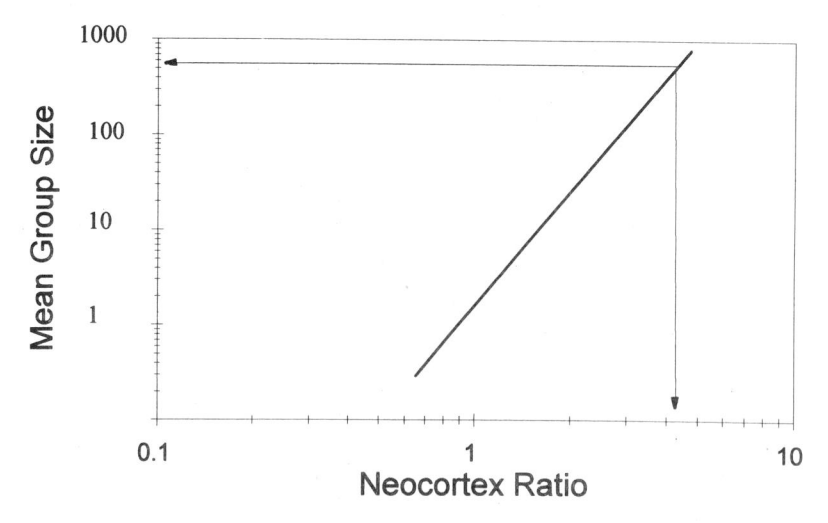

Figure 2. Predicting the group size for modern humans by interpolating the observed relative neocortex size for humans (4.2) into the regression equation for anthropoid primates.

viduals that one can ask a favour of without imposing. It is also a common size for villages or other local communities in traditional societies.

And herein lies the problem. Groups of 150 are approximately three times larger than the largest mean group sizes observed in other primates (50–55 in baboons and chimpanzees). If humans were to bond their groups using the same mechanism of social grooming as primates do, they would inevitably face an acute time budgetting crisis. There is a linear relationship between grooming time and group size in Old World monkeys and apes (Dunbar 1991) which we can use to predict the grooming time required to bond groups of 150. It turns out that something in the order of 40% of total day time would have to be devoted to intense social interaction.

Since no living primate species devotes more than 20% of its total time to grooming, this clearly imposes an enormous burden on any species that also has to earn its livelihood by foraging in the real world. The import of this for our ancestors would be the rather stark alternative between (1) attempting to increase brain size anyway in order to solve the ecological problem (but suffering a cost in terms of the cohesion of groups as a result) and (2) holding brain size constant and either going extinct or retreating to some more appropriate habitat where survival was possible in smaller groups. Neither of these solutions would have been especially helpful, since either way the species' ability to invade new habitats (or take advantage of new eciological conditions) would have been greatly restricted. Like the modern chimpanzees, humans would probably have been confined to the dwindling areas of forest and woodland.

The only alternative option would have been to evolve a more effective bonding mechanism that allowed time to be used more effectively. This leads directly to the suggestion that language cuts through this impasse by enabling us to use the time available for social interaction more efficiently.

There are two ways in which language allows us to do this. One is that conversations allow us to interact with several people simultaneously, whereas grooming is a strictly one-on-one affair. Even today, when humans groom each other, it remains a largely dyadic activity. In contrast, when we talk to each other, we often talk to several people at once. The other way that language increases the efficiency of social interaction is by allowing a wider dissemination of information. Nonhuman primates can only find out what is happening within their social world by direct observation. Humans, in contrast, can do so indirectly by word of mouth: I can tell you what your friends have been doing, and you can then update your mental social map accordingly. In addition, of course, we can use language for advertising: in this way, we can short-circuit the long drawn out business of finding out about each other by broadcasting a resume. I can tell you what kind of person I am, what I like and what I dislike, how I behave under different circumstances. You can then use this information to make an assessment of whether I would make a useful or reliable friend or ally. Of course, like all advertising, any such information needs to be carefully screened and checked. But even so, there must be considerable savings on time over what other monkeys and apes can do, given that they are restricted to direct personal experience to find these things out.

5. LANGUAGE AS A BONDING DEVICE

If this view of language is right, then we should at least be able to test some derivative predictions. One of these must be that human conversation groups are proportionately larger than nonhuman primate grooming groups. A second is that at least informal conver-

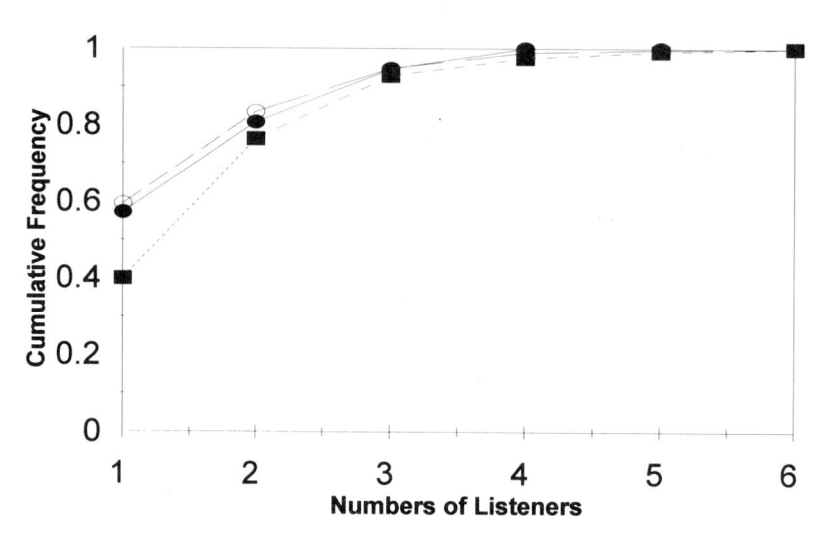

Figure 3. Cumulative frequency of audience size in naturally occurring conversations. Audience size is the number of individuals actively involved in a conversation by paying attention to the speaker. Data are shown for three separate samples, two in university canteens and one drawn from a series of public meetings. Source: Dunbar et al. 1995.

sations between acquaintances should be heavily biased towards what might loosely be called social gossip (in other words, information about oneself or one's acquaintances and their behaviour).

We have checked both of these predictions. We have sampled conversational groups in a number of different settings (including university canteens, receptions and fire practices). We have also undertaken detailed analyses of the topics of conversation (again, using both university canteens and other public places such as bars and trains).

The results from the first study suggest rather clearly that conversation group sizes are limited at about four individuals (one speaker and three listeners) (Dunbar et al. 1995). Fig. 3 plots the cumulative frequency distributions for the number of individuals that a speaker can reach (i.e. conversation group size less one, since there is always only one speaker at any given moment per conversation: Dunbar et al. 1995). All three datasets in the sample suggest that the number of listeners rapidly approaches an asymptotic value at around three.

If social grooming is a strictly dyadic activity, the number of individuals that an actor can reach is one by grooming and three by speaking, giving speaking an efficiency ratio of approximately 3.0 compared to grooming. The ratio of group sizes for humans and chimpanzees is 148 to 53.5, a ratio of 2.8:1. If one of language's functions is to upgrade the number of interactees by the amount needed to convert the maximum group sizes found in nonhuman primates to that predicted for (and found in) modern humans, then the efficiency of conversation turns out to be just about what it should be (roughly three times more efficient than grooming).

It is interesting to note at this point that this limit on conversation group sizes correlates rather nicely with the evidence from the study of psycho-acoustics suggesting that there may be an upper limit on the number of people who can interact in a conversational group. Cohen (1971) reviewed and reanalysed data from studies of voice attentuation in relation to both background ambient noise levels and the distance separating speaker and listeners and showed that, as group size increases, so the discriminability of speech sounds deteriorates. The rate of attentuation of speech sounds is such that, in an environment with

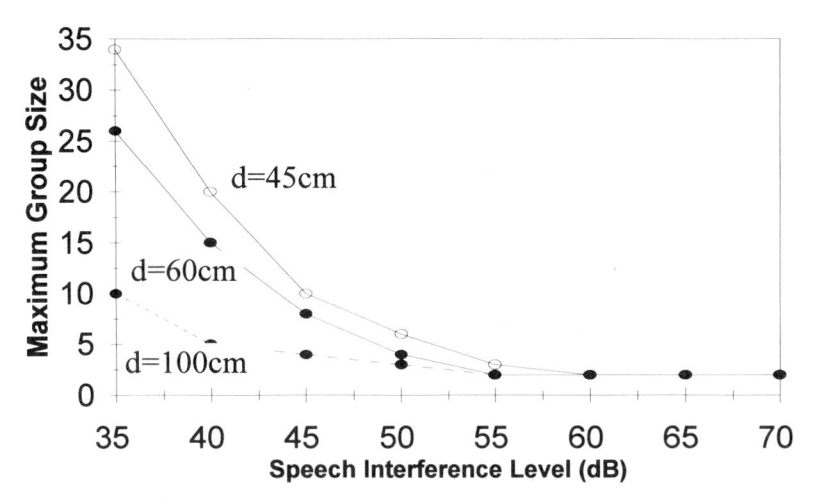

Figure 4. The maximum group size for a conversation in which all members are able to clearly discriminate speech sounds spoken at normal voice levels by all speakers, plotted against speech interference levels. The lower line (for $d=100$cm) is drawn by extrapolation from the other two lines. d is the inter-individual distance measured nose-to-nose between adjacent individuals on the circumference of the circle. Source: redrawn from Cohen 1971 (Fig. 7.1)

minimal background noise, it is possible to get only seven people in a circle that is small enough for everyone to hear clearly what the speaker is saying. As background noise levels increase, so the diameter of this circle (and hence the number of members) decreases, tending towards a limit at a group size of two (or even one!) as noise levels approach those that are physically damaging (such as at large receptions in confined spaces). At speech attenuation levels that are typical of normal everyday environments (e.g. a large office with a speech interference level of 47dB), the number of individuals that can form a circle without serious attentuation of speech sounds spoken at normal voice levels and with a nose-to-nose distance of 60cm between adjacent individuals is about 4.5 (Fig. 4). With larger inter-individual distances (e.g. 1m, with which many people would be more comfortable) or louder ambient noise levels (higher values of the speech interference level), the limiting number of members for the group would be proportionately lower.

The second prediction was tested by sampling conversational topics from natural groups in various settings (Dunbar et al. submitted). Fig. 5 shows the mean proportion of all speaking time devoted to different topics. Social topics included matters relating to personal relationships, personal experiences and social arrangements, including those of the speaker and the audience as well as those of third parties not present at the time. Topics in the politics category included matters relating to religion, ethics and morals as well as politics. The data clearly show that speakers in relaxed social settings talking informally with friends and/or acquaintances devote approximately two-thirds of their conversation time to topics relating to their personal social lives. The non-social world and even social matters that relate to the larger-scale aspects of society (politics, culture, etc) account for a surprisingly small proportion of total time. There are no significant differences between the sexes in the broad distribution of conversation time, though there are some interesting differences on a smaller scale.

I interpret these results as evidence in support of the hypothesis that language evolved as a bonding mechanism. Obviously, once such a feature is in place, it is perfectly possible for other functions to be brought into play such that it can be used to subserve

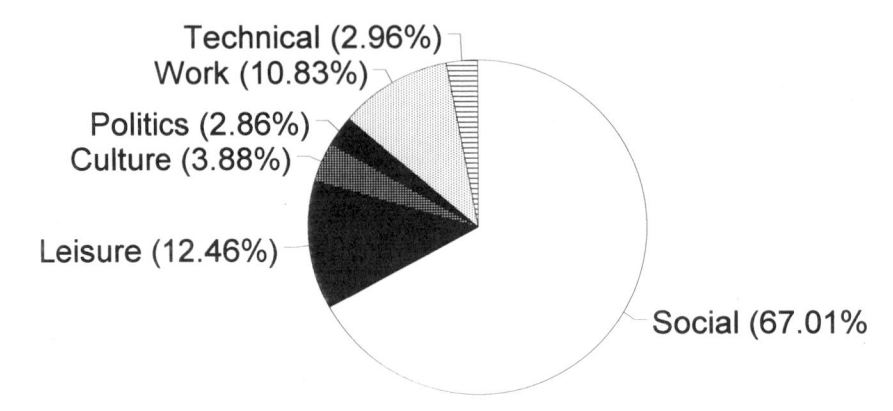

Figure 5. Proportion of conversation time devoted to different topics. The data are average values from three separate studies (two conducted in university canteens, one in public places). Social topics include anything concerned with personal relationships, personal experiences and arrangemenst for future social activities. The category politics includes all topics relating to religion, ethics and morals as well as politics. Source: Dunbat et al. submitted.

new mechanisms. Two likely possibilities in this respect are Machiavellian strategies of deception and the use of direct advertising (especially in the mate choice context). The former is obviously related to another important problem that human groups are likely to encounter, namely the problem of freeriders. Enquist & Leimar (1993) have pointed out that large dispersed groups are especially vulnerable to freeriders who exploit the cooperative nature of groups while failing to repay the debt. They suggested that gossip (the exchange of information on other's social failings may have evolved as a mechanism for controlling freeriders. Equally, the suggestion that sexual advertising may be an important function for language meshes well with Miller's (1994) ideas about the evolution of the human brain as a sexual signalling device.

The problem with both of these alternative suggestions is that they do not address the fundamental problem that large human groups would have faced—namely, how could such groups be kept functionally integrated? Freeriders only become a problem once groups have formed; they are not a problem that groups are formed to resolve. Similarly, mate advertising may be an important feature of everyday life, but it does not explain how large groups are kept together. Mate choice itself is not a sufficient function for large groups.

6. AN AFTERTHOUGHT

One final question remains unanswered. Primate social grooming works as well as it does because it is a highly pleasurable activity to the monkeys and apes concerned. It is pleasurable because it causes the release into the bloodstream of significant quantities of endogenous opiates (Keverne et al. 1989). It achieves this because the pinching and pummelling that is involved in grooming is just the kind of low intensity slightly painful activity that is particularly efficient at stimulating the release of opiates from the brain as part of the pain-control mechanism. This appears to provide the immediate reinforcement that makes grooming such a worthwhile activity to engage in. With speech, that immediate reinforcer was lost. This raises interesting questions about how this new bonding mechanism works at the proximate level.

One interesting possibility is that language actually mimics grooming by stimulating opiate production, such that the same proximate mechanism is used to reinforce social interaction (thereby providing the opportunity for the ultimate functional process of information transfer to take place). One way in which language might do this is by stimulating smiling and laughter. There is some evidence to suggest that both smiling and laughter are very good at stimulating opiate production. This is why we experience an opiate-like "high" after listening to a good comedian. It also explains why some conversations (invariably those in which we laugh a great deal) are particularly warming and make us feel good. A great deal of informal human conversation is designed to make us smile and laugh and laughter certainly plays a very important role in the dynamics of conversations (see for example Provine & Fischer 1989, Provine & Yong 1991, Provine 1993). My suggestion is that the use of laughter has replaced physical contact during grooming so that we can produce the same proximate effects by literally "grooming-at-a-distance."

This hypothesis is, at present, pure speculation. There is nothing other than circumstantial evidence to support it. However, it does provide an explanation for several unique features of human communication behaviour and, in doing so, it brings together into the same explanatory framework a number of hitherto unrelated facts. On these grounds alone, it clearly merits more detailed investigation.

7. CONCLUSION

In this paper, I have tried to spell out both the logic and the evidence relating to the hypothesis that language evolved as a bonding device designed to increase the efficiency of information transfer in humans in order to bond significantly larger groups than conventional primate mechanisms (social grooming and direct personal experience) are capable of doing.

It is, I think, important to bear in mind that during evolution processes and mechanisms are constantly taken over for new purposes. This applies as much to the transformation into speech of the original vocal behaviour of our primate ancestors as to language itself during subsequent evolutionary periods. In the latter respect, it is, I think, clear that the opportunities provided by language have been capitalised on to cope with other kinds of social problems faced by humans living in large groups (notably controlling the activities of freeriders).

It is clear that there is considerable work yet to do in testing the hypothesis and disentangling the details of the mechanisms that allow language to do the work we think it does. However, the main point for the moment is that this hypothesis offers an opportunity to test specific predictions in a way that no other previous alternative hypothesis has done. More importantly, it allows us to test between competing hypotheses. In one sense, all I have done is to formulate an alternative to the conventional wisdom that language evolved to enable early humans (hominids?) to organise and plan hunting activities or teach tool making.

REFERENCES

Aiello, L.C. & Dunbar, R.I.M. (1993). Neocortex size, group size and the evolution of language. *Current Anthropology* 34: 184–193.

Astington, J.W. (1995). *The Child's Discovery of the Mind*. Fontana, London.
Barton, R. (1996). Neocortex size and behavioural ecology in primates. *Proceedings of the Royal Society, B,* 263: 173–177.
Barton, R. & Dunbar, R.I.M. (1996). Evolution of the social brain. In: R.Byrne & A.Whiten (eds) *Machiavellian Intelligence*, Vol. 2. Cambridge University Press, Cambridge.
Barton, R. & Purvis, A. (1994). Primate brains and ecology: looking below the surface. In: J.Anderson, B.Theirry & N.Herrenschmidt (eds) *Current Primatology: Proceedings of XIVth Congress of the International Primatological Society*, pp. 1–11. University of Strasbourg Press, Strasbourg.
Brothers, L. (1990). The social brain: a project for integrating primate behaviour and neuropsychology in a new domain. *Concepts in Neuroscience* 1:27–51.
Byrne, R. (1995). *The Thinking Ape: Evolutionary Origins of Intelligence*. Oxford University Press, Oxford.
Byrne, R. & Whiten, A. (eds) (1988). *Machiavellian Intelligence*. Oxford University Press, Oxford.
Cheney, D. & Seyfarth, R. (1990). *How Monkeys See the World*. Chicago University Press, Chicago.
Clutton-Brock, T.H. & Harvey, P.H. (1980). Primates, brains and ecology. *Journal of Zoology, London,* 190: 309–323.
Cohen, J.E. (1971). *Casual Groups of Monkeys, Apes and Men*. Harvard University Press, Cambridge (Mass.).
Dunbar, R.I.M. (1988). *Primate Social Systems*. Chapman & Hall, London.
Dunbar, R.I.M. (1991). Functional significance of social grooming in primates. *Folia primatologica* 57: 121–131.
Dunbar, R.I.M. (1992). Neocortex size as a constraint on group size in primates. *Journal of Human Evolution* 20: 469–493.
Dunbar, R.I.M. (1993). Coevolution of neocortical size, group size and language in humans. *Behavioural & Brain Sciences* 16: 681–735.
Dunbar, R.I.M. (1995). Neocortex size and group size in primates: a test of the hypothesis. *Journal of Human Evolution* 28: 287–296.
Dunbar, R.I.M. & Bever, J. (submitted). Neocortex size determines group size in insectivores and carnivores. *Ethology*
Dunbar, R.I.M., Duncan, N. & Marriot, A. (submitted). Human conversational behaviour: a functional approach. *Ethology & Sociobiology.*
Dunbar, R.I.M., Duncan, N. & Nettle, D. (1995). Size and structure of freely-forming conversational groups. *Human Nature* 6: 67–78.
Enquist, M. & Leimar, O. (1993). The evolution of cooperation in mobile organisms. *Animal Behaviour* 45: 747–757.
Happe, F. (1994). *Autism: An Introduction to Psychological Theory*. University College London Press, London.
Harcourt, A.H. (1992). Coalitions and alliances: are primates more complex than non-primates? In: A.H.Harcourt & F. de Waal (eds) *Coalitions and Alliances in Humans and Other Animals*, pp.000–000. Oxford University Press, Oxford.
Jerison, H. (1973). *Evolution of Brain and Intelligence*. Academic Press, New York.
Keverne, E.B., N.D.Martinez & B.Tuite (1989). Beta-endorphin concentrations in
cerebrospinal fluid of monkeys are influenced by grooming relationships. *Psychoneuroendocrinology* 14: 155–161.
Kinderman, P., Dunbar, R. & Bentall, R. (in press). Theory of mind deficits and causal attributions. *British Journal of Psychology.*
Koyama, N. & Dunbar, R.I.M. (1996). Anticipation of conflict by chimpanzees. *Primates* 37:79–86.
Kudo, H., Lowen, S. & Dunbar, R. (submitted). Neocortex size as a constraint on grooming clique size in primates. *Behaviour.*
Leslie, A. M. 91987). Pretense and representation: the origins of "theory of mind." *Psychological Review* 94: 412–426.
Miller, G. *Evolution of the Human Brain Through Runaway Sexual Selection: The Mind as a Protean Courtship Device.* PhD thesis, University of California.
O'Connell, S. (1996). *Theory of Mind in Chimpanzees*. PhD thesis, University of Liverpool.
O'Keefe, J. & Nadel, L. (1978). *The Hippocampus as a Cognitive Map*. Clarendon Press, Oxford.
Pawlowski, B., Dunbar, R. & Lowen, C. (submitted). Neocortex size, social skill and mating success in male primates. *Behaviour.*
Povinelli, D., Nelson, K.E. & Boysen, S.T. (1990). Inferences about guessing and
knowing by chimpanzees (*Pan troglodytes*). *Journal of Comparative Psychology* 104: 203–210.
Premack, D., & Woodruff, G. (1978). Does the chimpanzee have a theory of mind? *Behavioural & Brain Sciences* 4: 515–526.
Provine, R.R. (1993). Laughter punctuates speech. *Ethology* 95: 291–298.

Provine, R.R. & Fischer, K.R. (1989). Laughing, smiling and talking: relation to sleeping and social contexts in humans. *Ethology* 83: 295–305.
Provine, R.R. & Yong, E. (1991). Laughter: a stereotyped human vocalisation. *Ethology* 89: 115–124.
Sawaguchi & Kudo, H. (1990). Neocortical development and social structure in primates. *Primates* 31: 283–290.
van Schaik, C.P. (1983). Why are diurnal primates living in groups? *Behaviour* 87: 120–144.
Stephan, H., Frahm, H., & Baron, G. (1981). New and revised data on the volumes of brain structures in insectivores and primates. *Folia primatologica* 35: 1–29.
Whiten, A. & Byrne, R. (1988). Tactical deception in primates. *Behavioural & Brain Sciences* 11: 233–244.

6

THE COMMUNICATION PARADOX AND POSSIBLE SOLUTIONS

Towards a Radical Empiricism

Karl Grammer,[1] Valentina Filova,[2] and Martin Fieder[1]

[1]Ludwig-Boltzmann-Institute for Urban Ethology
c/o Institute for Human Biology/University of Vienna
Althanstrasse 14 A-1090 Vienna/Austria
[2]Institute for Automation-Department for Pattern Recognition and Image
  Processing
Technical University Vienna
Treitlstrasse 3/1832 A-1040 Vienna/Austria

In the history of both animal and human ethology the direct observation of unstaged interactions in a natural habitat plays a critical role for methodological and theoretical considerations. Even when ethologists think that they already know much about adaptations and the ways in which they interact with the environment, the principles which have been involved in the evolution of increasingly complex human behaviour are still not very well understood.

A major reason for this lies in methodological problems connected with the observation and description and the nature of human behaviour itself. In order to asses causation and function of behaviour we rely on an "observational device." The process of information reduction which is applied to the study of behaviour results in highly variable observations. The assessment of meaning and function rarely produces reproducible results, and different signals especially in human communication seem to take many meanings which are context-specific. Partially this might be due to the observational approaches used for coding behaviour.

1. WHAT IS COMMUNICATION?

A straightforward definition of communication is not difficult. As a starting point we can define it as the transfer of information between two communicative units. Ethology has created many models for the process of information transfer. Basic to these approaches is the term "signal," an information carrier which is produced through encoding information in an

New Aspects of Human Ethology, edited by Schmitt *et al.*
Plenum Press, New York, 1997

communication channel by a sender. This signal is decoded by a receiver who adds information to the signal and then decodes its meaning. In a classical ethological approach many of the signals are a result of evolutive constraints and work in a quasi automatic way. Although many human signals have been isolated as cultural universals (Eibl-Eibesfeldt, 1972) a closer look reveals high variability. For instance Ekman and Friesen (1971) propose cross-cultural universal facial signals for emotions. An observational approach to the study of emotions in every day life reveals a high variability in the production of patterns and pure emotion patterns which occur rarely (Grammer et al., 1988).

The Lorenz-Tinbergen approach sees signals as discrete and deterministic: A sends a signal X and B decodes and returns Y. In this definition, visual, tactile, acoustical and verbal information are divided into units of meaning. Signals have a lexical structure: one entry in the lexicon has one specific meaning and one definite function. The basic assumption of this approach is that signals exist as independent units and sender and receiver share a common code. The signals themselves are considered as discrete units of movements each with a beginning and an end in time, which in turn can frame a static component like a posture. This is clearly demonstrated by the acoustic properties of laughter which separate it from speech (Provine and Young, 1991), or by bodily movements in interactions, like illustrating hand-movements (Ekman and Friesen, 1972). Thus, signals have a content which is different for each signal, and which makes them identifiable reliably. If so, signals have to show a certain form constancy which is necessary for identification and two different signs do not overlap in their meaning. Most of the times form constancy defines the relation of movements to each other, like the typical head movements connected with laughter (Grammer and Eibl-Eibesfeldt, 1989). If laughter occurs in interactions, the head is moved in a circular fashion away from the partner.

Another approach to signalling uses the same basic structure, but it takes the probabilistic nature of communication into account. It still assumes that signals are discrete: A sends X and B decodes and returns Y, or not (Argyle, 1988). These probabilistic models imply some law of summation over space and time. This means that A sends X and Z and T to B at the same time, or A sends X and Z and T sequentially. This model of communication is called "modulated communication" (Markl, 1985). Grammer (1995) proposed that "natural" signs sent in parallel, like age and sex, signals of dominance/ submission or emotions contain the decoding instruction for a signal. For instance age and sex of the sender can modulate the meaning of a smile from a "come-on" to a simple friendly gesture. A slightly different approach was suggested by Schleidt (1973) as tonic communication. He assumed that meaning could be encoded in the form of pulse rate modulation. The sender sends a signal of uniform height and duration repeatedly in distinctive intervals. The receiver then applies some kind of low pass filter in order to integrate the signals over time. The effect on the receiver then is a slowly accumulating, tonic one.

For the observation of communication behavioural categories are constructed as classes or prototypes of signals. These classes have to be stereotypical, homogenous and discrete in order to be reliable. In addition we want to avoid functional descriptions because the goal of ethological approaches is the description of function itself. Thus, in order to produce reliable results the reduction of information has to be enormous. An example for such a classical approach is provided by Grammer (1991) who paired strangers randomly and tried to find out if body movements and postures could predict self-reported interest in another person. The resulting clusters of movements proved to be inconsistent and unreliable in predicting interest in the other person.

Furthermore, single postures did not covary with reported interest in this study. But as soon as postures and vocal stimuli were combined, the situation changes dramatically.

The Communication Paradox and Possible Solutions

Figure 1. Body postures and laughter. The figure shows body poses 2 seconds before two strangers laughed in a waiting room experiment (Grammer, 1991). The postures where constructed by regression analysis from 4547 postures. In (a) the most frequent postures during laughter are shown. In the right picture the view from the female is shown, the middle picture shows the observers view and the left picture shows the view from the male. In (b) the postures are shown males and female take during laughter which signals aversion and in (c) postures during laughter which signals approval and interest in the other person are shown. Interestingly there is an additive effect for the single posture elements in the different body parts: the more of the single elements are present, the higher is the correlation with the self-reported intentions (Wire frame models by A. Jütte).

It was possible to show that postures which are taken during laughter might well transport the meaning of laughter. The acoustic event laughter does not alone delineate interest from no interest. Highly interested males or females do not laugh more often than persons with no interest. Moreover, people who are together with strangers of the same sex laugh more often. In sum, there is no contextual evidence that laughter alone is a sexual signal. When combined with postures however, laughter may take different meanings on a continuum from rejection to appraisal of the partner.

Evidence from the consequences of particular signals reveals many contradictions. The same behaviour can have different meanings. *Open legs* among females and *Hair Flip* (Fig. 2) where the hair is moved out of the face with the hand and the head tossed backward, indicates low female interest. If both behaviours are combined with laughter they covary with high interest. Other behaviours may take different meaning when they are static or dynamic. The *Head Akimbo*, a behaviour where the breast is pushed out and the hands are folded behind the neck, is associated with high interest when it occurs as a

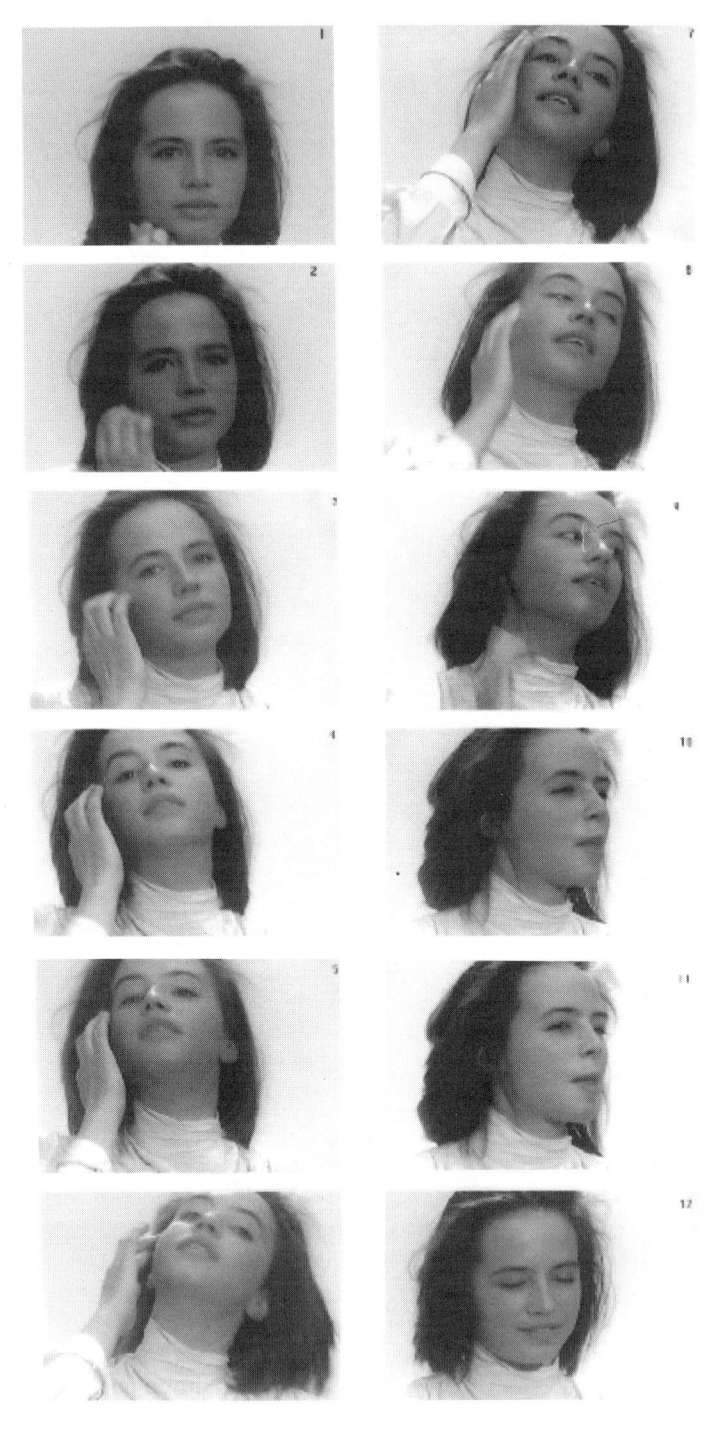

Figure 2. Hairflip—a female mannerism. The Hairflip consists of a typical movement sequence which starts with a slight head tilt, followed by a head up movement. The hand reaches out into the hair and the head turns back into the starting position with gaze aversion. This movement is performed more often by females (Grammer, 1991) than by males.

movement during laughter, but with low interest when it occurs as posture before laughter (Grammer, 1991).

The communicative situation presents itself as unclear and ambiguous although the receiver seems to be able to decode the senders intentions. Receivers are generally aware of what the sender wants to tell them. Thus the decoding of meaning in interactions can not be described with a simple signal oriented approach. As an alternative, we could speculate that meaning and intentions are communicated solely through the verbal channel by speech content itself. Krauss et al. (1981) showed that speech accompanying gestures were not related to the decoding of meaning. They assume that body and arm movements are results of speech production itself and propitiate speech production. In earlier rating experiment of politicians Krauss et al. (1981) had already shown that verbal information dominate visual and auditive information. If we agree with this approach it seems useless to search for signalling intentions in human non-verbal behaviour. In contrast to the above mentioned results, Mehrabian (1972) showed in series of experiments the relative role facial expression, vocal behaviour and speech content play in the perception of persons. Mehrabian comes to the general conclusion that non-verbal behaviour plays the main role for the decoding of meaning. He finds that the meaning of messages is determined to 55% by visual information, 38% by vocal information and only 7% by speech content. These relative relations have been replicated by Siddiqi et al. (1973) and Wallbott (1991).

If we look at the content of the information which is transferred between interactants, we find that facial information is used for decoding tendencies of dominance and positive affect (Rosenthal and Depaulo, 1979). Ekman et al. (1980) gave information from seven different communicative channels: only facial information, only bodily information, speech, filtered voice, transcribed speech content, and combinations of voice and speech, voice, body and speech. In this research none of the presented channels was dominating the others in the transfer of meaning.

1.1. What Is Communication for?

Communicative models are a description of a communicative process of information transfer. Most models will fail when it comes to explain the function of communication because of the nature of communication itself, and the tools which are applied to reduce the information, as we will show later. Social groups are complex structures and their main feature is that the goals of the members rarely are in accordance. Human groups can be seen as an agglomeration of conflicting interests. This fact ultimately may be the driving force behind the evolution of social intelligence. Proximately it may be the basic constraint for communication and thus the generation of signals in any channel of communication.

The probabilistic multi-meaning nature of human communication is present in verbal and in non-verbal communicative acts. Linguistic research shows that indirectness of verbalisation and verbal acts like "hedging" depend on the risk of the intended communicative act (Brown and Levinson, 1978). If the benefits for the sender are high and the costs for the receiver are also high, it is obvious that the risk of not reaching the pursued goal for the sender is also high. As a result, the sender has to use signals and actions which allow to manipulate the receiver in the sense of the sender. Evolutionary theorists have forwarded comparable ideas. Openly presenting intentions in communication might not pay (Dawkins and Krebs, 1981) because the signal receiver might act directly against the sender's intentions. The sender thus would not be able to reach his/her goals. In addition, as soon as the receiver recognises the intentions of the sender the probability of deception might rise. This situation drives any type of communication into manipulative efforts. The manipulative component of a signal has to force the receiver into a certain state where he is willingly accepting the goals of the sender, preferably without recognising that he was manipulated. This situation is the communicative paradoxon: showing intentions and not getting caught by a suspicious receiver. In this view the function of communication is manipulation and is used for risk dependent transfer of information. Thus, a prerequisite for any communicative model is the assessment of risk, which will be highly context dependent. Risk itself is created by the goal under quest, i.e. the imposition for the receiver, the relationship between the interactants and motivational factors. In our introductory example of the waiting room situation risk should be high for a person who develops interest in the other person. Risk is determined by the possible costs and benefits for both sender and receiver. Risk dependent communication allows the explanation of simple straightforward transfer of meanings (under low risk conditions) and highly ambiguous transfer of information in situations with high risk. In this view both verbal and non-verbal channels can be affected. So, any model of communication should take risk-assessment into account. Non-verbal behaviour may be an important tool in high risk situations because of its non-binding standard, when compared to verbal behaviour (Grammer et al., 1996). The contradiction in the results which show either dominance of verbal information over visual information or vice-versa lies in the fact that an independent rater of situations had no possible costs in such an experiment, nor had the sender of the signal. This trap forces research on communication to work under naturalistic conditions, that is to observe unstaged social interactions.

1.1.1. Direct Communication. How can the sender achieve the delicate task of risk dependent communication? The sender has to assess risk and to act accordingly. The production of signals then could be optimised. Signals used in a situation of high benefit and/or high costs for both the receiver and the sender should be easily decodable. In this case the only preconditions of effective signalling are low environmental noise, encoding error by the sender and decoding errors by the receiver. An important means to actively reducing these errors is to increase contrast in a signal. Another mechanism is to produce a signal repeatedly and constantly over time. Both processes lead to ritualisation of signals. A ritualised signal consists only of a few elements that are produced repeatedly and in a fixed sequence. The aim of ritualisation is to make a signal definite and unmistakable. Grammer and Eibl-Eibesfeldt (1989) have shown that laughter follows ritualisation principles. Under high risk conditions, female laughter becomes more stereotypic, the threshold for performance is lowered and it is accompanied by typical movement sequences. As soon as risk becomes more asymmetric and the possibility of deception rises, the communicative situation changes drastically.

1.1.2. Lying, Deception and Mind Reading. The first possibility is the use of deception in the sense of sending false information. There are some constraints connected to lying. An example are children who cry in conflicts (Grammer, 1992). A crying child who is engaged in a conflict with another child receives support by a third child in most cases. The risk of the supporter is high in such a situation, because he might get attacked. The probability of getting support depends on the frequency the child cried in the past. If it cries too often and uses crying in a deceptive way (i.e. to receive support) he/she won't receive any longer support. The receiver only engages in support if the honesty of the signal is guaranteed. Thus the use of deception will depend from the frequency in which it is used and the costs and benefits connected to the interaction. Harper (1992) pointed out that deception can only occur when the frequency of deception is low, the signal has little costs for the sender and high benefits for the receiver. This situation forces the receiver to apply "mind-reading" and try to find additional cues for possible detection of deception. If the sender tries to deceive the receiver, the sender will try to control his behaviour in order to avoid detection. Yet control is rarely complete. If the sender tries to control his emotions he for instance creates leaks in the rest of his non-verbal behaviour. The receiver then will be able to detect the deception (Ekman and Friesen, 1969). Therefore lying is not always a solution for the communicative paradox.

The second form of deception is withholding information. According to Harper (1992) this is the main form of deception. Even if its use is widespread, the signal sender has to clarify his intentions sooner or later, or he will not be able to reach his goals.

1.1.3. Direct Cognitive and Physiological Manipulation: A Smile Is Not Just a Smile. Direct manipulation of the cognitive apparatus or the physiology of the receiver can play a role. Lorenz (1973) and later Cosmides et al. (1992) proposed that our information processing apparatus was formed and optimized in the course of evolution. If our brains are optimized for adaptive information processing then these adaptive structures can be exploited. An example of this possibility is the perception of emotions. The signal receiver experiences the same physiological changes as the sender of an emotion (Ekman et al., 1983). Thus by sending signalling "emotion" the sender is able to influence the physiology of the receiver. In the case of a smile, this makes sense because emotions change the cognitive processing of social stimuli. Happy people process information less critically than sad people (Forgas, 1992). Thus, a smile does not only mean "I am Happy," it simply influences the information processing in the receiver in favour of the sender (Grammer, 1995). This also explains why smiling is not necessarily bound to emotions (Kraut and Johnston, 1979). Smiles are more reliably associated with social motivations than with emotional experience. Comparable physiological changes occur with the perception of olfactory stimuli. A female pheromone, i.e. copulin, which is produced in the vaginal secretion influences male processing of female attractiveness. Under the influence of female copuline, males judge female attractiveness more positively (Grammer et al., 1996).

Although direct manipulation of the receiver through signals might play a critical role for communication, it does not yet explain how senders can hide their intentions. This goal can not be achieved simply by doing nothing, because receivers will become suspicious. Again there are many possible solutions to such a goal.

1.1.4. Being Honestly Dishonest. The first solution is the sending of meaningful signals either out of context or with different motivational background. This motivational background migh contradict the actual goal of the sender. Grammer and Kruck (1996) showed that this solution is not uncommon. Females who are not interested in males do not send negative signals, they send a mixture of 60% positive and 40% negative signals. In this case the negative message is hidden in positive signals. In this case the deceptive attempt is the avoidance of face-loss by the male and thus potential aggression towards the female, which could occur when the male is bluntly rejected.

1.1.5. Multimodal Combinations: Signals from Different Channels and Metacommunication. A second solution to the problem of intention hiding lies in combining signals from different sensory channels. For instance in the case of laughter, body movements or postures, each with a determined meaning itself, can be added, but also laughter quality (i.e. amount of vacilsation, number of bouts etc.), odour and touching can co-occur. Such combinations have the potential to create an almost infinite number of meanings of laughter, from sexual enticement to mobbing (Grammer, 1991). Furthermore metacommunication, becomes possible through multimodal combinations. Laughter for instance might put everything else what happened in a "play-mode" which simply says "Look it is not serious what I am doing" (vanHooff, 1972). By doing so, multi-meaning combinations are possible, which allow almost endless combinations for communicative purposes.

In the case of multimodal combinations, communicative channels are interacting. If this occurs, there has to be cross-modal integration of information. By comparing the impact of the different parts of the signal the receiver has to arrive at a decision what actually is meant. This problem is well known in research on the interaction of non-verbal behaviour and verbal utterances. Eyebrow raises and slight vertical or lateral rotations of the head appear to serve both punctuating speech rhythm and the emphasizing utterances. Although eyebrow raises are a clearly identifiable and a cross-culturally constant and discrete signal (Eibl-Eibesfeldt, 1989, Grammer et al., 1988), it can take many different meanings when associated with speech. Thus we would reach an almost infinite number of possible combinations. This is also the case for combinations with of eye-brow raising with other facial muscle movements. Grammer et al. (1988) showed that the resulting patterns where highly variable. This makes it unlikely to reencounter the same combination of stimuli again and these could make decoding more difficult. The same could also account for the high variance in human non-verbal behaviour.

1.1.6. Higher Order Combinations: Signals from the Same Channels. Higher order combinations or the summation of signal combinations over time could also veil the senders intentions. Combinations can occur when simple or multimodally combined signals are combined again either sequentially or at the same time. It becomes even more complex when a a member of the combination can be substituted by another signal with the same meaning. As a result the same meaning can arise through different combinations, which are not identical. Moreover if in such combinations not only the presence of a signal plays a role, even the absence can be a signal itself. Moore's (1985) description of female solicitation may serve as illustration. She found 51 behavioural units in the validation of her repertoire and any combination of at least 10 units of female behaviour could predict male approaches. How could a male receiver have managed to "evaluate" or "recognise" female interest? Indeed there are $1.3*10^{10}$ possible combinations of 10 behaviour elements out of a repertoire of 51.

1.1.7. Manipulation of the Time Structure: Good Vibrations and the Generation of Noise. There is mixed evidence on the assessment of function when using discrete behavioural categories, thus syntactical rules could provide some cue for the interpretation of signals independent of the signal content. With a mathematically sophisticated method, Grammer et al. (in prep.) analysed behaviour sequences in male-female interactions. They found a highly complex interweaving of behavioural elements. Pairs created dance like movement patterns. The drawback is that these patterns are highly idiosyncratic—not one pattern occurred twice in almost 8000 patterns identified on the pair level. The temporal organisation of these patterns varies with interest: they become more stereotypical when interest in the partner is high. Even more, the female initiates the patterns and the more patterns are present the better the male feels himself in the interaction. This suggests that temporal organisation itself could predict male-female-interest. Manipulation of time structure and its perception can also by reached by trying to achieve synchronisation. Although most attempts to empirically describe synchronisation empirically have not been successful. In contrats to the impossibility of description subjects can rate the degree of synchronisation in interactions, and this ratings correspond to the subjective experience of interpersonal rapport among the interactants (Bernieri and Rosenthal, 1995).

Another solution to the hiding of intentions problem is the creation of "noise." That means sending many signals without attaching meaning to it, and hiding the meaningful signals in this "noise." The noise will make it almost impossible for the receiver to decode the real intentions of the sender. In this context, the concept of "protean behaviour" (Chance and Russell, 1959) could play a role. The concept proposes that behaving unpredictably and erratically will mask the signal sender's intentions like a prey trying to evade its predator. In the study cited above, the repeated and time constant patterns of behaviour are hidden in a continuous flow behaviour. A closer analysis of the data reveals that only the precise timing and not the content of the performed behaviour plays a role for the generation mutual understanding.

Another alternative is that rare events which are hidden in "noise" advertise intentions at "hot times" at "hot-spots." This means that one single signal with a distinctive and explicite meaning sent at the right time could signal the intentions of the sender. Thus one or two events which might differ from interaction to interaction can be enough for communicating intentions. This alternative is highly likely because individuals communicate stimulus information in a way which is adaptive for perceivers to detect, and perceivers detect this information, when they are attuned to it (MacArthur and Baron, 1983).

Mutual understanding can also be reached by sending information in a way the receiver can not consciously assess. This means sending unknown signals to the receiver or sending variants of signals the receiver is not likely to interpret. Thus, the receiver is forced to learn the shared code slowly. This hazard would lead to a highly variant shared code. The emphasis lies on the term "learning." The sender has to prepare the receiver slowly for receiving his actual intentions.

1.2. The Quality of Movements: A Neglected Dimension

Slow escalation and hiding intentions when at risk of failure on the sender's side is is paired with a cognitive apparatus which tries to unravel the intentions on the receivers side. This situation will force communication to a level where the receiver might not be able to asses consciously the manipulation which is underway. The receiver then has to look for honest and develop strategies for "mindreading."

This situation is contrasted with a scientific research apparatus which seems not at all adequate for the analysis of such a paradoxical situation. It is obvious for any observer that behaviour is distinctive and that there are many levels and methods of description for signals. Categories can embody muscle movements or groups of such movements like in Ekman's and Friesens Facial Action Coding system (Ekman and Friesen, 1978) or descriptions of limb movements like in the Berner System developed by Frey and Pool (1976) or even more complex units like walking. Muscle movement description is the most basic level of description. On the next level behaviour is already described by interpretative categories, even when the definition is highly operationalized. This interpretation sometimes can involve hypotheses on the function of a behaviour. The term "coy smile" is already a hypothesis on the function of behaviour although it describes a distinct motor pattern of head movements combined with a smile (Eibl-Eibesfeldt, 1989).

Common to all these approaches is that a continuous behaviour stream is forced into a series of event categories which might subsume comparable, but visually distinct behaviours. For example a non-verbal threat can be done in many ways. By raising an arm fast or slow, with fist clenched or not, the movement staying at the maximum flexion for a certain time and going back fast or slowly. Any of the possible combinations will produce a different type of "threat." We also can transfer the movement combination itself to a leg or even a head movement: moving the head fast towards somebody else then staring at him and finally look slowly away. In every case we will produce an event of "threat" by using a certain movment configuration. A solution would be to describe all these different types of threat, with the result that the numbers of each type of threat events will become small and useless for statistical treatment. An additional problem at this level is the reliability of observations: in order to be identified reliably and to avoid the development of an observer bias, categories have to be unmistakable, stereotypical, homogeneous and discrete. This can lead to oversimplification and broad categories.

Thus a dilemma arises—observers surely can interpret and understand the behaviour of others and their possible intentions correctly but we might not be able to identify them on the bases of communication theories which use discrete signals.

The solution we propose to this dilemma is that the categories which are used to break up the flow of behaviour are only a poor approximation to how the receiver processes information. Any behaviour is a change in a continuous information flow which could be noticed by a receiver. These changes in the information flow can be of various qualities. It is possible to encode information in the quality of a change. We propose that body movements and parts of body movements themselves are processed, i.e. the receiver does not perceive and summate behaviour in single categories "Legs Open" or "Hair Flips." In contrast the receiver could assess elementary dimensions of behaviour like speed, acceleration and amount of movement or motion quality. The brain might not use categorical perception in the same way as we classically analyse behaviour. An example for different possible meanings that arise through speed differences in onsets of the events was given by Grammer et al. (1988) for "eye-brow-flashes" in a cross-cultural analysis. In the most common event the pattern starts with the contraction of the M. corrugator supercilii. This contraction disappears and in a fast movement the brows are lifted and a smile appears. The duration of the contraction of the M. frontalis and the Pars palpebralis which causes the eyebrow lift is variable and disappears slowly, whereas the smile caused by a contraction of the M. zygomaticus major stays on the face. The second pattern also starts with a contraction of the M. corrugator supercilii. But then, there is a slow lift of the brows, and the contraction of the M. corrugator supercilii does not disappear. Rarely a smile is added (See figure 3 for details). In this case, meaning arises through the combination of elements present and the dynamic properties of the elements.

The solution to the dilemma between event oriented categorisation of behaviour and possible qualitative differences in the same behaviour with different meaning can be found in the way visual information is processed. The information is not only reduced, but parallel new information is created. Processing of visual signals takes place on two levels: one is low level processing where the perceived information is recoded: colour, motion, depth, time integration of movements. On this level information is detected which seems to be necessary for spatial navigation. During high level processing, where pattern recognition occurs, a "world-model" or a priori knowledge like form, size, schemes is added. On this level, the brain compares the results it has got so far and then tries to come to a coherent interpretation of the world (Arbib and Hansen, 1987). Traditional behaviour research is working on the second level and neglected the first level which basically describes the quality and not the content of behaviour.

In order to assess qualitative aspects of body movement Johansson (1973, 1976) used a point light display fixed to the joints of his subjects and filmed them in the dark. The resulting films appeared as a configuration of bright points against a dark background. If such point-light clips are shown to raters, they can recognise sex and age of the

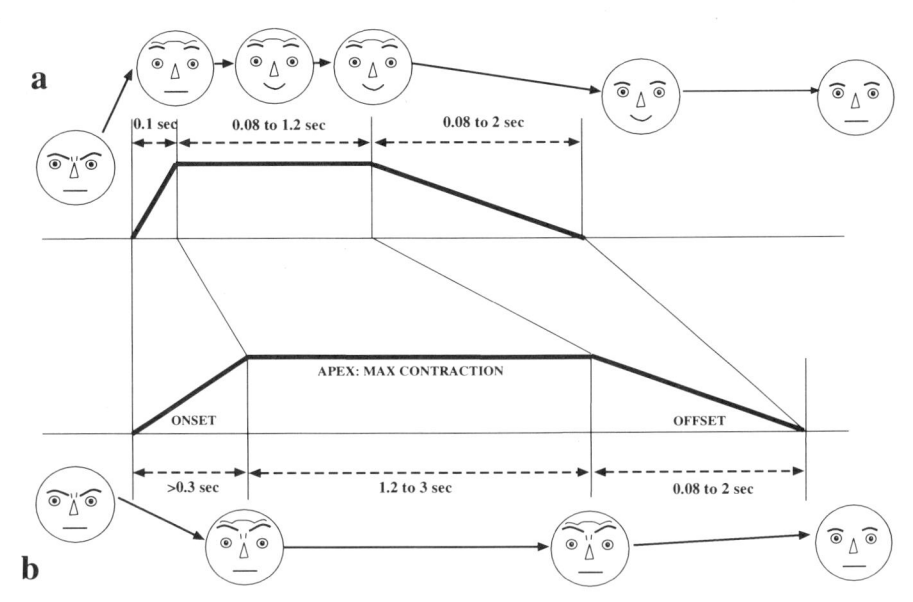

The Communication Paradox and Possible Solutions

Figure 3. Time structure and patterns in the eye-brow-flash. The figure shows two different patterns in a eye-brow-flash. The prototypical pattern in (a) was described by Eibl-Eibesfeldt (1972). This pattern starts with a frown which disappears. The brows are lifted quickly and a smile is added. The brow raise disappears, while the smile can stay on the face. The second pattern (b) is completely different: It also starts with a frown, which does not disappear whilst the brow raise appears on the face. The brow raise onset duration is three times as long as in the first pattern and the patternsduration is much longer than in (a) (Grammer et al., 1989). In addition to the difference in combinations of muscle movements there is also a difference in the time structure which changes the quality of the expression.

subject (Cutting and Proffitt, 1981). This effect has been replicated several times (e.g. Runeson and Frykholm, 1983). If the point light displays are presented statically they are rated as a random display of points. Observers are able to isolate abstract patterns of movements from such displays. Moreover, observers are able to detect effort, intentions and deception (Runeson and Frykholm, 1983), and if the point light displays are applicated in faces, perceivers are able to detect information connected to emotion (Bassili, 1979). Gait and movement analysis has a considerable tradition in medicine where it is used to describe movement disorders. It has rarely been used in behaviour research.

Motion quality is a powerful descriptor which is used for signal decoding by humans. Yet there is a serious methodological drawback: point light displays can not be used in unstaged interactions, because they are an obvious research device. They make any stimulus person rather self conscious and alerts them to the variables of interest to the researcher (Berry et al., 1991). We thus developed a digital image analysis system which can be used to decode automatically speed, acceleration and size of moving body parts from digitised video images.

2. AUTOMATIC MOVIE ANALYSIS (AMA) AND MOTION-ENERGY-DETECTION (MED)

Machine understanding of human action has become a fascinating topic in computer sciences through the last years. Unfortunately the researchers have mostly avoided dealing with people in their natural environments (Pentland, 1995). In contrast to this, several research groups have developed devices which are able to accomplish the basic task of making computers recognise who they are working with and to be sensitive to people's gestures and expressions. Computers who are able to track model human actions can be used for behaviour analysis and not only as new input devices. They can recognise faces (Moghaddam and Pentland, 1995), and facial expressions (Essa and Pentland, 1995). Unfortunately the approaches are made in a way that the dynamical component is lost and expressions are forced again in event categories, like surprise, happiness, disgust or anger with all the drawbacks of categorisation.

In a more general approach, Starner and Pentland (1995) and Pentland and Liu (1995) have developed a motion analysis system where the human body is modelled as a Markov device with a number of internal mental states, each with its own particular behaviour, and inter-state transition probabilities. Internal states are determined through an indirect estimation process, using the person's movement and vocalizations as measurements. These variables then are fed directly into a computer which then can observe a person's actions and respond accurately. Niyogi and Adelson (1995) developed a set of techniques which are capable of analysing the patterns which are generated by walking. These patterns are then translated into a stick-figure and the gait can be analysed.

Unfortunately these devices are quite sophisticated and expensive and they have not been used for behaviour research. In order to assess the quality of movements and human expression we developed a system which can be used even in labs with a low budget. We dispensed with real time operation capabilities and turned to the analysis of digitised video. These procedure is called Automatic Movie Analysis (AMA). The advantages are clear: it is possible to repeat any type of analysis and control for artefacts.

2.1. The Programming Platform

Automatic Movie Analysis is a programming platform which can apply a row of sophisticated filters to videoimages sequentially. In this article, we will restrict our analysis on Motion Energy Detection (MED). This is a simple but elegant filtering method for the determination of qualitative aspects of behaviour. The method relies on the fact that pictures of movies (frames) are time dependent distributions of grey-scales or colour values. A video-picture consists of an array of pixels which take different values. In a grey-scale these values usually range from 0 (white) to 255 (black). If the camera view is static, single pixels will change their values when movement occurs. If there is no movement, the pixels stay at the same value. The solution is to make the difference between two pictures—if there was no movement, we will get an all white picture, if there is movement we will get grey values in those regions where movement occurred. The amount of movement will become visible through the amount of pixels which are not white.

This image differencing (Sonka et al., 1993) works quite well if the camera position and the lighting conditions remain unchanged. The method can detect movements, not their direction. In this case AMA consists of the following steps (illustrated in Fig. 4):

1. *Digitising* of a movie in a range between 12.5 to 25 pictures a second (320 x 240 pixels frame size) in greyscales (Pixelvalue=[0..256] greys, where 0 is white and 256 is black).
2. *Making difference pictures* The arithmetic difference is calculated between two or more consecutive images of the movie (MED).
3. *Digital noise-reduction* Videonoise is related to tape quality, camera resolution and light conditions. In a video picture, pixels will change their colour ran-

domly. As a result "lonely pixels" will appear in the difference picture. The noise was reduced by the application of a median filter to the difference pictures, where a pixel received the median grey value of its nine surrounding neighbours.
4. *Error detection* Flashes result from poor videotape quality or technical problems of the recording equipment. If such flashes occur, there is an immediate high change in mean grey density in the overall picture. If this occurred the grey density for this frame (t) was replaced by the difference density between points t_{-1} and t_{+1}.
5. *Calculate mean grey values* for one or more predefined regions of the picture. These regions have to be determined by hand In a newer version of the program one or more persons can separated automatically from the background.
6. *Standardising of mean grey values* Different viewing angles of the person and changes of body postures during the experiment lead to different overall values of mean grey density, because visible contours change. Thus the mean grey values were transformed to z-scores.
7. *Smoothing* At this stage, noise was still present and resulted in small but very fast and short changes. The scores then were smoothed with a 5-point moving average.
8. *Thresholding* Further analysis is possible if the continuous recordings are collapsed into events. A threshold method was used. When the grey level change was under a certain threshold, the grey level change was set to zero. The threshold was determined with an optimal thresholding method (Sonka et al., 1993).

Figure 4a shows a videosequence of a female "Hair-Flip" and a male watching her. In Figure 4b all difference pictures are shown. Figure 4c shows the transformed z-scores of the male and female from Fig. 4a. The male shows no movement, three movement clusters were identified (Burst 1, 2, 3). First a hand movement together with a head toss (Burst 1, frame 4–17). Immediately after this hand movement the hair flip starts (Burst 2, frames 19–39). Finally, the female turns the head towards the male (Burst 3, frames 58–65).

Thus we have the following descriptors for movement quality: Number of bursts, their duration and the size of an event (the area enveloped by the burst). The number of elements in an event was calculated by counting the number of maxima and minima in the burst. A burst can have several elements which are produced by combining different movements. The number of elements describes the complexity of the movement. Finally we were able to record speed of movement change, which is the size of the burst divided by its duration.

2.2. Interactions between Strangers and Sublumineal Manipulation

This procedure was applied by Grammer et al. (1996) to social interactions between strangers in a waiting room situation. In an experiment, strangers of both sexes met and interacted for 10 minutes while they were videotaped with a hidden camera. The experiment was made in Japan and in Germany. This situation can be described in terms of high risk of social non-acceptance; thus communication should be forced into a manipulative level. After the experiment, the subjects made a self report on their interest in their partner and how pleasant they found the interaction.

The first two minutes of the videoclips were analysed first with traditional methods and then with AMA. On the one hand, traditional analyses yielded cultural differences be-

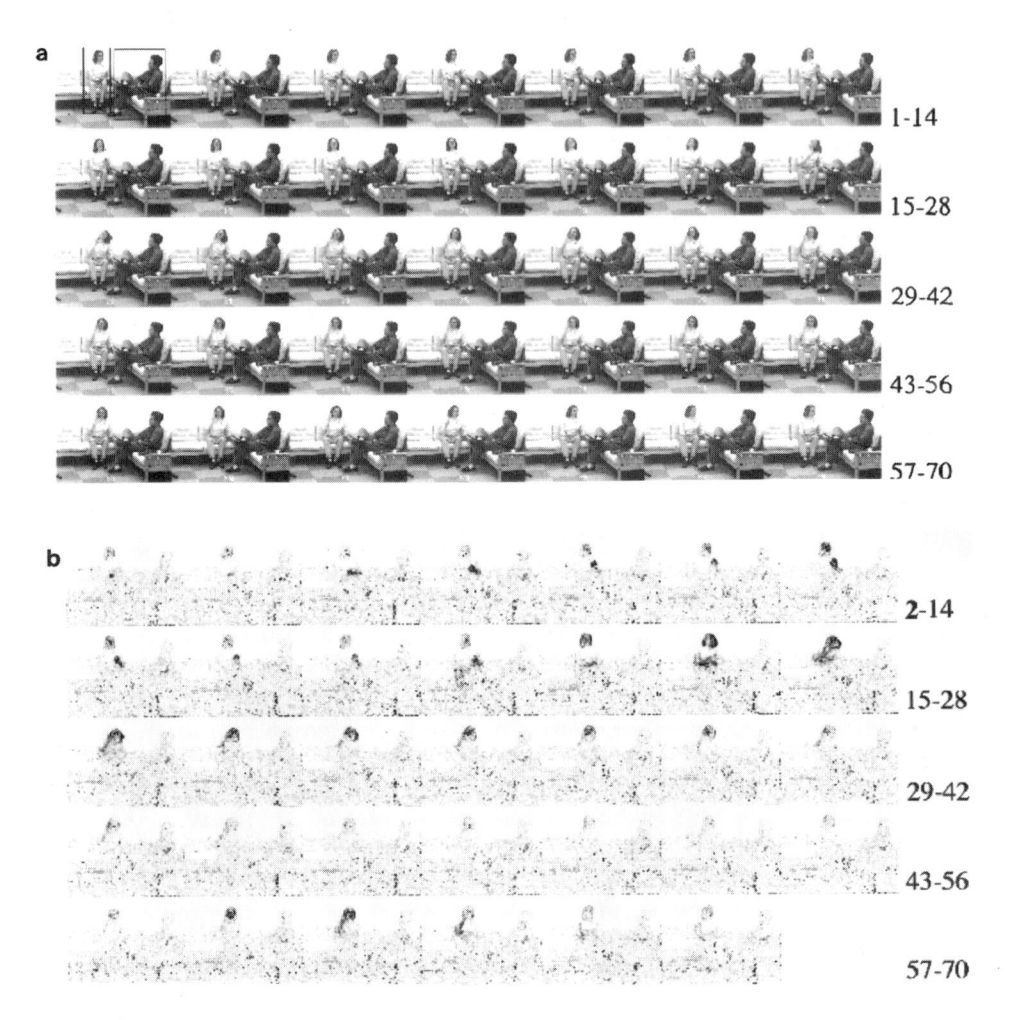

Figure 4. A female Hairflip analyzed with AMA. Instance of a female Hairflip preceded by hand movement. In (a), the original film sequence is shown (70 frames, every second frame skipped). The rectangles in frame 1 delimit the areas of interest for which difference-pictures were calculated (b). Note the changes in greydensity (e.g. head movement). In (c), terminology and the movement descriptors resulting from the automatic movie analysis (AMA) are shown: Number of bursts (total amount of movement), and duration, size and speed of bursts (size divided by duration, i.e. greydensity change per s). The number of elements (number of maxima) within a burst is considered to be a measure of complexity since elements normally are produced by different movements. The three movement bursts of the woman identified by the threshold method are in good agreement with the film sequence. The first burst depicts a hand movement accompanied by a head toss (frames 4–17), whereas the immediately following burst represents the hair flipping (frames 19–39). During the third burst, after a pause of about 1.5 s, she turns her head towards him (frames 58–67).

tween Japan and Germany. A typical Japanese behaviour is "nodding" which rarely occurs with comparable high frequencies in Germany. Moreover, when the frequencies of the generic behaviour codes were compared to male and female interest, no significant correlations were found. On the other hand, AMA reached similar results in Japan and in Germany. Females changed the quality of their behaviour when they had high interest in the male. These qualitative changes were not due to mere nervousness or excitement—the

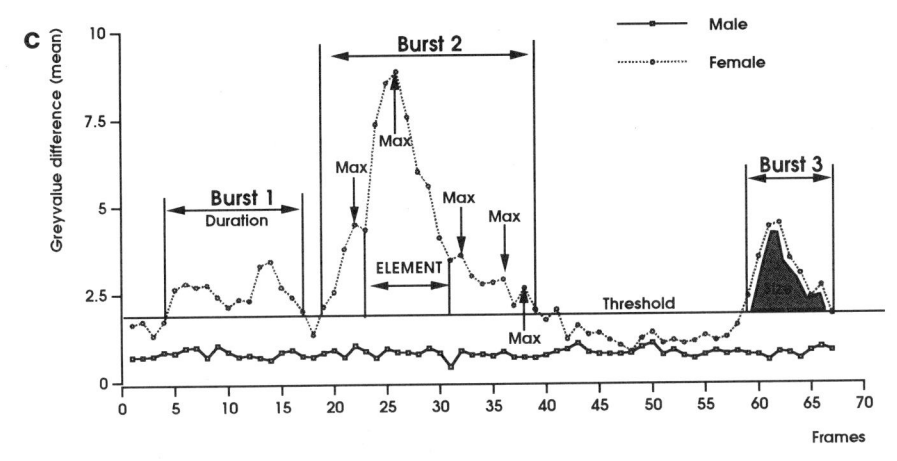

The Communication Paradox and Possible Solutions

Figure 4. (*Continued*)

females actually moved more, but showed smaller and slower movements. These qualitative changes give an impression of slow and determined movements where the single parts were accentuated. Males reacted to these qualitative modifications positively and experience the situation more pleasant although their interest in the partner is not affected. In addition, males who perceive the situation positively talk more. Thus we can conclude that it is not the content of a non-verbal behaviour, it is the quality of the movement which actually holds the information about the interest of the sender.

One objection could be that the movements are generated by speech and lively conversation between the interactants. This was not confirmed. The amount of speech did not correlate with qualitative changes in the females behaviour nor with female interest. When the correlations between movement data and interest where corrected for speech, it became clear that non-verbal behaviour is the means of communication in this situation. Thus in real life situations where risk is present, non-verbal behaviour plays the main role in communication. The evolutionary theory behind this explanation is the fact that females actually have a greater risk in male-female interactions than males (Trivers, 1972). But it is not only the risk of loosing investment: actually the risk of being deceived by a male is quite high. In a questionnaire study by Tooke and Camire (1991) 60% percent of the males reported that they had used deception in such interactions. Thus it seems logical that females would try to manipulate the males slowly without revealing their intentions in order to gather information about the male's behaviour tendencies. This is possible when the male feels pleasant and starts to talk and reveil information .

A second even more interesting result is the fact that behaviour is not a simple continuous flow of movements—it is definitely structured into single bouts (see Figure 5). Times of movement and non-movement alternate. Thus for a closer analysis we will look at the quality of the bouts themselves, and take a look which information qualitative changes might provide.

2.3. Showing Off: Simple Movements and Their Communicative Value

In order to accomplish this task, video-material from an observational study on female cycle and self-presentation (Grammer et al., 1996) was used. The starting point of

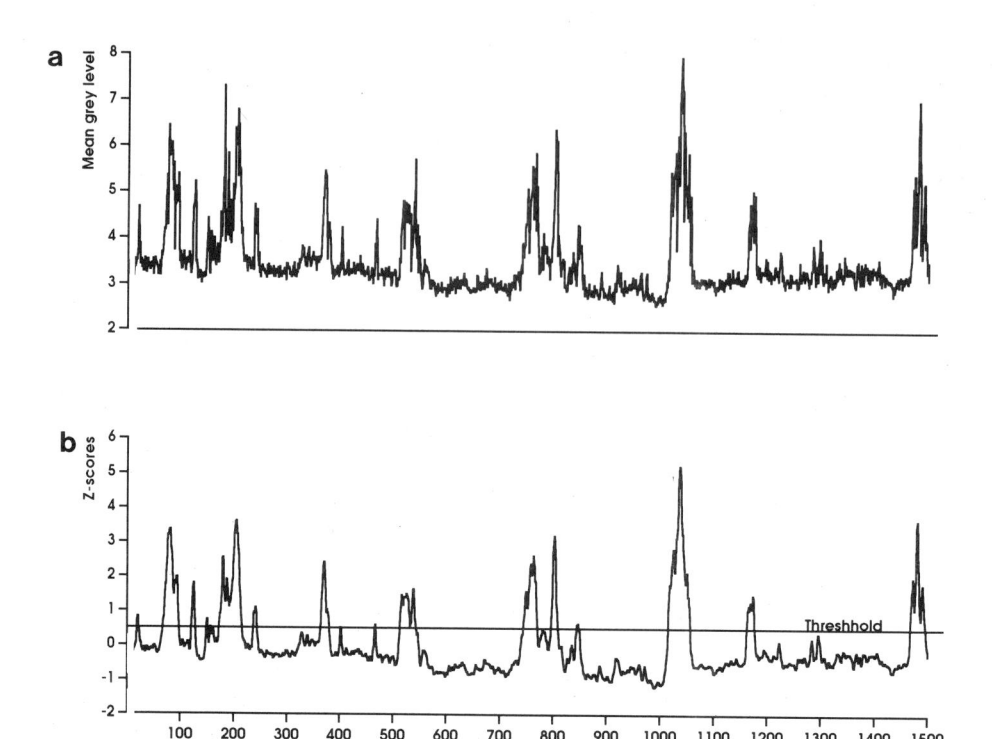

Figure 5. Movements from a two-minute interaction. Greydensity changes of a person throughout a whole film sequence of two minutes. In (a), greydensities obtained by averaging all greyvalue differences of a difference-picture are plotted (end of step 5, cf. text for details). In (b), the data are z-transformed (step 6), smoothed (step 7), and a threshold has been calculated (step 8). AMA identified 15 bursts.

this study was the fact that in humans we find almost complete female sexual cripsis. This means that in contrast to most non-human primates, humans show no obvious behavioural or visual sign for ovulation. There has been a lot of speculation about the function of female sexual cripsis and it is proposed that it leads to male-female bonding, that it promotes active female choice or that it may give a chance to females to induce sperm-competition (Alexander and Noonan, 1979, Benshoof and Thornhill, 1979, Baker and Bellis, 1995). By hiding the point where conception is quite likely, females would force the male to stay near them to ensure conception. The male would do so to be sure that the female conceives his child and not that of another male. The fact that the male has to stay near the female could promote the emergence of a male-female relationship and lead to male investment in the offspring. The induction of sperm-competition and gene-shopping only works if females use extra-pair copulations at the time of maximal likelihood of conception. This is actually the case and leads to relative paternity security of about 90% (Baker and Bellis, 1995) The function of sperm-competition is thought to be either the optimisation of offspring quality and, if a relationship already exists, to obtain genetically variable offspring. Grammer et al. (1996) showed that females who had a steady partner and who came alone to an anonymous discotheque, showed more skin and weared tighter clothes when oestrogen levels where high. Moreover, these females knew consciously that they were signalling availability through clothingstyle.

Non-verbal behaviour can be used strategically when somebody tries to impress annother person. In the context of self-presentation non-verbal behaviour is irrepressible and impactful. It is off-the-record and it is difficult to describe verbally (DePaulo, 1992). Many studies on self-presentation have shown that people can successfully make clear to others, using non-verbal cues, the internal state that they are actually experiencing and that they also can convey the impression of a state that they are not really experiencing. In this process many non-verbal cues can play a role, even body movements, postures (Cunningham, 1977) and gait (Montepare et al., 1987) can be used as signals. According to DePaulo (1992) women are more concerned with self-presentation than men, they are non-verbally more involved and spontaneously more expressive.

Out of these results we can hypothesise that the quality of behaviour may also change in order to signal sexual availability. The reason for unobtrusive signalling in this case would be enforced by the fact that active female choice has to exist for the induction of sperm-competition. Thus risk is also present in such a situation because signalling sexual availability will obviously attract all males. The risk is that one might attract the wrong males thus the signalling level should be subtle and not exaggerated.

2.3.1. Turning Around—An Instance of Self-Presentation. During this study 123 females (mean age 23) were filmed from front and back in order to determine the clothing style. The females had to turn around on a command issued by a female experimenter standing behind the camera. This turn-around movement has no communicative value on the first sight. In this situation however we had onlookers (male and female) and a videocamera. According to Goffman (1959) self-presentation will take place as soon as a stage for the presentation is present. In this type of self-presentation public self-conciousness plays also a role. This is defined as "an awareness of and a responsivity to the impressions that are being made on others" (Scheier and Carver, 1981, p.198).We thus assumed that the turning-around movements can be used as communicative tools. At the side of the subject, a male or female was placed for holding a blackboard with an identification number on it. The female subjects all were non-pilltakers, who either were singles or who had a steady partner but came alone to the discotheque. The result is a sex of experimenter (male/female) times type of female (single/paired but alone) design. Oestrogen levels which can give a hint on cycle state were determined by enzyme immuno assay from saliva. In addition skin exposure and tightness were analysed by Grammer et al. (1996) with digital methods and a rating scheme for tightness.

The turning movement was digitised in a 320x240 pixels size at 25 frames a second. Digitisation started with the beginning of the turning movement and ended when the turning movement came to a halt. These digital videos were analyzed with Automatic movie analysis, Motion energy detection was applied for each subject on the screen, the experimenter and the female separately. Duration of the movement, the number of bursts in the movement (the basic units of the movement), complexity (the number of single maxima), the maximum speed of the movement, and the information content (amount of grey value change per frame) were recorded (see figure 6). The resulting movement graphs then were played back parallel to the movies. This procedure allowed a frame exact determination of the beginnings and the endings of the movements by comparing the movement graph to the videopictures. The play back also showed that the movements could be divided into several parts (see figure 6 and table 1). Basically a turn-around can start with an intention movement prior to the actual turning. We find different types of movement initiation. The head can introduce the movement as can one foot, one knee or a turn of the upper body. Then the turning around itself follows. When the turning ends, the standing position can

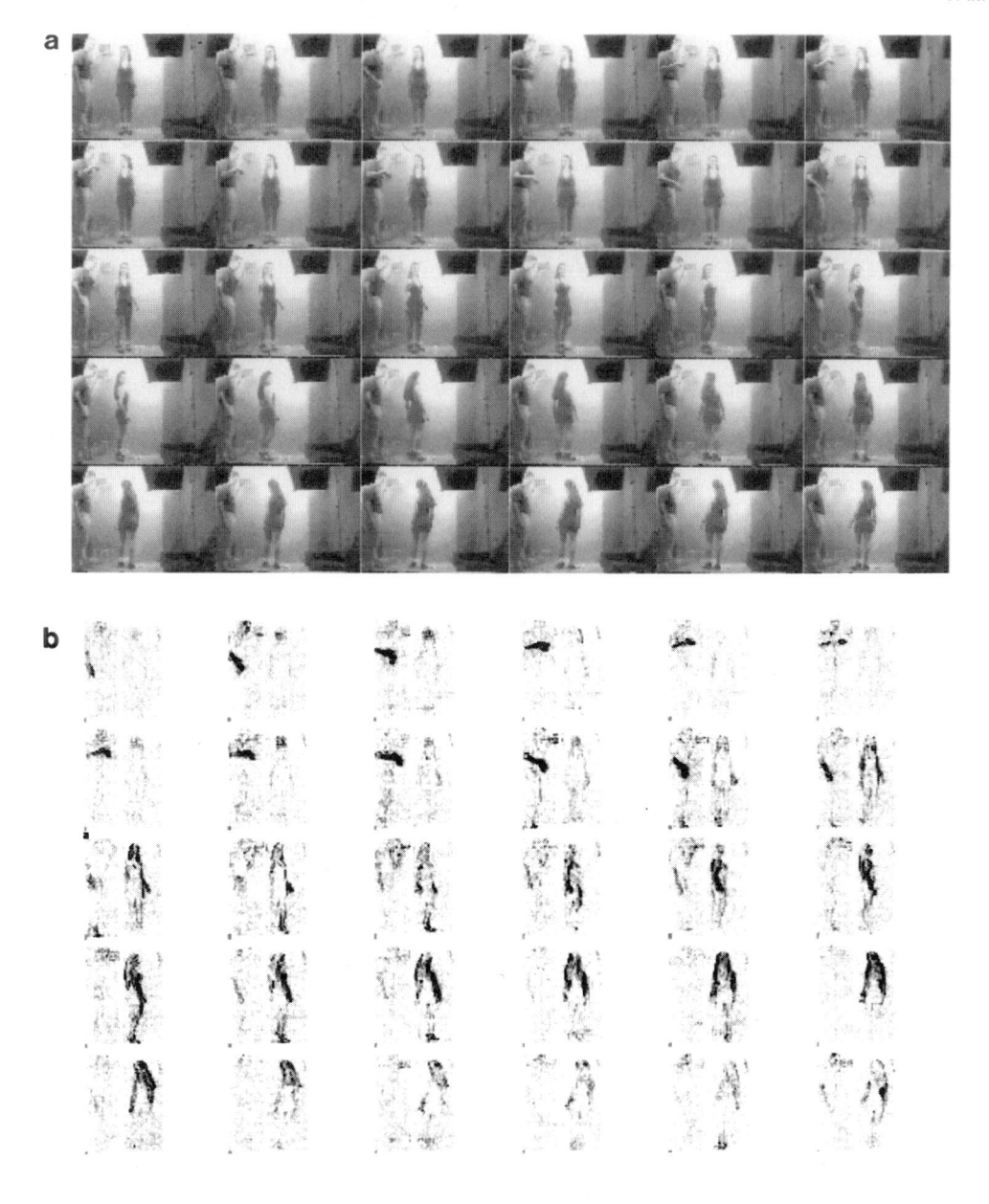

Figure 6. Female reaction to a command: Turning-around movements. Sequence (a) shows the digitized pictures for a reaction to a command. The female command giver is standing at the camera. On the command "Please turn around" the female turns around. On the left side of each picture the "stimulus male" can be seen. Sequence (b) shows the difference pictures with the regions for the stimulus and the subject which are used to calculate the mean grey density. In (c) these grey density levels are presented as z-scores. The movement has three phases: phase I where an initiation movement is made. Phase II where the actual turn around movement happens and phase III where an additional movement is added to the turn around (see also text). The description parameters for the movment are the same as in Fig.4c. The difference to Figure 4 is that the threshold is calculated dynamically for each 10 frames in order to deal with shifts of grey density changes.

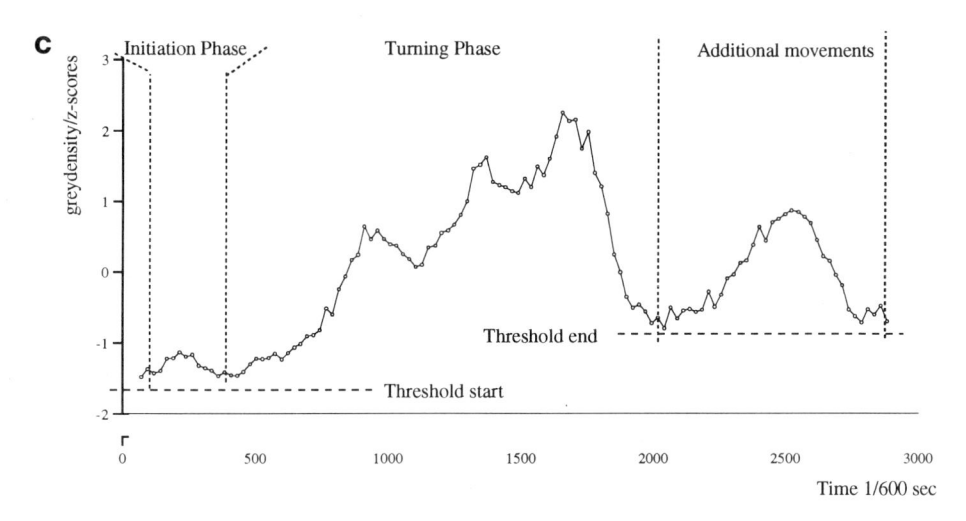

The Communication Paradox and Possible Solutions

Figure 6. (*Continued*)

be corrected or additional movements can be added. These movements may consist of hip swaying, hair-flip arm swaying or moving the body from the left to the right. Fig. 6 shows a breast-presentation movement, where the upper body turns over after the stop of the actual turning movement and the body silhouette becomes visible from the side. In addition a typical "breast out-shoulder-back movement" occurs. These four types form only the main classes, the additional movements can be done by any movable body part. The movements also were coded "traditionally" with the observation categories presented in Table 1.

Table 1. Behaviour occuring during turn-around movements

Movement phase	Behaviour type
Phase I: Intention movements	Head down
	Shoulders up
	Lift arms
	Turn head
	Others
Start of movement	Left/Right foot
	Upper body
	Arms
Phase II: Turn around	Number of steps needed
Pase III: Additional final movement	Position correction
	Hair Flip
	Hip Sway
	Move Shoulders
	Arms sway
	Arms lift
	Sway trunk
	Side presentation
	Head turn
	Head tilt
	Others

Table 2. Differences in movement quality between the experimental groups

		Duration		Bursts		Complexity		Information		Maximal speed	
Group	N	mean	std	mean	std	mean	std	mean	std	mean	std
Paired-FS	20	32.8	19.0	3.4	2.3	14.1	11.6	47.9	39.4	2.7	1.0
Paired-MS	14	19.3	6.4	1.5	0.7	6.3	3.2	29.5	8.7	3.3	1.1
Single-FS	57	30.1	17.9	2.8	2.3	12.5	10.7	37.0	14.9	2.8	1.1
Single-MS	29	28.0	15.0	2.5	2.1	12.4	10.1	31.9	10.4	2.8	1.2
K-W Anova		p=0.11		p=0.10		p=0.12		p=0.02		p=0.33	
Total	120	26.8	16.8	2.74	2.3	12.1	10.31	36.7	20.5	2.9	1.1

(FS: female as stimulus. MS: male as stimulus. Paired: female having partner but coming alone. Single: female with no partner).

2.3.2. Movement Quality and Oestrogen Levels. In a first step a conventional statistical analysis was performed in order to see if the four experimental groups are qualitatively different. Table 2 shows the results which actually propose that the four groups can be differentiated on the basis of partner status and stimulus person in only respect to information content of the movement. The highest amount of information is present when the female is confronted with a female stimulus. On the first glance this contradicts the showing off hypothesis.

If we go one step further, we find a high interrelation between the recorded movement parameters themselves. Duration, the number of bursts, complexity and information are positively correlated in all groups (n=120, r_s=0.65 to 0.84) and correlate negatively with speed (-0.45 to -0.87). That is short movements have fewer bursts, are less complex and have less information, and their speed is high. These interrelations suggest that there are physical constraints on movements.

In a next step, we correlated oestrogen levels with the movement parameters. The Bonferroni corrected correlations show that only females confronted with a male stimulus change their behaviour quality. Single females make slower and more complex movements. Both single and paired females show more information per time in their movements with increasing oestrogen levels (Table 3).

2.3.3. Movement Quality and Stimulus Reaction. These results suggest that single females react toward a male stimulus by changing the quality of their behaviour. But how does the stimulus male perceive it? When we correlate the stimulus male's behaviour quality with those of the subjects, we do not find any significant correlation. Thus it seems that males do not notice the female's qualitative behaviour changes directly. It is also possible to directly test if the male reacts to female oestrogen levels. The stimulus male's behaviour depends on the paired female's oestrogen levels. He changes the duration of his

Table 3. Female movement quality and oestrogen levels

		Duration		Bursts		Complexity		Information		Maximal speed	
Group	N	r_s	p	r_s	p	r_s	p	r_s	p	r_s	p
Paired-FS	20	−0.14	n.s.	−0.19	n.s.	−0.19	n.s.	−0.13	n.s.	0.25	n.s.
Paired-MS	14	0.27	n.s.	0.34	n.s.	0.19	n.s.	0.61	0.05	0.04	n.s.
Single-FS	57	0.05	n.s.	0.04	n.s.	0.01	n.s.	0.14	n.s.	−0.09	n.s.
Single-MS	29	0.21	n.s.	0.33	n.s.	0.52	0.02	0.52	0.02	−0.45	0.04

(FS: female as stimulus. MS: male as stimulus. Paired: female having partner but coming alone. Single: female with no partner).

movements (n=14, r_s=0.68) makes more bursts (r_s=0.66) and makes more complex movements (r_s=0.57). The stimulus female does not show any significant reaction to other females oestrogen levels.

But as we know, there is no direct coupling between the male's and the female's qualitative behaviour changes, i.e. stimulus males' movement quality does not change parallel to subjects qualitative change. Maybe the males use other sources for information, like skin showing or tightness of clothes.

If this is the case, the reaction of the stimulus males might not be due to qualitative changes in female behaviour at all. The reaction of the males could be a reaction to the exalted sexual signalling through clothing style. Thus we controlled the correlations between stimulus behaviour and oestrogen levels for skin with partial correlations. The correlations for the male stimulus reactions to unpaired females oestrogen levels do not disappear (df=11, duration: $r_{spartialized}$=0.64, p=0.02, complexity: 0.62, p=0.02).

So far the results suggest that the behavioural changes are actually changes which occur together with high oestrogen levels. We may conclude that females who develop interest in a male signal high oestrogen levels. If this assumption is true, then we should expect that these changes are present in all females with high oestrogen levels, and that it is impossible to suppress these changes completely. Nevertheless, under the right stimulus conditions, females could either fake or superelevate them suggesting a cognitive accessibility.

Females with higher oestrogen levels show higher information content in their movements when they are confronted with the stimulus male but only in the case of paired females the male reacts. We have found the highest values of information content when a female stimulus is present but there is no difference between paired and single females when a male stimulus is present (See Table 2).

If we look back at the considerations about possible communicative mechanisms, we find many possible solutions for this. The most obvious one is that the simple physical measures we used for movement description are not adequate, or there is a summation of different features over time. So far we can exclude at least multimodal communication where skin showing and movement quality add up.

2.3.4. A Neural Network Approach for the Analysis of Movement Quality: Parallel Distributed Processing. In recent years, connectionism has become a focus of research in a number of disciplines. Neural networks represent a special kind of information processing: connectionistic systems simulated by a computer consist of many primitive cells which are working in parallel and are connected via directed links. This forms an analogy to the human brain: the cells are analogous to neurones and the links are the connections between those neurones. The main processing principle of these cells is the distribution of activation patterns across the links similar to the basic mechanisms of the brain. Information processing in the brain is based on the transfer of activation from one group of neurones to others through synapses. In analogy to activation passing in biological neurones each unit receives a net input that is computed from the weighted output of prior units with connections leading to this unit. However, the most current neural networks do not try to closely imitate biological reality.

In neural networks "knowledge" is distributed through the activation of cells and the weighting of the links. The networks are organised by training. In supervised training, the network "learns" a set of patterns together with their classification by repeated presentation. Through this "learning process," classical logical conclusions are replaced by vague and associative recalls. This is of advantage in all cases where no set of clear logical rules

Table 4. Network classification of movements according to oestrogen levels

Pattern classified as	Original oestrogen level associated with pattern		
	High	Middle	Low
High	25	0	1
Middle	0	16	16
Low	0	7	11
None	0	3	0
% correct	100%	61.5	57.1

can be given. After learning, the neural network can be able to classify unlearned patterns correctly or not. In the first case we then can assume that in the patterns is at least some information present which is common to certain classes. Unfortunately it is very difficult to recall the information the network has used for classification.

Neural networks thus can be used to look if information is present in a pattern which then can be used to classify these patterns. A neural network analysis was applied to the raw data from the "showing-off" study above. The network was constructed as a time delayed network (TDNN, Waibel, 1989) on the SNNS-Simulator (Zell, 1994). The network embodied 10 (features) times 24 (total delay length) input units, 120 hidden units (receptive field) and 3 output units for low, middle and high oestrogen levels. Time delayed networks do not use a static presentation of patterns and they can be used for the independent recognition of features within a larger pattern. The update algorithm forces the network to train on time/position independent detection of sub-patterns. However, there is no specific set of rules on how to construct a network, and building networks heavily relies on trial and error. Thus, the fact that it is not possible to train a network does not mean that there is no information to learn.

We applied two basic training methods. First the network was trained with data from single females with male stimulus. The validation was done with single females with female stimulus. The data from paired females with male and female stimulus were then tested for classification analysis. Second, the training was done with data from paired females with male stimulus, the validation was done with paired females with female stimulus, and the testing with single females with male or female stimulus.

The classification results showed astonishing stability: 66% of cases from method One and 70% of the cases from method Two were classified correctly. A closer look reveals that the wrong classifications were due to the fact that only low and middle oestrogen levels were classified sometimes incorrectly as either low or middle but never as high. With both methods combined high oestrogen levels were classified 100% correct. Thus the TDNN was able to discriminate between high and middle/low oestrogen levels correctly (See Table 4) using MED data from videopictures processed through AMA.

In order to isolate the movement prototypes, we calculated the mean movement curves for the three classes of oestrogen levels. Figure 7 shows the results. The three curves for high, middle and low oestrogen look different—but when tested the only significant difference is in the information content. Lowest oestrogen shows lowest information content in movement (Median:34). Middle oestrogen shows middle information (Median:36) and highest oestrogen level shows the highest information content (Median:39, K-W 1-Way Anova, p=0.018). This result finally brings us back to coding with discrete categories.

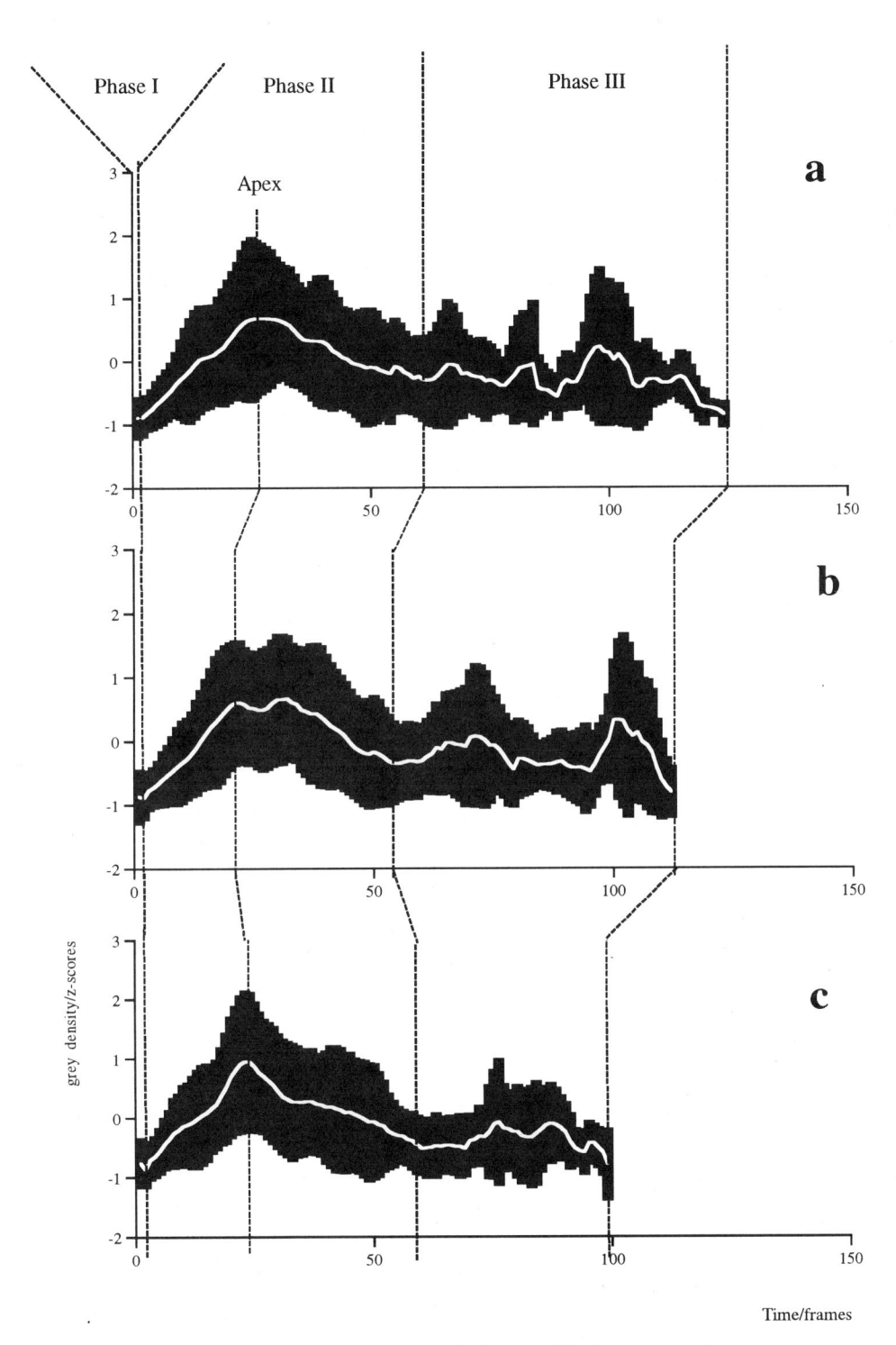

Figure 7. Movement prototypes for oestrogene levels in females. This figure shows the mean movement curves for the three oestrogen levels: (high (a), middle (b) and low (c)). The black bars indicate the standard deviation for each frame. The three movement phases and the apex (See Table 1) are indicated by dashed lines.

2.3.5. Discrete Coding of Movement Patterns. In order to find out if certain discrete patterns as described in Table 1 are connected to oestrogen levels a traditional coding was applied to the digitised videos. It turned out that in none of the three phases discrete codes could separate between oestrogen levels. The exception was an additional movement in phase three. The presence of one or more additional movements occurred significantly more often under high oestrogen levels. This relation was independent from the content of the behaviour (Median-test, p=0.02).

So far we can conclude that it is possible to describe intentions in communicative acts with the help of qualitative changes in movements. Yet it is still unclear which changes are present, because MED only crudely describes qualitative changes on a holistic level. Single movement features are not captured by this method. Yet our hypothesis is confirmed that under high risk conditions, communicative acts are forced to a level where it is only difficult to assess them with generic coding methods. This situation has lead us to a series of new developments which we currently pursue.

3. FUTURE DEVELOPMENTS—ALYSIS

The description of the whole human body and its moving parts seems to be an unsolvable endeavour. In recent years however, digital analysis of human movements and bodies has moved far away from simple MED approaches. Basically, all methods which have been used up to now are derivatives from two approaches. Either the contours of moving or non-moving objects are separated from a background, or the displacement of pixels or groups of pixels are calculated as optical flow (Sonka et al, 1993). For instance, when we want to look at emotions, a method for surface analysis of the face has to be developed in contrast to a three dimensional tracking method for an arm. The assessment of movements will differ from the assessment of postures and a method to translate the static states will also be necessary. Basically a body has to be separated from its background and then divided into its segments. This means that the body has to be dissolved into head, face, arms, body and legs. Each of these parts can then be described separately. Interestingly enough, there are many approaches to solve the task of body movement tracking. The isolation of body parts including head and face does not pose a problem. This task has been solved repeatedly (Pentland, 1995). Kakadiaris, Metaxas and Bajcsy (1994) for instance proposed an integrated approach to segmentation, shape and motion estimation of complex articulated objects which can also be used for human bodies.

The best results in the human body tracking and action recognition are achieved at MIT Media Lab (Maes et al., 1995). The ALIVE "Artificial Life Interactive Video Interface" allows wireless full-body interaction between the human participants and a rich graphical world inhabited by autonomous agents. Agents are modelled as autonomous behaving entities that have their own sensors and goals and that can interpret the actions of the human and react to them in "interactive time." Vision routines compute figure/ground segmentation and analyse the user's silhouette to determine the location of the head, hands, and other parts of the body in a colour image. This self-calibrating stereo person tracker can recover the 3D shape and motion of the hands and head of the moving person (Pentland, 1995).

Our new developments work on a model base. The human body can be modelled as a system of objects connected together by joints with one or more degrees of freedom. Tracking motion of human body can be formulated as the real-time visual tracking of kinematic chains. A kinematic object is a collection of objects connected by joints. With each object a local co-ordinate system can be associated to specify its 3D position and orientation. Since the objects are connected instead of using a sixdimensional vector for each object to describe its 3D position and orientation, the joint parameters can be used to define the mutual relationships of the objects and degree of freedom. We refer to these parameters as kinematic parameters. For modelling the shape of the objects different models can be used ranging from line, plane, to more sophisticated surface models. We refer to this parameters as shape parameters. As the objects project onto the image plane, the image data may reflect the texture of the objects surface, the object contour, the optical flow if the objects are in motion, etc. We refer to this as image features. The problem of tracking articulated objects can be viewed as an estimation of the objects kinematic and shape parameters from image features. Our aim is to isolate the (movement) vectors for the joints of the model. This first will give a position vector for all body parts and second we will get a movement vector for the head, which allows to gather data for the assessment of gaze direction, the shoulders, the elbows, the wrists, the lower body, the thighs, the knees and the ankles. These movement vectors then will be applied to a rendered simultaneously moving model. Speech will be processed simultaneously for loudness and frequency, thus allowing comparisons between movements and speech.

This analysis will produce a continuous data-stream that can be analysed in various ways. Each posture is defined as an unique set of vectors and each movement through unique changes of these vectors. Its advantage is that it does not need any interpretation on a higher level through an observer nor a computer based expert system which tries to reinterpret the movements. The vector data can be fed directly to neural nets for pattern recognition, and the patterns can be verified through rating studies. Future applications are person-recognition from gait, the monitoring of therapy success and the comparison of quality of signals in different species, under different contexts and physiological conditions.

4. COMMUNICATION THEORY AND PHYSIOLOGICAL STATES

Although it seems that we are just proposing a new method, this approach has consequences not only for the observation but also for the explanation of behaviour.

The main advantage of this approach, when it is compared to those using conventional coding methods, is that no presupposition on the structure and content of behaviour is made. This frees us from restrictions of conventional coding methods. With the use of conventional codes we can only find what we have put in the codes; behavioural codes are already hypotheses about behaviour. Behavioural codes are categories which represent many, sometimes different behavioural events. Although these categories correspond to the basic construction principles of our brain, which uses prototypes in order to reduce environmental information (Rosch, 1978), the assumption that communication works on the same level may be wrong. Indeed signals can be organised as prototypes but this is not necessarily so. We have shown that communication between humans can work on a level where no categorisation exists. The fallacy of looking at communication with the principles of the apparatus involved in it leads to a false and incomplete understanding of the nature of communication.

With this approach we possess the almost complete data-stream and we can look at how the brains of both the receiver and sender actually construct communicative reality. Our approach allows the manipulation of stimuli which can be tested against reality. We will propose a new communication theory which is in its nature multi-modal and multi-layered with different channels and many possible communicative principles.

The starting point for such a theory is that like the evolution of intelligence, the evolution of human communication has its basic constraints in machiavellism. Human brains are devices for processing information. We can suppose that there was differential survival and reproduction connected to optimal information processing (Lorenz, 1973, Cosmides et al. 1992). If there are adaptations to optimal information processing, these adaptations can be exploited. Thus, communication research has to deal with the constraints and possibilities presented by theses adaptations. Future studies should look for adaptive information processing structures which could be exploitable through communication. We suppose that low-level processing of information is at least one possibility. This means that levels where the basic information is extracted from visual stimuli could be exploited. Comparable approaches could be made with the complexity of a stimulus. The less complex a stimulus is, the easier it could be decoded, producing higher levels of excitation in the brain. This is basically an open field but it can not be mastered with traditional coding and research methods. The problems which are connected to any communication theory we have shown in the introduction starting from the possibility of deception and ending with the possibilty that noise is used tactically to veil intentions are avoided by our methodological approach. The method can even deal with repeated meaning encoded in pulse rate modulation when small changes or movements are repeated in time as shown in the waiting room study.

We propose that there are multi-layered processing mechanisms. The top layer for processing holds consciously accessible information. The bottom layers can not be assessed directly and controlled. On the top layer communication is actually an accessible information exchange about the real world with its social and ecological aspects. We can tell each other what we think about each other or gossip about others (Dunbar, 1993), we can create and use non-verbal signals like gestures (Morris et al., 1979) differently in different cultures and we are able to lie and detect lies.

On the other hand we are able to veil our intentions by many measures like the creation of noise. We try to manipulate each other's physiological states and influence information processing in our social environment. This is the basic assumption of a new communication theory: brains are able to exploit others brain's functions and structures in order to manipulate them. These manipulations are intended and planned, and conscious access to these plans is not necessary for their realisation. Qualitative changes of behaviour which are present under different oestrogen levels can be used intentionally when male brains have been selected for detection of qualitative changes in behaviour caused by oestrogen levels which promise stable female cycles and successful reproduction. Such an approach does not need necessarily innate behaviours—just basic construction principles for "what brains like." If our brain perceives "approaching speed" as danger then any fast movement toward another will be interpreted as threat. In such a way individuals could learn to use the same behaviours again and again. This would explain the wide variety and individual differences.

We are able to show that on the sender's side information might be encoded in the quality of behaviour. Females seem to do so under high risk conditions. This corresponds to the fact that females are more sensitive to the production and decoding of non-verbal behaviour (Rosenthal and Depaulo, 1979). The fact that communication is goal directed and depends on the pursued goals and possible risk of not achieving the goal has been neglected so far.

The results from qualitative movement changes in both studies have theoretical consequences. So far it is the first time where it has been shown that females try to manipulate male perception directly. Moreover, there should be at least some conscious assessment of cycle state, because females signals can become more obvious when a stimulus is present and when the female is at an ovulatory stage. This underlines the hypotheses that female sexual cripsis has indeed the function of promoting active female choice and thus can be used to induce sperm competition.

Showing off and sublumineal manipulation is a means to manipulate the perception of one's self through others, and there is no need to assume that this is done consciously. In this article we have shown that the principles are comparable: changing the quality of behaviour, so that the receiver actually can not access the changes directly. This brings up another principle of manipulative communication. The sender has to avoid that the receiver might be able to learn. We can suppose that there is pressure on learning signals very fast in order to assess other's intentions early and reliably. Thus, communication should be variant and use different means in the same situations constantly. This leads to a model of parallel distributed processing for the decoding of meaning. The results on the classification of movements through neural nets propose such a model, although the model is only a poor approximation of human parallel processing.

Classical communication theories also do not account for the fact that signalling is not only about external information, but also about internal states and the manipulation of internal states which are encoded in behaviour quality. An exception to these hypotheses seems to be emotions which can be produced as signals. The problem in this is that the nature and signal value of emotions are unclear and it is not known to what extent qualitative changes in facial muscle movements affect emotional interpretation. There are some hints that actual movement quality is the cue which could be used for decoding information and not the actual configuration of muscle movements. Emotions among expressive actors are recognised easier than emotions among non-expressive actors (Wallbott, 1990). The solution to the communicative paradox thus lies in the possibility to observe the actual nature of communication with the help of new methods. Only behaviour recordings which are free from interpretation and which produce direct data are useful for the detection of communicative principles.

ACKNOWLEDGMENT

Funded by the Jubiliäumsfond of the Austrian National Bank, P5676.

REFERENCES

Alexander, R. D. & Noonan, K. M. 1979. Concealment of ovulation, parental care, and human social evolution. In: *Evolutionary biology and human social organization* (Ed. by N. A. Chagnon & W. G. Irons), pp. 436–453. Duxbury: North Scituate.

Arbib, M. A. & Hansen, A. R. 1987. Vision, Brain and Cooperative Computation: an overview. In: *Vision, Brain and cooperative computation* (Ed. by M. A. Arbib & A. R. Hansen), pp. 1–86. Cambridge MA: The MIT Press.

Argyle, M. 1988. *Bodily communication*. London: Methuen.

Baker, R. R. & Bellis, M. A. 1995. *Human sperm competition. Copulation, masturbation and infidelity*. London: Chapman and Hall.

Bassili, J. N.1979. Emotion recognition: The role of facial movement and the relative importance of upper and lower areas of the face. *J. Personality. Soc. Psych.*, **37**, 2049–2058.

Benshoof, L. & Thornhill, R. 1979. The evolution of monogamy and concealed ovulation in humans. *J. Soc. Biol. Struct.*, **2**, 95–106.

Bernieri, F. J. & Rosenthal, R. 1991. Interpersonal coordination: behaviour matching and interactional synchrony. In: *Fundamentals of Nonverbal Behavior Part V. Interpersonal Processes* (Ed by Feldman and Rime), pp. 401- 431. Harvard: Harvard University Press.

Berry, D. S., Kean, K. J., Misovich, S. J. & Baron, R. M.1991. Quantized displays of human movement: a methodological alternaive to the point light display. *J. Nonverb. Behav.*, **15**, 1–97.

Brown, P. & Levinson, S. 1978. Universals in Language Usage: politeness phenomena. In: *Questions and Politeness. Strategies in Social Interaction.* (Ed. by E. Goody), pp. 56–289. Cambridge: Cambridge Univ.Press.

Chance, M. R. A. & Russel, W. M. S. 1959. Protean displays: a form of allaesthetic behaviour. *Proc. Zool. Soc. London,* **132**, 65–70.

Cosmides, L., Tooby, J. & Barkow, J. H. 1992. Evolutionary psychologyand conceptual integration. In: *The adapted mind* (Ed. by L. Cosmides, J. Tooby & J. H. Barkow), pp. 3–18. Oxford: Oxford University Press.

Cunningham, M. R. 1977. Personality and the structure of the nonverbal communication of emotion. *J. Personality*, **45**, 564–584.

Cutting, J. E. & Proffitt, D. E. 1981. Gait perception as an example of how we may perceive events. In: *Intersensory perception and sensory integration* (Ed. by R. D. Walk, & D. E. Proffitt), pp. 249–273. New York: Plenum Press.

Dawkins, R. & Krebs, J. R. 1981. Signale der Tiere: Information oder Manipulation. In: *Öko-Ethologie.* (Ed. by J. R. Krebs & N. B. Davies), pp. 222–242. Berlin und Hamburg: Parey.

De Paulo, B. M. 1992. Nonverbal Behavior and Self-Presentation. *Psych. Bull.,* **111/2**, 203–243.

Dunbar, R. I. M. 1993. Coevolution of neocortical size, group size and language in humans. *Behav.Brain.Sci.*, **16**, 681–735.

Eibl-Eibesfeldt, I. 1972. Similarities and differences between cultures in expressive movements. In: *Nonverbal communication.* (Ed. by R. A. Hinde), pp. 297–312. Cambridge: Cambridge University Press.

Eibl-Eibesfeldt, I. 1989. *Human Ethology.* New York: Aldine de Gruyter.

Ekman, P. & Friesen, W. V. 1969. Nonverbal leakage and clues to deception. *Psychiatry*, **32**, 88–106.

Ekman, P. & Friesen, W. V. 1971. Constants across cultures in the face of emotion. *J. Personality Soc. Psych.*, **17**, 124–129.

Ekman, P. & Friesen, W. 1972. Hand Movements. *J. Communication*, **22**, 353–374.

Ekman, P. & Friesen, W. 1978. *Facial Action Coding system.* Palo Alto, CA: Consulting Psychologists Press.

Ekman, P., Friesen, W. V., O'Sullivan, M. & Scherer, K. R. 1980. Relative importance of face, body, and speech in judgements of personality and affect. *J. Personality Soc. Psych.*, **38**, 270–277.

Ekman, P., Levenson, R. W. & Friesen, W. V. 1983. Autonomous nervous activity distinguishes among emotions. *Science*, **221**, 1208–1209.

Essa, I. & Pentland, A. 1995. Facial expression recognition using a dynamic model and motion energy. Int'l Conference on Computer Vision, Cambridge, MA, June 20–23, 1995.

Forgas, J. P. 1992. Affective Influences on Partner Choice—Role of Mood in Social Decisions. *J. Personality Soc. Psych.*, **61/5**, 708.

Frey, S. & Pool, J. 1976. A New Approach to the Analysis of Visible Behaviour. *Forschungsberichte aus dem Psychologischen Institut der Universität Bern.* Bern.

Goffman, E. 1959. *The presentation of self in everyday life.* NewYork: Doubleday.

Grammer, K. 1989. Human Courtship: Biological Bases and Cognitive Processing. In: *The sociobiology of sexual and reproductive Strategies* (Ed. by A. Rasa, C. Vogel & E. Volland), pp. 147–169. London: Chapman and Hall.

Grammer, K. 1992. Intervention in cconflicts among children: context and consequences. In: *Coalitions and alliances in humans and other animals.* (Ed. by A. Harcourt & F. deWaal), pp. 259–283. Oxford: Oxford University Press.

Grammer, K. 1991. Strangers meet: laughter and nonverbal signs of interest in opposite-sex encounters. *J. Nonverb. Behav.*, **14**, 209–236.

Grammer, K. 1995. *Signale der Liebe* . 3., neu überarbeitete Auflage. München: dtv-Wissenschaft.

Grammer, K. & Eibl-Eibesfeldt, I. 1989. The ritualisation of laughter. In: *Natürlichkeit der Sprache und der Kultur* (Ed. by W. A. Koch), pp. 192–214. Bochum: Brockmeyer.

Grammer, K., Honda, M. & Schmitt, A. 1996. Human courtship: digital image analysis of body movements. *J Personality Soc. Psych.*, under revision.

Grammer, K., Jütte, A. & Fischmann, B. 1996. Der Kampf der Geschlechter und der Krieg der Signale. In: *Sexualität im Spiegel der Wissenschaft*. Edition Universitas, Stuttgart:Hirzel. In press.

Grammer, K. & Kruck, K. 1991. Decision making in opposite sex-encounters: love at first sight ?. Kyoto, 22nd International Ethological Conference.

Grammer, K. & Kruck, K. 1996. Female control and female choice. In: *When women want sex: perspectives on female sexual initiation and aggression.* (Ed. by B. Anderson & C. Struckmann-Johnson) New York: Guilford Press.

Grammer, K., Kruck, K. & Magnusson, M. 1996. The courtship dance: mathematical algorithms for pattern detection in non-verbal behaviour. *J. nonverb. Behav.* (under revision).

Grammer, K., Schiefenhövel, W., Schleidt, M., Lorenz, B. & Eibl-Eibesfeldt, I. 1988. Patterns on the Face: brow movements in a crosscultural comparison. *Ethology*, 77, 279–299.

Harper, D. G. C. 1992. Communication. In: *Behavioural ecology. An evolutionary approach.* (Ed. by J. R. Krebs & N. B. Davies), pp. 347–398. Oxford: Blackwell.

Johansson, G. 1973. Visual perception of biological motion and a model of its analysis. *Perception & Psychophysics*, 14, 201–211.

Johansson, G. 1976. Spatio-temporal differentiation and integration in visual motion perception. *Psychol. Res.*, 38, 379–393.

Kakadiaris, I. A., Metaxas, D. & Bajcsy, R. 1994. Active part-decomposition, shape and motioon estimation of articulated objects: A physics-based approach. *Proc. of IEEE Conference on Computer Vision and Pattern Recognition*, pp. 980–984. Seattle, Washington.

Krauss, R. M., Apple, W., Morency, N. L., Wenzel, C. & Winton, W. 1981. Verbal, vocal, and visible factors in judgements of another's affect. *J. Personality Soc. Psychol.*, 40, 312–320.

Kraut, R. E. & Johnston, R. E. 1979. Social and emotional messages of smiling: An ethological approach. *J. Personality Soc. Psych.*, 37, 1539–1553.

Lorenz, K. 1973. *Die Rückseite des Spiegels.* München: Piper.

MacArthur, L. Z. & Baron, R. M. 1983. Toward an ecological theory of social perception. *Psychol. Rev.*, 90, 215–238.

Maes, P., Darrell, T., Blumberg, B. & Pentland, A. 1995. The ALIVE system: wireless, full-body interaction with autonomous agents. *Proc. Computer Animation*, IEEE Press, April 1995.

Malatesta, C. A. & Izard, C. E. 1984. The facial expression of emotion: Young, middle-aged, and other adult expressions. In: *Emotion in adult developmaent* (Ed. by C. Z. Malatesta & C. E. Izard), pp. 253–273. Beverly Hills, CA: Sage.

Markl, H. 1985. Manipulation, modulation, information, cognition: some of the riddles of comunication. *Fortschritte der Zoologie*, 31, 163–194.

Mehrabian, A. 1972. *Nonverbal communication.* Chicago: Aldine.

Moghaddam, B. & Pentland, A. 1995. Probabilistic visual learning for object detection. Int'l Conference on Computer Vision, Cambridge, MA, June 20–23 1995.

Montepare, J. P., Goldstein, S. B. & Clausen, A. 1987. The identification of emotions from gait information. *J. Nonverb. Behav.*, 11, 33–42.

Moore, M. M. 1985. Nonverbal courtship patterns in women: context and consequences. *Ethol. Sociobiol.*, 6, 237–247.

Morris, D., Collett, B., Marsh, P. & O'Shaugnessy, M. 1979. *Gestures, their origins and distribution.* London: Jonathan Cape.

Pentland, A. 1995. Machine understanding of human action. M.I.T. Media Laboratory Perceptual Computing Section Technical Report No.350, Sept.1995. Appeared: *7th Int'l Forum on Frontier of Telecommunication Technology*, Nov. 1995, Tokyo, Japan.

Pentland, A. & Liu, A. 1995. Toward augmented control systems. IEEE Intelligent Vehicle Symposium 95, September 25–26, Detroit, MI.

Provine, R. R. & Young, Y. L. 1991. Laughter: a stereotyped human vocalization. *Ethology*, 89, 115–124.

Rosch, E. H. 1978. Principles of Categorization. In: *Cognition and Categorization* (Ed. by E. Rosch & D. Lloyd), pp. 27–48. Hillsdale: Erlenbaum.

Rosenthal, R. & Depaulo B. M. 1979. Sex differences in eavesdropping on non-verbal cues. *J. Personality. Soc. Psychol.*, 37, 273–285.

Runeson, S. & Frykholm, G. 1983. Kinematic specification of dynamics as an informational basis for person-and-action perception: Expectation, gender recognition, and deceptive intention. *J. Exp. Psychol.*, 112, 585–615.

Scheier, M. F. & Carver, C. S. 1981. Private and public aspects of self. In: *Review of personality and social psychology*, Vol. 2 (Ed. by L. Wheeler), pp. 189–216. Beverly Hills, CA: Sage.

Schleidt, W. M. 1973. Tonic communication:contionous effects of discrete signs in animal communication systems. *J. Theoret. Biol.*, 42, 369–386.

Siddiqi, J.A., Schwind, H.L. & Voss, H.G. 1973. Irrelevanz des Inhalts—Relevanz des Ausdrucks. *Z. Experimentielle und Angewandte Psychologie*, 20, 472–488.

Sonka, M., Hlavac, V. & Boyle R. 1993. *Image Processing, Analysis and Machine Vision.* London: Chapman and Hall.

Starner, T. & Pentland, A. 1995. Visual recognition of american sign language using hidden Markov models. *Proc Int'l Workshop on Automatic Face- and Gesture-Recognition*, Zurich, Switzerland, June 26–28, 1995.

Tooby, J. & Cosmides, L. 1990. On the universality of human nature and the uniqueness of the individual: the role of genetics and adapation. *J. Personality*, **58**, 1.

Tooke, W. & Camire, L. 1991. Patterns of Deception in Intersexual and Intrasexual Mating Strategies. *Ethol. Sociobiol.*, **12**, 345–345.

Trivers, R. L. 1972. Parental investment and sexual selection. In: *Sexual selection and the descent of man 1871–1971.* (Ed. by B. Campbell), pp. 136–179. Chicago: Aldine.

Van Hooff, J. A. R. A. M. 1972. A Comparative Approach to the Phylogeny of Laughter and Smile. In: *Non-Verbal Communication.* (Ed. by R. A. Hinde), pp. 209–241. Cambridge: Cambridge University Press.

Waibel, A., Hanazawa, T., Hinton, G., Shikano, K. & Lang K.J. 1989. Phoneme recognition using time-delay neural networks. *IEEE Transactions On Acoustics, Speech, and Signal Processing*, **37/3**, 328–339.

Wallbott, H. G. 1990. *Mimik im Kontext.* Göttingen: Verlag für Psychologie Dr.C.J.Hogrefe.

Wallbott, H. G. 1991. The Emotional in Social Psychology and the Social in Emotion Psychology—An Overview Concerning the Intersection Between Social Psychology and Emotion Psychology. *Z. für Sozialpsychologie*, **22/1**, 53–65.

Zell, A. 1994. *Simulation neuronaler Netze.* Bonn: Addison-Wesley.

7

TWIN STUDIES OF BEHAVIOR

New and Old Findings

Thomas J. Bouchard, Jr

Department of Psychology and Institute of Human Genetics
University of Minnesota
Minneapolis, Minnesota 55455-0344

1. INTRODUCTION

In the age of molecular genetics some commentators have questioned whether twin studies will continue to be a useful tool for studying genetic influences on behavior. Many of us think the answer is yes (Bouchard & Propping, 1993). Work with monozygotic twins reared apart provides an imperfect, but nevertheless powerful window on the direct influence of genes on whole organisms. Extended behavior genetics designs that include twins also provide important information about more complex modes of inheritance that will be very difficult to implement with molecular techniques (Lykken, McGue, Tellegen, & Bouchard, 1992). In this presentation I will attempt to illustrate these points and bring you up-to-date on new findings and new twists on old findings in the field of behavior genetics. In order to illustrate the breath and implications of these findings, first, I will sample the major human individual differences—psychological interests, mental abilities, personality, social attitudes and psychopathology—and illustrate the breath and magnitude of genetic influence in each of these domains. Secondly, I will illustrate the much less well-known fact that measures of the environment that are often purported to be "causal agents" may also be a manifestation of genetic influences rather than causal agents in their own right.

2. METHOD

Before turning to specific content, I would like to take a moment to review briefly the logic of behavior genetic methods. This is most easily accomplished by showing how path models are used to represent the latent constructs underlying correlations and covariances. My examples all utilize correlations, however, the models are typically fit to covariances.

New Aspects of Human Ethology, edited by Schmitt *et al.*
Plenum Press, New York, 1997

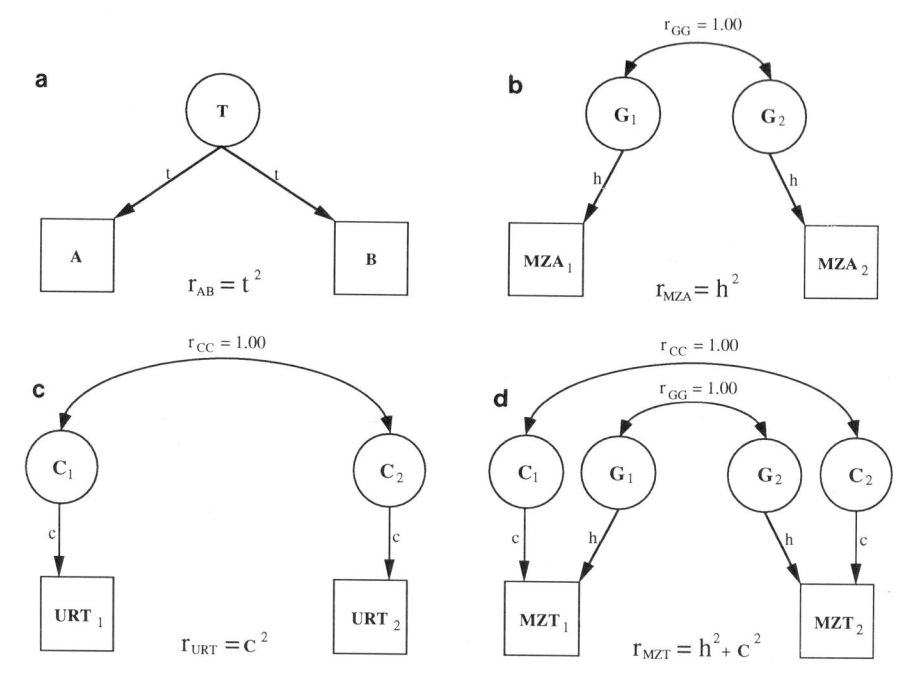

Figure 1. Path diagrams for (a) parallel form reliability, (b) monozygotic twins reared apart, (c) unrelated individuals reared together, and (d) monozygotic twins reared together.

Let us begin with a well known example. Figure 1a shows, in path diagram form, the intraclass correlation between alternate forms of a test, observed scores represented as boxes. As the diagram shows this correlation can be interpreted as representing the variance accounted for by the influence of the true score (T)—a latent trait shown in the circle (Hayes, 1973, p. 535). The rules of path analysis tell us that the correlation is made up of the sum of the multiplication of the terms of each path that connect the trait. Thus the equation;

$$r_{AB} = t^2.$$

The correlation is not squared. Figure 1b illustrates the correlation for a trait, represented by the boxes, between monozygotic twins reared apart (MZAs).

The similarity is assumed to be caused by the latent genetic factor shown in the circle (G). The causal paths are labeled with path coefficients. Again the rules of path analysis tell us that the correlation is made up of the sum of the multiplication of the terms of each path that connect the trait. Thus the equation;

$$r_{MZA} = h^2.$$

This value represents what is called the broad heritability. The model makes the usual assumptions; no placement effects, no genotype x environment interaction and no genotype x environment correlations, adequate representation of genotypes and adequate representation of environments. The MZA correlation alone directly estimates the broad heritability. The correlation is not squared.

Figure 1c shows the path model for unrelated individuals reared together (URT). There is a single latent factor (C) and the equation reads;

$$r_{URT} = c^2.$$

This design is, in essence the inverse, of the MZA design, and the correlation directly estimates the influence of common or shared environmental factors. Again the correlation is not squared.

Figure 1d shows the correlation between monozygotic twins reared together. We have simply added a latent variable to represent common or shared environmental influence (C). This model represents the obvious fact that monozygotic twins reared together (MZTs) can be similar for both genetic and environmental reasons and that the correlation is confounded. Thus the equation;

$$r_{MZT} = h^2 + c^2.$$

The difference between the MZA correlation and the MZT correlation estimates the magnitude of common or shared environmental influence. Models for dizygotic twins reared apart (DZAs) and dizygotic twins reared together (DZTs) would make use of a correlation of .5 between the additive genetic components because such twins share half their segregating genes in common by descent.

Behavior genetic designs, contrary to widespread belief, are not limited to assessing genetic influences. The URT design, for example, estimates only environmental influence. Kinships that estimate a single parameter are almost always used in conjunction with other kinships in order to provide independent confirmation of the findings although they are fully informative in their own right.

More information about fitting genetic models can be found in Neale and Cardon (1992) and Neale (1995) and references cited therein. Let's now turn to some actual psychological traits and representative findings.

3. RESULTS

3.1. Psychological Interests

I begin with psychological interests because in my opinion the measurement of psychological interests is an important and seriously under-appreciated contribution of differential psychology to human welfare. Interest measures relate to important facets of everyone's daily life—the worlds of work and leisure. There are three widespread but false beliefs about psychological interests that deserve brief attention. First, contrary to the opinions of many psychologists, particularly introductory psychology textbook authors, the domain of psychological interests is not simply a reflection of personality—the correlations between the two domains are, with a few exceptions, trivial (Waller, Lykken, & Tellegen, 1995). Psychological interests are a discrete domain of human traits worthy of study in their own right. Second, interests are not ephemeral. They stabilize in late adolescence and are quite resistant to change after that time (Campbell, 1966). Consider a study by Rohe (1980) of 14 young males who suffered spinal cord damage. Pre-injury Strong Vocational Interest Blank scores were recovered from their personal files and compared to their current scores. There was little change in interests following injury and what change did exist was comparable to that of a control group that did not undergo spinal cord damage. It does not come as a surprise to differential psychologists that young male paraplegics play basketball and compete in wheel chair races. This robustness of psychological interests in the face of powerful environmental influences has been well-known for many years. On the other hand some psychologists are surprised to find that the interest profiles of American Brigadier Generals are remarkably like those of high-level corporate executives, except for the former's specific interest in military activities (Campbell, 1995). Finally, consistent with the above findings, and contrary to a widespread belief among psychologists, psychological interests are very significantly influenced by genetic factors. Gati (1991) in a comprehensive review of the structure of interests in the Psychological Bulletin, the main review journal of the American Psychological Association, claimed that "although there is evidence for the contribution of genetic factors to interests, these factors account for less than 5% of interest variance" (p. 312). I will demonstrate that Gati is off by a factor of about 8 to 10. Even more interesting is that fact that common family environmental influence on psychological interests is quite small. The belief that children follow in the occupational footsteps of their parents for environmental reasons is largely incorrect because the inference is based on an observed correlation that is most often interpreted as causal rather than confounded.

I have also chosen the interest domain to illustrate a simple but important methodological principle—the importance of measurement error and specificity. Measurement error and specificity saturate all psychological instruments and failure to take them into account results in theoretically misleading conclusions[1].

Table 1 shows the twin correlations and model fitting heritability from Minnesota Study of Twins Reared Apart (MISTRA) for ten factors derived from a factor analysis of two widely used inventories, the Strong Vocational Interest Blank (Hansen & Campbell, 1985) and the Jackson Vocational Interest Survey (Jackson, 1977).

The MZA correlations which directly estimate the broad heritability clearly suggest strong genetic influence on these measures of interests. More interestingly the very low DZA correlations suggest that the source of genetic variance is non-additive or what my colleague David Lykken has called "emergenic." Emergenic genetic influences are those that are genetic in origin but do not run in families (Lykken, McGue, Tellegen, and Bouchard, 1992). A few years after we began the Minnesota Study of Twins Reared Apart (MISTRA), we realized that very few studies of adult twins reared together, with whom we hoped to compare our findings, had ever been carried out. Most twin studies up to that point had been carried out on children. Consequently, we launched the Minnesota Twin Registry (Lykken, Bouchard, McGue, & Tellegen, 1990), a large adult twin registry where twins were ascertained from birth records. Because the cost of using the commercial instruments included in the MISTRA assessment were prohibitive (tens of thousands of dollars), we created our own vocational interest inventory as well as a leisure time interest inventory and scales to measure self-ratings of talent and personal qualities (e.g., abstract intelligence, self-discipline, etc.). These instruments were then administered to a sample of 512 pairs of MZTs and 390 pairs of DZTs. The same instruments were mailed to MZAs

---

[1] I distinguish between theoretical and practical conclusions because theoretically we are interested in the degree of genetic influence on a latent trait or theoretical construct. Practically we may be interested in the heritability of a specific measure (e.g., IQ as measures by the Wechler Adult Intelligence Scale). An excellent discussion of measurement eror in psychological research can be found in Schmidt and Hunter (1996).

Table 1. Intraclass correlations for MZA and DZA twins
for ten factor scales derived from two interest inventories
and heritabilities estimated by model-fitting

	Intraclass correlation		
Scale	MZA (N=52)	DZA (N=27)	Model-fitting heritability
Realistic	.44	−.09	.41
Investigative	.68	.02	.66
Artistic	.52	−.07	.50
Social	.54	.02	.52
Enterprising	.41	.30	.50
Conventional	.38	.17	.38
Academic orientation	.73	.19	.82
Work Style	.25	−.03	.22
Adventure	.54	.04	.53
Medical	.47	.14	.48
Mean	.50	.07	.50

Note: Data compiled from Moloney et al., (1991)

who had previously participated in our study as well as newly recruited twins yielding 54 pairs.

A factor analysis of the results yielded 39 first-order and 11 second order factors. The mean intraclass correlations for the three groups of twins are shown in Table 2. The heritability estimate for the 50 factors is highly consistent across methods; the MZA direct estimate is (.42) and the estimate based on reared together twins is .52. In this instance we found evidence for non additive genetic variance for only a few measures.

In order to pursue the question of non additive genetic variance further we collaborated with colleagues at a number of other Universities and were able to combine family data gathered on the SVIB from eight different kinships. They included MZ and DZ twins reared apart and together, adopted parents and their offspring, biological parents and their offspring, adopted siblings and biological siblings. Because different versions of this inventory had been used in the various studies we had to use rather short scales (Hansen, 1982) that contained items common to all versions of the instrument. With this combination of kinships it was possible to estimate additive genetic effects, nonadditive genetic effects, shared environmental effects and nonshared environmental effects plus error. The

Table 2. Mean intraclass correlations for MZA, DZT and MZT twins
and Falconer heritability estimated from DZT and MZT twins
for 50 first and second-order interest factors

Intraclass correlation			
MZA (N=54)	DZT (N=390)	MZT (N=512)	Heritability based on reared together twins
.42	.23	.49	.52

Note: Data compiled from Lykken (1993).

Table 3. Median proportions of variance attributed to additive genetic, nonadditive genetic, shared environmental, and nonshared environmental effects plus error and broad heritability, derived from eight kinships, for seventeen brief strong vocational interest blank scales

Additive genetic	Non additive genetic	Shared environmental	Nonshared environmental plus error	Heritability
.14	.22	.08	.54	.36

Note. Data compiled from Betsworth et al., (1993).

median results for the 17 SVIB scales that can be scored in this manner are shown in Table 3.

The results of this study strongly confirm our initial findings of nonadditive genetic variance for psychological interests.

Notice the remarkably small component of shared or common family environmental influence. As I mentioned earlier the most direct estimate of this value is the correlation for unrelated siblings reared together. They share a common family environment but no genes. That correlation is also .08 corresponding precisely to the overall estimate. Common family environmental influence can also be estimated from Table 2. It is simply the difference between the MZT and MZA correlations as the MZTs live together and experience a common family environmental influence and the MZA's do not. In that instance the value is .07.

The heritabilities found in this last study are smaller than in the previous studies. The reason for this is simple—unreliability of measurement. Table 4 shows the MZA correlations for six common higher-order factors that can be estimated in two of the three studies and comparable factors derived from two instruments. As expected the correlations increase with the psychometric quality of the instruments. We have demonstrated the same effect in the twin registry sample (Lykken, Bouchard, McGue, & Tellegen, 1993).

The failure to find non additive genetic effects in the large twin registry sample remains a puzzle and will be resolved only when we are able to test this sample with instruments comparable to those used in the other studies. As an aside it is worth mentioning that the two instruments used to derive the factor scales make use of quite different methods of measurement (paired comparisons vs. a Like, Indifferent, Dislike format) and utilize different content. Each instrument, however, yielded the same results.

Table 4. MZA Intraclass correlations for the Holland General Occupational Themes for the Hansen Brief Scales, the SCII Full-Length Scales and for factor measures based on two instruments

Scale	Hansen brief scales	SCII full-length scales	Factor scales
Realistic	.20	.30	.44
Investigative	.39	.48	.68
Artistic	.23	.38	.52
Social	.42	.42	.54
Enterprising	.41	.42	.41
Conventional	.24	.25	.38
Mean	.32	.38	.50

Note: Data compiled from Betsworth et al., (1993) and Moloney et al., (1991)

Table 5. Intraclass correlations, confidence intervals, samples sizes, and test utilized for IQ in five studies of monozygotic twins reared apart

Study and test used (primary/secondary/tertiary)	N for each test	Primary test	Secondary test	Tertiary test	Mean of multiple tests
Newman et al. (1937) (Stanford-Binet/Otis)	19/19	.68 ± .12	.74 ± .10		.71
Juel-Nielsen (1980) (Wechsler-Bellevue /Raven)	12/12	.64 ± .17	.73 ± .13		.69
Shields (1962) (Mill-Hill/Dominoes)	38/37	.74 ± .07	.76 ± .07		.75
Bouchard et al. (1990) WAIS/Raven-Mill-Hill first principal component	48/42/43	.69 ± .07	.78 ± .07	.78 ± .07	.75
Pedersen et al (1992) first principal component	45	.78 ± .06			.78
Weighted average					.75

Note: From Bouchard (1996).

3.2. Intelligence

Another psychological trait that is known to be important in everyday life is intelligence (Barrett & Depinet, 1991). The nature-nurture question has been explored in the domain of intelligence in more depth than any other (Bouchard, 1996a; Bouchard, 1996c). The recent publication of the *Bell Curve* (Herrnstein & Murray, 1994), and the controversy surrounding it, brought out once again the huge discrepancy in views between most (not all) of the scientific researchers in the domain and the media. According to Staples (1994), in a signed editorial in the New York Times: "Despite the impression that there is something new in 'The Bell Curve' its authors, Charles Murray and Richard Herrnstein, have merely reasserted the long-unproven claim that I.Q. is mainly inherited. The language is calmer, the statistical gimmicks slicker, but the truth remains the same: There exists no plausible data to make the case" (p. A18). Specialists in the domain have taken a quite different view. While they recognize that there are many unknowns regarding the nature of intelligence (Neisser, Boodoo, Bouchard, Boykin, Brody, Ceci, et al., 1996), they agree that it is an important and substantive construct (Gottfredson, 1994; Neisser, et al., 1996) and virtually all agree that genetic factors underlie variation in IQ to an important degree (Snyderman & Rothman, 1988).

Contrary to Staples' claim, the evidence for genetic influence on IQ has become much stronger in recent years and a number of important discoveries have been made. I summarize the key findings in a few tables and figures. Table 5 shows the most direct evidence regarding the heritability of IQ, the correlations for monozygotic twins reared apart.[2]

The first three studies by Newman, Freeman and Holzinger (Newman, Freeman, & Holzinger, 1937), Juel-Nielsen (Juel-Nielsen, 1980), and Shields (Shields, 1962) have been savaged by critics (Taylor, 1980). The data, however, have stood up to these criticisms quite well (Bouchard, 1982; Bouchard, 1983; Bouchard, 1996c). This is not to say they are without flaws. No experiment of nature is. The more interesting fact is that they

---

[2] This table omits the controversial findings of Cyril Burt (Macintosh, 1995).

now have been fully replicated with much better instrumentation, diagnostic techniques, larger sample sizes and in one case far superior sampling. The Minnesota Study of Twins Reared Apart (MISTRA) (Bouchard, Lykken, McGue, Segal, & Tellegen, 1990), with few exceptions, recruited twins as they came to light throughout the English speaking world. The sample is thus one of convenience, and the justification for its use lies in the adequacy of the descriptive statistics (Bouchard, et al., 1990). The MISTRA researchers have shown that the similarity between the twins in IQ (as well as other traits) cannot be explained by such factors as amount of contact, age of separation, or measurable background characteristics of the rearing families. The Swedish Adoptive Twin Study (SATSA) based on a birth registry sample has also replicated the previous studies (Pedersen, Plomin, Nesselroade, & McClearn, 1992; Plomin, Pedersen, Lichtenstein, & McClearn, 1994). The SATSA researchers have also shown that their results cannot be explained by known and measured artifacts.

In spite of empirical demonstrations to the contrary, the argument that some kind of similarity in environmental treatment makes reared apart identical twins similar in IQ continues to lurk in the background (Lewontin, Rose, & Kamin, 1984). This hypothesis can be directly tested in a number of ways. A simple test is to examine the correlation in IQ for cousins. These individuals are reared in ordinary homes by their biological parents where one parent in each pair grew up in the same home. The median IQ correlation for cousins based on four studies (1,176 pairings) is .15, a figure considerably smaller than that found for MZA twins. Part of this correlation, of course, is genetic in origin.

A more powerful test consists of examining the correlations for unrelated individuals reared together. This correlation directly estimates the influence of common family environmental influence. The results of all such studies are shown in Figure 2.

Notice that when measured in childhood such pairings reveal a common family environmental influence of about 30%. When measured in adulthood, however, the correlation drops precipitously. Three of the data points in the five adult samples report longitudinal data. That is the correlations are based on samples also tested in childhood. In the study with a sample of 108 (Scarr, Weinberg, & Waldman, 1993), the drop was from .31 to .19. In the group with a sample of 107 the drop was from .20 to −.03, and, in the group with a sample of 75 the drop was from .11 to −.02. Both of the last two groups were from the Texas Adoption Study (Loehlin, Horn, & Willerman, 1996). The correlations given above are based on participants tested at both times and do not necessarily correspond with numbers in the figure for children as there is always a loss of participants in follow-up studies. Almost all of the "adults" in these samples are between 16 and 21 years of age. It would be very desirable to have data on an older sample.

Critics of the MZA data who argue that various forms of bias (placement, rearing by relatives rather than random placement, etc.) probably lead to serious overestimates of genetic influence seldom address both the MZA and URT data sets simultaneously (Dorfman, 1995; Fancher, 1995). Since URTs live in the same home and are "matched" on far more variables than MZA twins, the whole array of commonly cited biases in MZA studies is brought into question. I used to call this kind of selective reporting "pseudo-analysis" (Bouchard, 1982) until I discovered it already had a name. It is called the Neglected Aspect Fallacy (Castell, 1935) and it violates Carnap's Total Evidence Rule (1950).

The data in Figure 2 raise an interesting question. Does the heritability of IQ really change over time? I and my colleague Matt McGue failed to take account of this possibility when we reviewed the IQ literature in 1981 (Bouchard & McGue, 1981). Figure 3 shows what happens when the twin reared together literature is organized by age (McGue, Bouchard, Iacono, & Lykken, 1993).

Twin Studies of Behavior 129

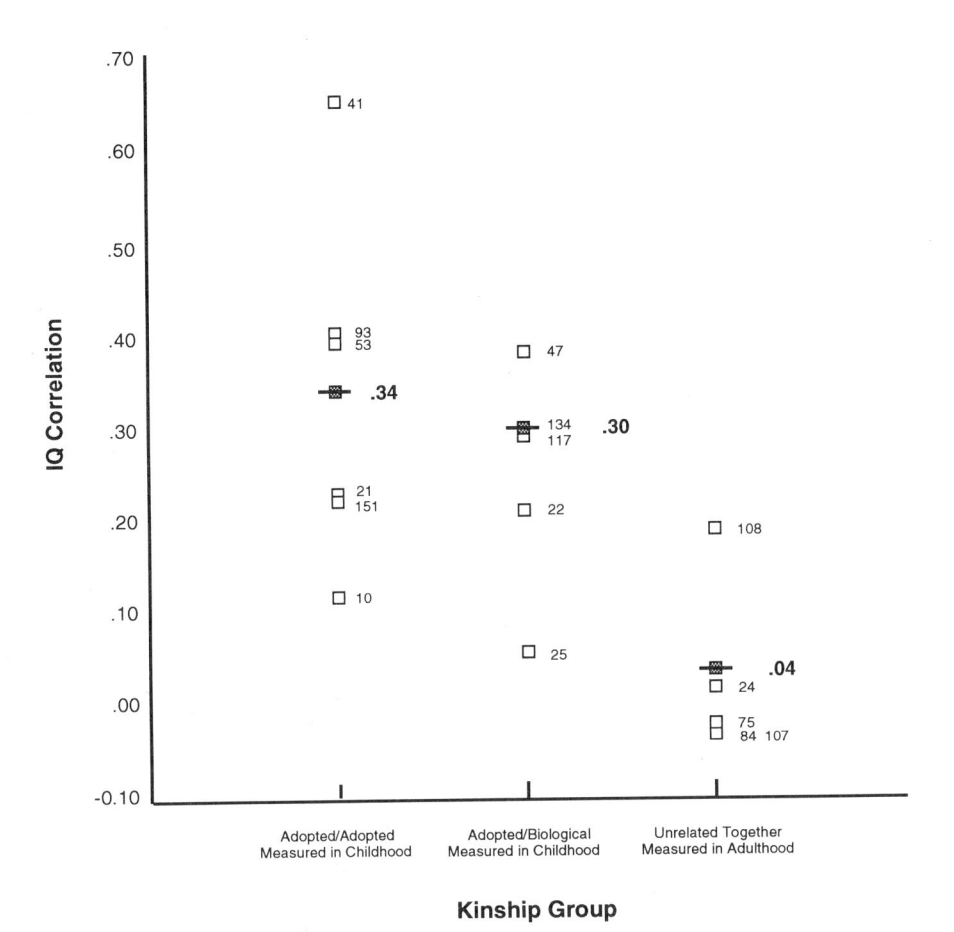

Figure 2. IQ correlations, sample sizes and weighted mean correlations (shaded square) for unrelated individuals reared together organized as adopted/adopted pairs measured in childhood, adopted/biological pairs measured in childhood and both types of pairs measured in adulthood.

There is a clear increase in heritability and a decrease in common family environmental influence with age. More recent studies also demonstrate roughly the same effect both in the U.S. and Sweden (Finkel, Pedersen, McGue, & McClearn, 1995), although there is some evidence for a decline in heritability in old age.

3.3. Personality

There is a widespread consensus that five major factors (Big Five) describe a significant part of variance in the domain of personality. They are Extraversion, Neuroticism, Conscientiousness, Agreeableness and Openness. The behavior genetic findings for these factors are summarized in Table 6 which presents the intraclass correlations for MZA, DZA, MZT and DZT twins and the genetic and environmental parameter estimates for the Multidimensional Personality Questionnaire (MPQ) markers of the Big Five.

The model fitting for this data was based on an analysis of multiple kinships gathered from the entire world literature by Loehlin (1992). His analysis required both additive and non additive variance and we allowed for them in our model. Our results mirrored

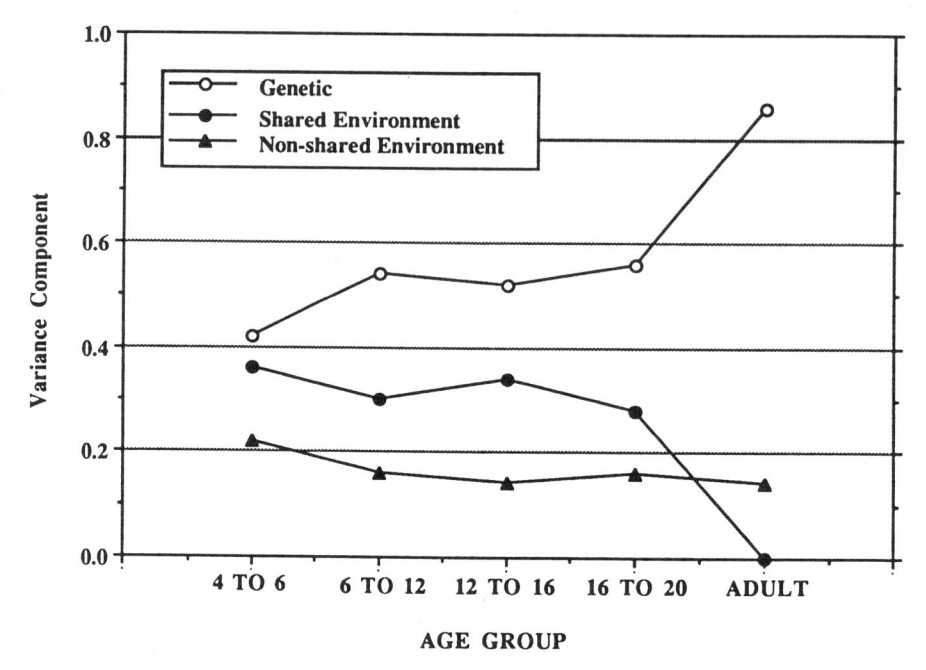

Figure 3. IQ variance component estimates for five age groups, derived from published IQ correlations using the Falconer heritability formula.

his very closely (Bouchard, 1994). The broad heritability of these traits is around .40, there is significant nonadditive variance for most of them, and the degree of common environmental influence is modest for most and zero for some.

Another example of how twin data can help illuminate interesting issues deals with the more detailed analysis of single variables. Recently Diener and Diener (1996) published an article entitled "Most People are Happy." They noted that most people reported Subjective Well-Being (SWB or Happiness) with marriage, work and leisure as well above the neutral point. This finding holds for disadvantaged groups as well as advantaged groups and is true in 86% of the 43 nations for which representative samples are available. They also demonstrated that many factors influence SWB but that their influence is transitory.

Table 6. Intraclass correlations for MZA, DZA, MZT and DZT twins and model-fitting results for Multidimensional Personality Questionnaire big five markers

Trait	MZA (59)	DZA (47)	MZT (522)	DZT (408)	Additive genetic	Non-additive genetic	Common environment	Broad heritability
Extraversion	.41	−.03	.54	.19	.09	.29	.15	.38
Neuroticism	.49	.44	.48	.19	.41	.09	.00	.50
Conscientiousness	.54	.07	.54	.29	.29	.13	.13	.42
Agreeableness	.24	.09	.39	.11	.05	.25	.09	.30
Openness	.57	.27	.43	.14	.29	.15	.00	.44
Mean	.45	.17	.48	.18	.23	.18	.07	.41

Note: Data compiled from Bouchard (1996b).

Table 7. Intraclass correlations on the well-being scale of the Multidimensional Personality Questionnaire for middle-aged twins reared together and reared apart

Type of twin pair	Number of pairs	Intraclass R
Twins reared together		
Monozygotic	647	.44
Dizygotic	733	.08
Twins reared apart		
Monozygotic	75	.52
Dizygotic	36	−.02

Note: Data compiled from Lykken and Tellegen (1996).

Similar results obtain for the Well-Being scale on the MPQ in a large Minnesota Twin Registry sample (Lykken & Tellegen, 1996). Important and powerful measures of worldly success such as educational level, income, socioeconomic status only account for 2–3% of the variance in Well-Being. On the other hand a very significant amount of the variance in Well-being is due to genetic factors. Table 7 shows the intraclass correlations for Well-Being for four twin groups.

These figures should be compared with the reliability of the test. The MPQ was given to 245 twins tested at ages 20 and 30. This test-retest correlation for Well-Being was .50. The cross-twin MZ correlation (twin 1 time 1 with twin 2 time 2 and similarly twin 2 at time 1 with twin 1 at time 2) was .40 and the cross twin DZ correlation was .07 or nearly zero. Thus heredity accounts for about 80% of the reliable variance over 10 years in young adulthood. As my colleagues David Lykken and Auke Tellegen conclude: "If the transitory variations of well-being are largely due to fortune's favors, whereas the midpoint of these variations is determined by the great genetic lottery that occurs at conception, then we are led to conclude that individual differences in human happiness—how one feels at the moment and also how happy one feels on average over time—are primarily a matter of chance" (p. 189).

Diener and Diener discussed happiness with work and leisure. They failed to note that twin research has been published on these topics as well. Job satisfaction—happiness in the world of work—has also been shown to be unrelated to numerous factors that would be expected to correlate with it. On the basis of this evidence, Staw (1986; 1985) argued that job attitudes should be viewed in the same way as personality dispositions, that is, as relatively enduring structures. We have shown using both samples of twins reared apart and twins reared together that job satisfaction and work values do have very significant heritabilities (Arvey, Bouchard, Segal, & Abraham, 1989; Bouchard, Arvey, Keller, & Segal, 1992; Keller, Arvey, Bouchard, Segal, & Dawis, 1992). These results have been replicated in an independent twin sample (Arvey, McCall, Bouchard, & Taubman, 1994).

3.4. Social Attitudes

The domain of social attitudes is one to which psychologists bring a strong preference for environmental explanations of variance. I was misled by this bias when I initially set up the assessment battery for MISTRA in 1979. Having been trained in Social Psychology and having taught Social Psychology courses, I had a number compilations of social attitude measures on my office bookshelf (Robinson & Shaver, 1969; Shaw & Wright, 1967). I consulted them at that time and quickly concluded that it was implausible that

Table 8. Intraclass correlations for five twin groups, spouse correlations and heritabilities for conservatism as measured by the Wilson-Patterson Scales in the Australian study and radicalism and tough-mindedness as measured by the Public Opinion Inventory in the London study

	Monozygotic		Dizygotic			Spouse	Heritability
	Male	Female	Male	Female	Unlike		
Australian twin study							
Number of pairs	565	1,232	351	750	905	103	
Conservatism	.60	.64	.47	.46	.41	.68[a]	.62
London twin study							
Number of pairs	120	325	59	194	127	562	
Radicalism	.75	.60	.52	.51	.48	.51	.76 (M)
							.62 (F)
Tough-mindedness	.49	.69	.18	.41	.28	.55	.42 (M)
							.69 (F)

Note: Data compiled from Martin et al. (1986).
a. Spouse correlation taken from Feather (1978).

such measures would demonstrate genetic variance. Attitudes, I was certain, were shaped entirely by family and social circumstances. Seven years latter Martin, Eaves, Heath, Jardine, Feingold and Eysenck (1986) demonstrated significant levels of heritability for social attitude measures using two different instruments in two different twin studies conducted in London and Australia. A summary of their results is shown in Table 8.

The high spouse correlations in this table are of interest. Assortative mating is higher for attitudes than any other domain of psychological traits. Assortative mating, if based on genetic sources of variance, increases the correlation between first degree relatives. Thus the DZ correlations are high relative to the MZ correlations. The model-fitting estimates of heritability in Table 8, which take account of assortative mating, are given in the last column of the table. It should be noted that sex had to be taken into account in the London study. The authors are aware that these estimates do not agree with our intuition "that cultural factors derived from parents are major determinants of family resemblance in attitudes" (Eaves, Eysenck, & Martin, 1989. p. 387), and they urge more extensive data gathering to test their models. It will be interesting to see if these results can be constructively replicated using other instruments and kinships.

Somewhat shocked by these findings, and seven years late, we inserted some attitude measures in our assessment. We also turned to our assessment battery and asked if there were any measured that might be "attitudinal like" in nature. We had, by virtue of their inclusion in some of our standard instruments, measures of religious interests, religious attitudes and religious values. In addition, as a consequence of not being able to administer a commercial test to the Minnesota Twin Registry, we created our own measures of Occupational and Leisure Time Interests. We had measures of religious interests on each of these instruments that had been administered to both twins reared apart and twins reared together. We also had a measure of religious interests from the SVIB (Hansen & Campbell, 1985) a measures of religiousness that could be scored from the Minnesota Multiphasic Personality Inventory (MMPI) (Dahlstrom, Welsh, & Dahlstrom, 1975) and a measure of religious values that could be scored from the Allport-Vernon-Lindzey Study of Values (Allport, Vernon, & Lindzey, 1960). The results of our analysis of shown in Table 9.

Table 9. Intraclass correlations for MZT, DZT, MZA and DZA twins, number of pairs of twins and model-fitting heritability estimates for various religiousness measures

	MZT	DZT	MZA	DZA	Heritability
Religious leisure time interests					
Number of pairs	458	363	32	24	
Correlations	.60	.64	.39	.04	.47
Religious occupational interests					
Number of pairs	458	363	32	25	
Correlations	.41	.19	.59	.20	.41
MMPI Religious Fundamentalism Scale					
Number of pairs			50	30	
Correlations	no data	no data	.55	−.22	.46
SVIB religious interest					
Number of pairs			52	31	
Correlations	no data	no data	.49	.15	.48
AVL religious values					
Number of pairs			38	21	
Correlations	no data	no data	.55	−.08	.52

Note: Data compiled from Waller et al. (1990).

While our heritabilities are not quite as large as those of Martin et al., our results do strongly suggest an important degree of genetic influence on social-attitude like measures. Nevertheless, the findings raise some interesting questions as well. For example, the very low DZA correlations are not consistent with the hypothesis that the high DZT correlations in the London and Australia Studies are due to assortative mating! Some critics have asserted that these results are "preposterous." We, on the other hand, believe that they seriously question our understanding of how human being come to hold the social attitudes. At least one social psychologist has begun a research program to explore the implications of these findings (Tesser, 1993).

3.5. Psychopathology

The domain of psychopathology is very large and cannot be reviewed in any simple manner. What I will do here is simply present the summary results of two representative studies using continuous measures. The first is a twin study of personality disorders using a recently developed 18 scale instrument (Livesley, Jang, Jackson, & Vernon, 1993). The sample consists of ordinary Canadian twins recruited in Vancouver Canada. The second is a recently published study of the MISTRA twins using the MMPI (DiLalla, Carey, Gottesman, & Bouchard, in press). The findings are shown in Table 10.

The results are remarkably similar given that two kinship groups and two different instruments were used. They are also very much like the personality results in general.

4. THE BEHAVIOR GENETICS OF "ENVIRONMENTAL MEASURES"

One of the most provocative "new areas" in behavior genetics might be labeled the genetics of environmental measures (Plomin, 1994a). This domain has two facets. The first is external dealing with objective characteristics of the environment, for example, the

Table 10. Mean intraclass correlations and broad heritabilities of measures of psychopathology in two studies, one using twins reared together and the other using twins reared apart

Measures	MZ	DZ	Broad heritability
Vancouver Study—twins reared together			
Mean correlation for 18 scales	.54	.26	.44
MISTRA—twins reared apart			
Mean correlation for 11 clinical scales	.46	.12	.44

Note: Data complied from Livesley et al., (1993) and Dillala et al. (in press).

socioeconomic characteristics of a child's parents (SES). The second is more subjective and experiential dealing with features of the environment created by the individual him- or herself. As with many new topics it is not new at all, merely a domain ignored and largely forgotten, only to be revived by a new generation of investigators.

Consider the use of SES as an environmental (causal) variable. Developmental psychologists are particularly fond of using parental SES, often indexed by income level, to explain low school achievement in children (Blau, 1981), anti-social behavior, and race differences in IQ (Brooks-Gunn & Klebanov, 1996). The problem with this approach is that it assumes causation rather than testing it. Sir Francis Galton introduced the adoption method into psychology well over 100 years ago to test the influence of relative social advantage (Galton, 1869/1914) rather than assuming it. Barbara Burks formulated the question in mathematical terms 1928 in a paper entitled "Statistical hazards in nature-nurture investigations" (Burks, 1928b). Ten years later she published the results of an adoption study that led her to conclude that about 2/3rds to 3/4 of the average IQ difference among children born into these social classes were due to genetic differences (Burks, 1938). Burks, of course, did not study the truly disadvantaged so it is not appropriate to generalize beyond what might be considered the ordinary range of social classes. Forty years later, Scarr and Weinberg (1978) replicated Burks work with a Minnesota sample.

In his famous 1969 monograph entitled "How much can we boost IQ and scholastic achievement?," Arthur Jensen (1969) pointed out that "Few if any students of this field today would regard socioeconomic status per se as an environmental variable that primarily causes IQ differences. Intellectual differences between SES groups have hereditary, environmental and interaction components" (p. 75). In a definitive treatment of this topic Meehl (1970) argued that "the commonest error in handling nuisance variables of the 'status' sort (e.g., income, education, locale, marriage) is the error of suppressing statistically components of variance that, being genetic, ought not to be thus arbitrarily relegated to the 'spurious influence' category" (pp. 393–394). Reviewing this problem again in 1973, Jensen referred to it as "The sociologists fallacy" (Jensen, 1973). I thought the characterization somewhat too strong and unfair, but consider the following two exemplars. The first is a quote from a 1981 book dealing with both between and within race (Blacks and Whites) IQ differences. The second example involves a meta-analysis of the literature dealing with the correlations between parental SES and children's achievement and the correlations between parental SES and children's IQ.

According the Blau, "The basic object of this book is to identify the social processes that influence the development of intellectual competence in black children and white children in order to account for their differences in measured ability in the early years of schooling. The results of my study provide strong evidence that the sources are social, not

Table 11. Intraclass correlations, sample sizes for MZA, MZT, DZA and DZT twins and model fitting parameter estimates for measures of occupational status by sex

Sex	Twin types				Parameter estimates			
	MZA (24)	MZT (42)	DZA (38)	DZT (50)	Additive genetic	Non shared Environment	MZA correlated environment	Common environment
Men	.50	.84	.66	.36	.60	.18	.12	.10
	(24)	(42)	(38)	(50)				
Women	.13	.56	.18	.47	.12	.46	.33	.10
	(16)	(36)	(52)	(40)				

Note: Data compiled from Lichtenstein et al. (1995).

genetic, in origin" (1981, p XV). The entire book, however, relegates SES and a series of related variables to the 'spurious influence' category while explicitly purporting to test and refute Jensen! The research does not include any twins nor adoptees and consequently cannot even begin to address questions dealing with genetic variance.

My second example is just as egregious in its own way. White (1982), in what is otherwise an exemplary and still valuable meta-analysis, discusses the use of SES both as a spurious influence (covariate or nuisance variable) and as a causal agent (p. 462). He completely fails to warn the reader, however, that it is mandatory to sort correlations into those based on relationships between biological relatives and those based on relationships between adopted parents and their adopted children. His failure to discuss such a critical issue is perplexing as he does cite two classic adoption studies (Burks, 1928a; Freeman, Holzinger, & Mitchell, 1928) but not that of Scarr and Weinberg published in 1978.

Neither the Blau quote nor the White literature review are isolated examples. The entire April 1994 issue of the journal *Child Development* commits the same error as does the 1995 presidential address of the President of the Division of Developmental Psychology of the American Psychological Association (Houston, Fall 1995). Both sources purport to deal with poverty, but conflate it with ordinary SES differences.

The best evidence for SES as a variable that reflects genetic influence when measured within families comes from adoption studies such as those of Burks (1938) and Scarr and Weinberg (1978). There are other lines of evidence as well. Waller (1971) has shown that if one tracks brothers over the course of their lifetime, sons with IQs higher than their father move up in SES relative to him and those with IQs lower than their father move down in SES. Variations on this design have also been carried out in England (Gibson & Light, 1967; Gibson, 1973). The issue can be addressed using twins as well. Table 11 shows the twin correlations and genetic and environmental parameters for occupational status using the four group design.

There are a number of interesting findings in this table. First, and not unexpectedly, there are large differences in genetic effects for men and women. Second, the amount of genetic influence on men's occupational status is large. Third, the placement bias effect (correlated environments) for the MZA twins is quite large. For both genders, however, the genetic effect is large.

The study of stressful life events is a nice example of the type of variable studied in the second domain of environmental measures mentioned previously. Psychologists have for years studied the influences of a wide variety of life events, family interaction patterns, etc., under the hypothesis that they were proximal causal variables with regard to a variety of psychological outcomes. The now discredited hypothesis of the "refrigerator mother" is a classic example.

Stressful influences both at home and in the work place have often been proposed as proximal causal variables for both physical mental health outcomes (Barling, 1990; Cooper & Kirkcaldy, 1994; Weiner, 1992; Zahn-Waxler, 1995), and numerous measures of stress have been devised (Cohen, Kessler, & Gordon, 1995). The importance of such variables on long-term outcomes has always been questioned (Hudgens, 1974; Pearlstone, Russell, & Wells, 1994), and the direction of causation question has been raised repeatedly. Only recently has the question of genetic influence on such measures been addressed. Total score on the well-known Social Readjustment Rating Scale (Holmes & Rahe, 1967) yield heritabilities of about .30 to .40 (Moster, 1991; Plomin, Lichtenstein, Pedersen, McClearn, & Nesselroade, 1990). This is a very high figure given the inherently unstable nature of such scales, and it probably accounts for much of the reliable variance in these measures. More importantly, dividing up these events into meaningful categories such as "controllable" vs. "uncontrollable" and "to self" vs. "to others," etc. results in strong differential heritability with uncontrollable events showing, as expected, zero heritabilities (Kendler, Neale, Kessler, Heath, & Eaves, 1993; Lichtenstein & Pedersen, 1995; Moster, 1991; Plomin, 1994b; Plomin, et al., 1990)

Finally a few comments on one additional facet of the domain of personality which overlaps with stressful life events—divorce. Divorce is associated with significant negative psychological outcomes in children and consequently has been interpreted by most psychologists as an environmental stressor (negative life event). One problem with this "causal" interpretation is that adjustment difficulties in these children predate the time of parental separation (Block, Block, & Gjerde, 1986). One can argue that this finding simply reflects problems in a marriage prior to divorce. It is possible, however, to ask a different question. Do genetic factors underlie divorce and might these factors be expressing themselves in the children of divorced parents? We do not know the full answer to these questions yet, but twin studies show a significantly higher concordance for divorce among MZ twins than DZ twins, and family background, involving divorce of both spouses and both spouses' parents, contributes independently to risk of divorce (McGue & Lykken, 1992). This suggests that characteristics each spouse brings to the relationship contribute to divorce. These characteristics may, of course, be transmitted genetically to their children. There may also be an interaction between the environmental stress created by the transactions in these families and the underlying characteristics transmitted to some of the children. In any event the possibility exists that what has been purported to be an environmental stressor actually reflects an underlying genetic disposition. This example also highlights the fact that a "phenotype" that is socially defined and may not even exist in some societies—those that do not allow divorce—may nevertheless be influenced by genetic factors. Another obvious example is alcoholism—a trait that could not exist in a population that was unfamiliar with or banned alcoholic substances.

5. CONCLUSIONS

At the end of a brief review of the genetics of personality published a few years ago, I drew some conclusion that I believe apply just as well to all psychological traits. Consequently, I believe it is appropriate to repeat it here.

"Current thinking holds that each individual picks and chooses from a range of stimuli and events largely on the basis of his or her genotype and creates a unique set of experiences — that is, people help to create their own environments (Scarr, 1992).This view of human development does not deny the existence of inadequate and debilitating environments nor does it minimize the role of learning. Rather, it views humans as dynamic creative organisms for whom the opportunity to learn and to experience new environments amplifies the effects of the genotype on the phenotype. It also reminds us of links to the biological world and to our evolutionary history. This brings us to the core problem of the genetics of personality — the function of variation in personality traits. The purpose of this variation is undoubtedly rooted in the fact that humans have adapted to life in face-to-face groups (sociality). Unraveling the role human individual differences play in evolution is the next big hurtle (Buss, 1993), and its solution will turn the behavior genetics of human personality from a descriptive discipline to an explanatory one" (Bouchard, 1994, p. 1701).

Since the above statement was written, we have begun to develop such a theory—the theory of Experience Producing Drives-Revised or EPD-R theory (Bouchard, Lykken, Tellegen, & McGue, 1996; Bouchard, under review). It is based on a theory originally proposed by Keith Hayes (1962), the American Comparative Psychologist who with his wife raised a chimpanzee in their home and who struggled mightily to disentangle the puzzle of how genes and environment drive behavior. But that is a story for another day.

REFERENCES

Allport, G. W., Vernon, P. E., & Lindzey, G. (1960). *Manual for the study of values (3rd ed.)*. Boston: Houghton Mifflin.

Arvey, R. D., Bouchard, T. J., Jr., Segal, N. L., & Abraham, L. M. (1989). Job satisfaction: environmental and genetic components. *Journal of Applied Psychology, 74*, 187–192.

Arvey, R. D., McCall, B., Bouchard, T. J., Jr., & Taubman, P. (1994). Genetic influence on job satisfaction and work values. *Personality and Individual Differences, 17*, 21–33.

Barling, J. (1990). *Employment, stress and family functioning*. New York: Wiley.

Barrett, G. V., & Depinet, R. L. (1991). A reconsideration of testing for competence rather than for intelligence. *American Psychologist, 46*(10), 1012–1024.

Betsworth, D. G., Bouchard, T. J., Jr., Cooper, C. R., Grotevant, H. D., Hansen, J. C., Scarr, S., & Weinberg, R. A. (1993). Genetic and environmental influences on vocational interests assessed using adoptive and biological families and twins reared apart and together. *Journal of Vocational Behavior, 44*, 263–278.

Blau, Z. S. (1981). *Back children/White children: Competence, socialization, and social structure*. New York: Free Press.

Block, J. H., Block, J., & Gjerde, P. S. (1986). The personality of children prior to divorce: A prospective study. *Child Development, 57*, 827–840.

Bouchard, T. J., Jr. (1982). [Review of Identical twins reared apart: A reanalysis]. *Contemporary Psychology, 27*, 190–191.

Bouchard, T. J., Jr. (1983). Do environmental similarities explain the similarity in intelligence of identical twins reared apart? *Intelligence, 7*, 175–184.

Bouchard, T. J., Jr. (1994). Genes, environment and personality. *Science, 264*, 1700–1701.

Bouchard, T. J., Jr. (1996a). Behavior Genetic Studies of Intelligence, Yesterday and Today: The Long Journey from Plausibility to Proof. In C. N. G. Mascie-Taylor & C. R. Brand (Eds.), *Biological and social aspects of intelligence* London: Macmillan.

Bouchard, T. J., Jr. (1996b). The genetics of personality. In K. Blum & E. P. Noble (Eds.), *Handbook of psychoeurogenetics* Boca Raton, FL: CRC Press.

Bouchard, T. J., Jr., Arvey, R. D., Keller, L. M., & Segal, N. L. (1992). Genetic influences on Job Satisfaction: A reply to Cropanzano and James. *Journal of Applied Psychology, 77*, 89–93.

Bouchard, T. J., Jr., Lykken, D. T., McGue, M., Segal, N. L., & Tellegen, A. (1990). Sources of human psychological differences: The Minnesota study of twins reared apart. *Science, 250*, 223–228.

Bouchard, T. J., Jr., Lykken, D. T., Tellegen, A. T., & McGue, M. (1996). Genes, drives, environment and experience: EPD theory - Revised. In C. P. Benbow & D. Lubinski (Eds.), *Psychometrics and social issues concerning intellectual talent*. Baltimore: John Hopkins University Press.

Bouchard, T. J., Jr., & McGue, M. (1981). Familial studies of intelligence: A review. *Science, 212*, 1055–1059.

Bouchard, T. J., Jr., & Propping, P. (Ed.). (1993). *Twins as a tool of behavior genetics*. Chichester, England: Wiley & Sons, Ltd.

Bouchard, T. J. J. (1996c). IQ similarity in twins reared apart: Findings and response to critics. In R. J. Sternberg & E. L. Grigorenko (Eds.), *Intelligence: Heredity and environment* (pp. ***-***). New York: Cambridge University Press.

Bouchard, T. J. J. (under review). Exprience producing drive theory: How genes drive exprience and shape personality. *Acta Paediatrica, Paper presented as part of the Nobel Symposium on Genetic vs. Environmental Determination of Human Behavior and Health, Stockholm Sweden, January 22–24, 1996.*

Brooks-Gunn, J., & Klebanov, P. K. (1996). Ethnic differences in children's intelligence test scores: Role of economic deprivation, home environment, and maternal characteristics. *Child Development, 67,* 396–408.

Burks, B. S. (1928a). Chapter X. The relative influence of nature and nurture upon mental development: A comparative study of foster parent-offspring child resemblance and true parent-true child resemblance. *Yearbook of the National Society for the Study of Education, 27,* 219–316.

Burks, B. S. (Ed.). (1928b). *Statistical hazards in nature-nurture investigations.* Blomington, Il: Public School Publishing Company.

Burks, B. S. (1938). On the relative contributions of nature and nurture to average group differences in intelligence. *Proceedings of the National Academy of Sciences, 24,* 276–282.

Buss, D. M. (1993). Strategic individual differences: The evolutionary psychology of selection, evocation, and manipulation. In T. J. Bouchard, Jr. & P. Propping (Eds.), *Twins as a tool of behavioral genetics* Chichester: Wiley.

Campbell, D. P. (1966). Stability of interests within an occupation over thirty years. *Journal of Applied Psychology, 50,* 51–56.

Campbell, D. P. (1995). The psychological test profiles of Brigadier Generals: Warmongers or decisive warriors? In D. Lubinski & R. V. Dawis (Eds.), *Assessing individual differences in human behavior* (pp. 145–175). Palo Alto, CA: Davies-Black.

Carnap, R. (1950). *Logical foundations of probability.* Chicago: University of Chicago Press.

Castell, A. (1935). *A college logic.* New York: Macmillan.

Cohen, S., Kessler, R. C., & Gordon, L. U. (1995). *Measuring stress: A guide for health and social scientists.* New York: Oxford University Press.

Cooper, C. L., & Kirkcaldy, B. D. (1994). A model of job stress and physical health: The role of individual differences. *Personality and Individual Differences, 16,* 653–657.

Dahlstrom, W. G., Welsh, G. S., & Dahlstrom, L. E. (1975). *An MMPI handbook, Volume II: Research Applications.* Minneapolis: University of Minnesota Press.

Diener, E., & Diener, C. (1996). Most people are happy. *Psychological Science, 7,* 181–185.

DiLalla, D. L., Carey, G., Gottesman, I. I., & Bouchard, T. J., Jr. (in press). Personality indicators of psychopathology via MMPI in twins reared apart. *Journal of Abnormal Psychology.*

Dorfman, D. D. (1995). Soft Science with a neoconservative agenda. Review of "The Bell Curve." *Contemporary Psychology, 40,* 418–421.

Eaves, L. J., Eysenck, H. J., & Martin, N. G. (1989). *Genes, culture and personality: An empirical approach.* New York: Academic Press.

Fancher, R. E. (1995). The Bell Curve on separated twins. *The Alberta Journal of Educational Research, 41,* 265–270.

Feather, N. (1978). Family resemblance in conservatism: Are daughters more similar to parents than sons? *Journal of Personality, 46,* 260–278.

Finkel, D., Pedersen, N. L., McGue, M., & McClearn, G. E. (1995). Heritability of cognitive abilities in adult twins: Comparison of Minnesota and Swedish Data. *Behavior Genetics, 25,* 421–431.

Freeman, F. N., Holzinger, K. J., & Mitchell, B. C. (1928). The influence of environment on the intelligence, school achievement and conduct of foster children. In G. M. Whipple (Eds.), *Nature and nurture: Their influence on intelligence (The 27th yearbook of the National Society for the Study of Education, Pt. 1)* Bloomington, Ill: Public School Publishing.

Galton, F. (1869/1914). *Hereditary genius: An inquiry into its laws and consequences.* London: Macmillan.

Gati, I. (1991). The structure of vocational interests. *Psychological Bulletin, 109,* 309–324.

Gibson, J., & Light, P. (1967). Intelligence among university scientists. *Nature, ?,* 441–443.

Gibson, J. B. (1973). Social mobility and the genetic structure of populations. *Journal of Biosocial Science, 5*(251–259).

Gottfredson, L. S. (1994, Tuesday, December 3, 1994). Mainstream science on intelligence. *Wall Street Journal,* p.

Hansen, J. C. (1982). Hansen combined form scales for the SII. In Minneapolis: University of Minnesota Center for Interest Measurement Research.

Hansen, J. C., & Campbell, D. P. (1985). *Manual for the SVIB-SCII (4th ed.).* Stanford, CA: Stanford University Press.

Hayes, K. J. (1962). Genes, drives, and intellect. *Psychological Reports, 10,* 299–342.

Hayes, W. L. (1973). *Statistics for the social sciences* (2 ed.). New York: Holt, Rinehart & Winston.

Herrnstein, R. J., & Murray, C. (1994). *The bell curve: Intelligence and class structure in American life.* New York: Free Press.

Holmes, T. H., & Rahe, R. M. (1967). The social readjustment rating scale. *Journal of Psychosomatic Medicine, 11,* 213–218.

Houston, A. C. (Fall 1995, Children in poverty and public policy. *Developmental Psychology Newsletter, American Psychological Association,* p. 1–8.

Hudgens, R. W. (1974). Personal catastrophe and depression: A consideration of the subject with respect to medically ill adolescents, and a requiem for retrospective life-event studies. In B. S. Dohrenwend & B. P. Dohrenwend (Eds.), *Stressful life events: Their nature and effects* (pp. 119–134). New York: Wiley.

Jackson, D. J. (1977). *Jackson vocational interest survey manual.* Port Huron, MI: Research Psychologists Press.

Jensen, A. R. (1969). How much can we boost IQ and scholastic achievement? *Harvard Educational Review, 39,* 1–123.

Jensen, A. R. (1973). Equating for socioeconomic variables. In A. R. Jensen (Eds.), *Educability and group differences* New York: Harper and Row.

Juel-Nielsen, N. (1980). *Individual and Environment: Monozygotic twins reared apart (revised edition of 1965 monograph).* New York: International Universities Press.

Keller, L. M., Arvey, R. D., Bouchard, T. J., Jr., Segal, N. L., & Dawis, R. V. (1992). Work values: Genetic and environmental influences. *Journal of Applied Psychology, 77,* 79–88.

Kendler, K. S., Neale, M., Kessler, R., Heath, A., & Eaves, L. (1993). A twin study of recent life events and difficulties. *Archives of General Psychiatry, 50,* 789–796.

Lewontin, R. C., Rose, S., & Kamin, L. J. (1984). *Not in our genes: Biology ideology, and human nature.* Pantheon: New York.

Lichtenstein, P., Hersberger, S. L., & Pedersen, N. L. (1995). Dimensions of occupations: Genetic and environmental influences. *Journal of Biosocial Science, 27,* 193–206.

Lichtenstein, P., & Pedersen, N. L. (1995). Social relationships, stressful life events, and self-reported physical health: Genetic and environmental influences. *Psychology and Health, 10,* 295–319.

Livesley, W. J., Jang, K. L., Jackson, D. N., & Vernon, P. A. (1993). Genetic and environmental contributions to dimensions of personality disorders. *American Journal of Psychiatry, 150,* 1826–1831.

Loehlin, J. C. (1992). *Genes and environment in personality development.* Newbury Park, CA: Sage Publications.

Loehlin, J. C., Horn, J. M., & Willerman, L. (1996). Heredity, environment and IQ in the Texas adoption study. In R. J. Sternberg & E. L. Grigorenko (Eds.), *Heredity, environment and intelligence* (pp. ***-***). New York: Cambridge University Press.

Lykken, D., & Tellegen, A. (1996). Happiness is a stocastic phenomenon. *Psychological Science, 7,* 186–189.

Lykken, D. T., Bouchard, T. J., Jr., McGue, M., & Tellegen, A. (1990). The Minnesota twin family registry: Some initial findings. *Acta Geneticae Medicae et Gemellogiae, 39,* 35–70.

Lykken, D. T., Bouchard, T. J., Jr., McGue, M., & Tellegen, A. (1993). Heritability of interests: A twin study. *Journal of Applied Psychology, 78,* 649–661.

Lykken, D. T., McGue, M., Tellegen, A., & Bouchard, T. J., Jr. (1992). Emergenesis: Genetic traits that may not run in families. *American Psychologist, 47,* 1565–1577.

Macintosh, N. J. (Ed.). (1995). *Cyril Burt: Fraud or framed.* Oxford: Oxford University Press.

Martin, N. G., Eaves, L. J., Heath, A. C., Jardine, R., Feingold, L. M., & Eysenck, H. J. (1986). Transmission of social attitudes. *Proceedings of the National Academy of Sciences USA, 83,* 4364–4368.

McGue, M., Bouchard, T. J., Jr., Iacono, W. G., & Lykken, D. T. (1993). Behavior genetics of cognitive ability: A life-span perspective. In R. Plomin & G. E. McClearn (Eds.), *Nature, nurture and psychology* (pp. 59–76). Washington, D. C.: American Psychological Association.

McGue, M., & Lykken, D. T. (1992). Genetic influence on risk of divorce. *PS, 3,* 368–372.

Meehl, P. E. (1970). Nuisance variables and the ex post facto design. In M. Radner & S. Winokur (Eds.), *Minnesota Studies in the Philosophy of Science IV* Minneapolis: University of Minnesota Press.

Moloney, D. P., Bouchard, T. J., Jr., & Segal, N. L. (1991). A genetic and environmental analysis of the vocational interests of monozygotic and dizygotic twins reared apart. *Journal of Vocational Behavior, 39,* 76–109.

Moster, M. (1991) *Stressful life events: Genetic and environmental components and their relationship to affective symptomatology.* Ph.D. Dissertation, University of Minnesota.

Neale, M. C. (1995). *MX: Statistical modeling.* Richmond, VA: Department of Human Genetics, Box 3 MCV.

Neale, M. C., & Cardon, L. R. (Ed.). (1992). *Methodology for genetic studies of twins and families.* Dordrecht: Kluwer Academic Publishers.

Neisser, U., Boodoo, G., Bouchard, T. J., Jr., Boykin, A. W., Brody, N., Ceci, S. J., Halpern, D. F., Loehlin, J. C., Perloff, R., Sternberg, R. J., & Urbina, S. (1996). Intelligence: Knowns and unknows. *American Psychologist, 51,* 77–101.

Newman, H. H., Freeman, F. N., & Holzinger, K. J. (1937). *Twins: A study of heredity and environment.* Chicago: University of Chicago Press.
Pearlstone, A., Russell, R. J. H., & Wells, P. A. (1994). A re-examination of the stress/illness relationship: How useful is the concept of stress. *Personality and Individual Differences, 17,* 577–580.
Pedersen, N., L., Plomin, R., Nesselroade, J. R., & McClearn, G. E. (1992). A quantitative genetic analysis of cognitive abilities during the second half of the life span. *Psychological Science, 3,* 346–353.
Plomin, R. (1994a). *Genetics and experience: The interplay between nature and nurture.* Thousand Oaks, CA: Sage.
Plomin, R. (1994b). The nature of nurture: The environment beyond the family. In R. Plomin (Eds.), *Genetics and experience: The interplay between nature and nurture* (pp. 82–101). Thousand Oaks: Sage.
Plomin, R., Lichtenstein, P., Pedersen, N. L., McClearn, G. E., & Nesselroade, J. R. (1990). Genetic influences on life events during the last half of the life span. *Psychology and Aging, 5,* 25–30.
Plomin, R., Pedersen, N. L., Lichtenstein, P., & McClearn, G. E. (1994). Variability and stability in cognitive abilities are largely genetic later in life. *Behavior Genetics, 24,* 207–215.
Robinson, J. P., & Shaver, P. R. (1969). *Measures of social psychological attitudes.* Ann Arbor, MI: Publications Division, Institute for Social Research.
Rohe, D. (1980) *Change in vocational interests after disability.* Ph.D., Minnesota.
Scarr, S. (1992). Developmental theories for the 1990's: Development and individual differences. *Child Development, 63,* 1–19.
Scarr, S., & Weinberg, R. A. (1978). The influence of family background on intellectual attainment. *American Sociological Review, 43,* 674–692.
Scarr, S., Weinberg, R. A., & Waldman, I. D. (1993). IQ correlations in transracial adoptive families. *Intelligence, 17,* 541–555.
Schmidt, F. L., & Hunter, J. E. (1996). Measurement error in psychological research: Lessons from 26 research scenarios. *Psychological Methods, 1,* 199–223.
Shaw, M. E., & Wright, J. M. (1967). *Scales for the measurement of attitudes.* New York: McGraw-Hill.
Shields, J. (1962). *Monozygotic twins: Brought up apart and brought up together.* London: Oxford University Press.
Snyderman, M., & Rothman, S. (1988). *The IQ controversy: The media and public policy.* New Brunswick, NJ: Transaction Books.
Staples, B. (1994, Friday, October 28, 1994). The 'scientific' war on the poor. *New York Times: Editorial,* p. A18.
Staw, B. M., Bell, N. E., & Clausen, J. A. (1986). The dispositional approach to job attitudes: A lifetime longitudinal test. *American Science Quarterly, 31,* 56–77.
Staw, B. M., & Ross, J. (1985). Stability in the midst of change: A dispositional approach to job attitudes. *Journal of Applied Psychology, 70,* 469–480.
Taylor, H. F. (1980). *The IQ game: A methodological inquiry into the heredity environment controversy.* New Brunswick, N. J.: Rutgers University Press.
Tesser, A. (1993). The importance of heritability in psychological research: The case of attitudes. *Psychological Review, 100,* 129–142.
Waller, J. H. (1971). Achievement and social mobility: Relationships among IQ score, education, and occupation in two generations. *Social Biology, 18,* 252–259.
Waller, N. G., Kojetin, B. A., Bouchard, T. J., Jr., Lykken, D. T., & Tellegen, A. (1990). Genetic and environmental influences on religious interests, attitudes, and values: A study of twins reared apart and together. *Psychological Science, 1*(2), 1–5.
Waller, N. G., Lykken, D. T., & Tellegen, A. (1995). Occupational interests, leisure time interests, and personality: Three domains or one? Findings from the Minnesota Twin Registry. In R. Dawis & D. Lubinski (Eds.), *Assessing individual differences in human behavior: New concepts, methods, and findings* (pp. 233–259). Palo Alto: Davies-Black.
Weiner, H. (1992). *Perturbing the organism: The biology of stressful experience.* Chicago: University of Chicago Press.
White, R. K. (1982). The relation between socioeconomic status and academic achievement. *Psychological Bulletin, 91,* 461–481.
Zahn-Waxler, C. (1995). Introduction to special section: Parental depression and distress: Implications for development infancy, childhood, and adolescence. *Developmental Psychology, 31,* 347–348.

8

HORMONAL INFLUENCES ON HUMAN BEHAVIOR

C. Sue Carter

Department of Zoology
University of Maryland
College Park, Maryland 20742

1. OVERVIEW

Human behavior is influenced by many factors; among these are hormones or other chemicals produced by various tissues or glands throughout the body. This paper reviews the current status of research relating hormones to human behavior. To provide background for this discussion, a brief history of this field and a primer in behavioral endocrinology are provided.

Although, reproductive hormones, including androgens and estrogens, have been studied in some detail, remarkably few strong conclusions are possible. However, new studies implicating neuropeptides, such as oxytocin and vasopressin, in human behavior suggest physiological substrates that may offer insights into hormonal influences on human behavior.

2. HISTORY OF THE STUDY OF HORMONAL INFLUENCES ON HUMAN BEHAVIOR

Among the endocrine organs, the testes are unique because they are suspended in a tissue pouch outside of the body cavity. The testes can be felt and these organs are vulnerable to physical insults, either deliberate or accidental. The effects of castration were described by Aristotle over three hundred years B.C. Removal of the testes or castration as a form of punishment or tribute dates to antiquity. Domestic animals and some cases humans were castrated to make them more docile. Castrata were valued as harem keepers. In addition, seasonal changes in behavior and the dramatic anatomical and behavioral events associated with puberty were components of the natural world (Bronson and Heideman, 1994). Thus, an awareness of a relationship between the testes and human behavior predates written history.

New Aspects of Human Ethology, edited by Schmitt *et al.*
Plenum Press, New York, 1997

In 1849 Arnold Berthold conducted the first systematic experiment in this field - now recognized as the beginning of experimental endocrinology. Berthold removed the testes of chickens, noting that following castration males no longer strutted, crowed or attempted to mate. Berthold also discovered that he could restore these behaviors by transplanting tissue from an intact male. The realization that a particular substance from the testes might influence behavior had to await the invention of the hypodermic needle. Claude Brown-Sequard (1889), a renowned physiologist, used himself as a subject in the first published experiment on the effects of testicular extracts. Brown-Sequard, motivated by his own experiences with aging, reasoned that the testes were involved in producing traits associated with youthful vigor. To test this hypothesis he injected himself with aqueous extracts from the testes of guinea pigs and dogs, keeping meticulous notes on the changes he observed in his body and behavior. The remarkable rejuvenation heralded by Brown-Sequard was probably both short-lasting and primarily due to placebo effects, because the major hormones of the testes are not water soluble. However, this experimental flaw was not recognized for several decades and word spread quickly of Brown-Sequard's apparent success in reversing the aging process. A new science and industry of organ therapies emerged, and modern interest in the capacity of "glandular" secretions to influence various behaviors can be traced to these early experiments.

3. A BRIEF PRIMER OF BEHAVIORAL ENDOCRINOLOGY

3.1. General Issues

The twentieth century brought awareness that specific chemicals, now known as hormones, were produced in the testes or other organs and could travel via the blood stream to target tissues and specific receptors throughout the body. In addition, the importance of local or "paracrine" chemical communication among neighboring cells was articulated. Neurotransmitters were defined as chemicals that were synthesized in nerve cells, crossing the space between two neurons (a synapse) to act on receptors on the postsynaptic cells. In addition, chemicals also can act as neuromodulators, altering receptor sensitivity, or neurotransmitter availability or action. Behavioral effects for particular hormones and neurotransmitters generally are assumed to reflect tissue specific receptors. For example, steroid hormones may act on a variety of protein-based "steroid" receptors which are distributed throughout the nervous system. Steroid hormones diffuse into cells, bind to intracellular receptors, and are transported to the cell nucleus. In the nucleus the hormone-receptor complex is capable of activating genetic transcription, and the subsequent production of other compounds, including enzymes and receptors for other molecules. Steroid hormones also are capable of acting directly on cell membranes and in some cases may have rapid non-genomic behavioral effects. Amino-acid based hormones and neurotransmitters generally act on cell membrane receptors, allowing rapid and reversible effects, which are compatible with comparatively quick and transient behavioral changes. Many hormones, especially steroids and peptides, circulate in conjunction with binding proteins. Hormone actions generally depend on the availability of free, i.e. not protein bound, hormone.

Various experimental criteria are used to formally differentiate among hormones, neurotransmitters and neuromodulators. However, for behavior it often is difficult to discriminate among these categories of function, since a given chemical compound may act at several locations and have more than one mode of action (Brown, 1994). For the purposes of this review a particular compound will be described as acting either at a distance from its site of origin (as a hormone which may require hours or even days to effect its action) or in a paracrine fashion (acting locally and usually more quickly as a neurotransmitter or neuromodulator).

The first half of this century saw the beginning of a new era of medicine dominated by biochemistry. Various hormones and neurotransmitters were identified, synthesized and became available for research or medical applications. Several generations of behavioral endocrinologists have taken advantage of these breakthroughs to begin the tedious task of identifying the behavioral effects of hormones in humans and other animals (reviewed Becker, et al., 1992; Brown, 1994; Nelson, 1995).

Practical and ethical issues have impeded the analysis of biochemical influences on human behavior, and at present questions are more common than answers. However, in conjunction with more readily controlled animal studies, patterns of relationships between hormones and behavior have begun to emerge.

In order to review these putative relationships it is first useful to define a subset of well-characterized hormones and neurotransmitters that have been implicated in behavior. The chemicals selected for discussion here are among those for which a robust relationship with behavior has been proposed, including steroids (estrogens, progestins, androgens and glucocorticoids), proteins (prolactin) and the neuropeptides (oxytocin and vasopressin). All of these chemicals may act as hormones, neurotransmitters and/or neuromodulators. In addition, to understand the action of these hormones, it is helpful to be familiar with some of the more common neurotransmitters (described below). Space does not permit a discussion of the behavioral effects of many additional compounds with endocrine or paracrine properties.

This present paper will emphasize the activational effects of hormones in adults. However, many hormones, and especially steroid hormones, also have developmental effects which have long-term consequences for the organization of both brain structure and function. Information regarding the role of hormones in human brain development is rare, and at present difficult to interpret. However, the animal literature on the organizational effects of hormones suggests that both steroid and peptide hormones have major effects on neural development. The developmental consequences of hormonal manipulations differ among species, and probably are an important source of individual differences in adult behavioral responses to either endogenous or exogenous hormones.

3.2. Types of Hormones

3.2.1. Gonadal Steroids. Gonadal steroids, including those produced by the testes and ovaries, were among the first to be identified and implicated in behavior. In addition, the adrenal cortex also produces steroid hormones. Steroids are derived from fatty acids and are composed of carbon chains, configured in a unique series of four interlocking rings. Cholesterol, which has 27 carbons, is the precursor molecule for steroid biosynthesis and biologically active steroids are metabolic derivatives of cholesterol. The production of specific steroids is regulated by tissue-specific enzymes. The major gonadal steroids are progestins, androgens, and estrogens.

Progestins are very abundant during pregnancy, and were first identified as hormones that could "promote gestation." The ovary and placenta secrete high levels of a specific progestin, known as progesterone, which is necessary for pregnancy. Progestins of adrenal origin also are common in both sexes. Progestins have 21 carbons and serve as precursors for the smaller sex steroids, including androgens and estrogens. Androgens,

Table 1. Steroid hormones and tissues of origin

Steroid hormone	# of carbons	Major tissue source
Cholesterol	27	steroid precursor (fatty acid derivative)
Cortisol	21	Adrenal cortex
Progesterone	21	Corpus luteum of ovary
		Placenta (in pregnancy)
Testosterone	19	Testes (male)
		Ovary (female)
Dehydroepiandrosterone (DHEA)	19	Adrenal cortex
		CNS?
Estradiol	18	Follicle of ovary
		Placenta (in pregnancy)
		Aromatization of testosterone

and especially testosterone, are made in abundance in males by the testes and to a less extent in females by the ovaries. Additional androgens, often with somewhat lower androgenic potency, are made by the adrenal cortex in both sexes. Dehydroepiandrosterone (DHEA) is the most abundant steroid in the human body, acting as a precursor for other hormones. DHEA also is found in neural tissue and is considered a "weak" androgen. Androgens have been described as "masculine" hormones, because of their capacity to produce male-like secondary sex traits, including the growth of the penis or semen-producing secretory glands, such as the prostate or seminal vesicles. Androgens are defined by the fact that they have 19 carbons and also by their capacity to induce growth in tissues of the male reproductive tract.

Estrogen is a general category of steroid hormone which occurs in humans in one of three forms; estradiol, estrone and estriol. Ovarian tissues, and especially the follicle of the ovary, synthesize comparatively high levels of estrogens. Estrogens are derived biochemically from androgenic compounds through a process known as aromatization which involves the removal of a carbon atom and the formation of an aromatic ring. Aromatization occurs in the presence of a specific enzyme known as aromatase. Aromatization can occur in many tissues including the testes and brain.

Estradiol is the most potent and best understood of the sex steroids. Estrone and estriol are weaker estrogens. Estrone, in a sulfated form known as "premarin" is derived from the urine of pregnant mares or stallions, and is commonly used as a hormone replacement therapy (HRT) in postmenopausal women. The estrogens have 18 carbon atoms and were originally defined by their capacity to induce growth in immature uterine tissue. Plants also produce estrogenic compounds (phytoestrogens); these plant hormones, as well as various synthetic compounds, such as diethylstilbestrol (DES) can stimulate estrogen receptors and thus have estrogenic effects.

3.2.2. Adrenal Steroids. Hormones synthesized by the adrenal cortex are typically characterized as either glucocorticoids (regulating glucose metabolism) or mineralocorticoids (regulating mineral balance). Adrenal corticoid hormones share many properties with the progestins and also are composed of 21 carbons. Among the glucocorticoids, cortisol is most abundant in humans, while many rodents rely primarily on a similar compound known as corticosterone. The major mineralocorticoid hormone is aldosterone. Adrenal corticoids are studied most frequently in the context of their metabolic effects, but also can have behavioral consequences. Weak androgens, including DHEA, also are produced by the adrenal.

3.3. Peptide Hormones and Neurotransmitters

A vast number of hormones and neurotransmitters are synthesized from simple structural units including chains of amino acids. Among the amino acid based hormones are comparatively large "protein" molecules such as prolactin, growth hormone and insulin, and shorter chains of amino acids which may form "peptide" hormones, such as oxytocin and vasopressin. Protein based hormones are essential for metabolism, growth and some components of reproduction. However, evidence for specific influences of these compounds on human behavior is limited.

The present review will focus on the behavioral consequences of two peptidergic compounds, oxytocin and vasopressin. A variety of other peptides, such as corticotropin releasing hormone (CRH), gonadotropin-hormone releasing hormone (GNRH), and the endogenous opiates, such as beta-endorphin (to name only a few) also influence behavior.

Because they have been extensively implicated in human behavior, it is helpful to describe a subset of the better known neurotransmitters. The catecholamines, including dopamine, norepinephrine and epinephrine are modified amino acids, manufactured in various parts of the central and peripheral nervous system. Epinephrine (also known as adrenaline) is produced primarily by the adrenal medulla. Dopamine and norepinephrine have functions throughout the central nervous system (CNS including the brain and spinal cord) and autonomic nervous systems (ANS or peripheral nervous system). In general catecholamines have been associated with behavioral activation and stress and various functions of the sympathetic branch of the ANS. Acetylcholine is released throughout the body, is involved with inhibitory and "vegetative" states, and is important for neuromuscular transmission, cortical function and the many activities of the parasympathetic (PNS) branch of the ANS. Serotonin and, its pineal gland derivative, melatonin are described structurally as indolamines. Both have been associated with sleep and rhythms. Serotonin is produced in the brain stem, but acts throughout the brain, including at sites in the hypothalamus and cortex. Serotonin plays a critical role in the inhibition of emotional reactivity.

Many psychoactive drugs act to alter neurotransmitter functions either through effects on their synthesis, metabolism or reuptake or by directly affecting the receptors for naturally occurring compounds. For example, drugs such as "prozac" increase serotoninergic activity by selective serotonin reuptake inhibition (SSRI).

3.4. Types of Behaviors That Can Be Influenced by Hormones

Behavioral endocrinologists have focused their attention on sexual, parental and aggressive behaviors, describing various relationships between the gonadal steroids and these behaviors. Most research in this area has been done in laboratory rodents, and especially in domestic rats.

The selection of rodents as the primary model for behavioral endocrinology has influenced our expectations regarding the nature of hormonal influences on human behavior. For example, domestic strains of rats were derived from *Rattus norvegicus*. Wild rats are highly versatile ecological generalists, and share the human habitat throughout the world. Domestic rats were selected for various traits including docility and the capacity to reproduce under laboratory conditions. Different breeders have selected for different physical and behavioral traits, with varying underlying physiological substrates. The rodents which are now used as models for endocrine research are the products of both natural and artificial selection. The degree to which these models are appropriate to advance our understanding of human behavior remains uncertain.

4. FEMALE SEXUAL BEHAVIOR

In rodents the most commonly studied female sexual behavior is the lordosis posture, in which the female remains immobile and concavely arches her back. Lordosis, usually in response to mounting, has been used as an index of female sexual receptivity. In addition, a variety of other measures such as elective proximity to a male have been used to index sexual proceptivity. Studies of these and related behaviors have shown that in rats ovarian secretions are essential for the expression of lordosis and can increase the expression of a variety of sociosexual behaviors (Pfaff, et al., 1994). Surgical removal of the ovary (ovariectomy) eliminates female sexual behavior in most rodents, and treatments with estrogen and progesterone can produce levels of sexual receptivity that closely resemble those seen in a gonadally-intact estrous female.

4.1. Ovarian Cycles

Studies of female sexual behavior in both humans and nonhuman primates have tended to examine changes in behavior as a function of the menstrual cycle or as a function of treatments with estrogen and/or progesterone. In most primates, including humans, ovarian hormones are not essential for the expression of female sexual behavior. Experimental studies of sexual behavior have most typically involved testing with a single male in a small test area. Tested under these conditions, female rhesus monkeys will mate throughout the entire menstrual cycle (Michael and Zumpe, 1993). In addition unlike the situation in rodents, ovariectomy does not abolish sexual behavior in most female primates. However, hormonal influences on female sexual behavior in primates are detected, especially when animals are tested under more naturalistic conditions that permit the establishment of social systems (Wallen, 1990). Under conditions in which female rhesus monkeys are at liberty to express their social and sexual preferences or "desires," mating occurs primarily around midcycle, near the time of ovulation (Wallen and Tannenbaum, 1996). Midcycle increases in sexual interest, and in some cases in sexual activity, also have been reported in human females. Thus, as expressed by Wallen (1990), in rodents both the "ability" and "desire" to mate are hormone dependent, while in primates, females have the ability to engage in sexual behavior in the absence of hormones, although sexual "motivation" or "desire" tends to vary as a function of the menstrual cycle.

4.2. Estrogen

Estrogen, either of ovarian origin or given as an exogenous treatment apparently enhances female sexual motivation, at least in monkeys (Wallen, 1990). The follicular stage of the menstrual cycle is typically associated with comparatively high levels of sexual behavior. Estrogen production declines during the menopause, and some, but not all, women experience postmenopausal declines in sexual activity (McCoy, 1992). Estrogen replacement therapy is associated with enhanced sexual activity and sexual interest. Estrogen also increases vaginal lubrication, inhibits hot flashes, normalizes sleep patterns and may indirectly influence sexual behavior (Sherwin, 1996).

Table 2. Possible hormonal effects on sexual function

	Androgen	Estrogen	Progesterone	Oxytocin	Vasopressin
Sexual interest	INC	INC female ? male	DEC[d]	INC (rats [e])	INC ?[e]
Vasocongestion or erection	INC (but not essential)	INC female ? male			INC ?[b]
Ejaculation (male)	INC			INC ?[e]	
Orgasm				INC [e]	
Gamete formation	Essential[a]	Essential[a]		INC ?[e]	

INC = increase; DEC = decrease
[a] Very high levels estrogen or androgen may interfere with sperm production, and possibly other normal functions of these steroids.
[b] Vasopressin plays a major role in the control of blood pressure, which is probably essential for vasocongestion or erection.
[c] Oxytocin gene expression occurs in ovary and testes, but the exact functions remain unknown.
[d] Progestins, such as "cyproterone acetate," are used clinically as anti-androgens (Kravitz, et al., 1995).
[e] Vasopressin release is associated with sexual arousal and oxytocin release with orgasm. Oxytocin also may play a role in sexual satiety. (Murphy, et al., 1987; reviewed Carter, 1992.)

4.3 Progesterone

During the menstrual cycle progesterone is secreted at the time of ovulation and continues to be elevated throughout most of the luteal (postovulatory) phase of the menstrual cycle. Under some conditions a short-term exposure to progesterone may facilitate sexual activity in rhesus (Michael and Zumpe, 1993). However, longer exposures to progesterone are primarily inhibitory, and there are reports of inhibited sexual activity and desire during the luteal phase of the menstrual cycle. The use of oral contraceptives, which are primarily synthetic progestins, may be associated with a facilitation of sexual activity in some women, probably because of a freedom from unwanted pregnancy. However, other women have reported inhibited sexual desire while taking oral contraceptives, which may be due to the inhibitory effects of the progestins.

4.4. Androgens

Removal of the adrenal gland, and thus adrenal androgens, has been associated with an inhibition of sexual behavior in both humans and primates. In addition, in primates exogenous androgens, but not estrogens, reportedly reversed the effects of adrenalectomy. Exogenous androgen treatments reportedly can facilitate sexual responses in women (Sherwin, 1996).

4.5. Oxytocin

Oxytocin is released during orgasm in both females and males of several species (reviewed Carter, 1992; Walburton, et al., 1994). Oxytocin can facilitate sexual receptivity in rats, but in prairie voles, which are monogamous rodents, the effects on female sexual behavior are inhibitory. Levels of oxytocin have been correlated with the orgasmic intensity in women. There is at present no direct evidence for a role for oxytocin in female sexual behavior in humans or other primates, although it has been proposed that oxytocin may have a role in both the phenomenology of orgasm and in sexual satiety (Carter, 1992).

5. MALE SEXUAL BEHAVIOR

5.1. The Role of the Testes

In most mammals intact gonads are a prerequisite to the onset of male sexual behavior. The expression of male sexual behavior is rare prior to puberty. However, once a male has become sexually active, removal of the testes does not produce an immediate cessation of mounting or intromission. However, even in laboratory rodents, male sexual behavior is not tightly linked to gonadal function. Following castration or inhibition of testicular function, ejaculatory reflexes, intromission frequency and later mounting usually declines over a period of weeks (Burris, et al., 1992; Bagatell, et al, 1994), although some individuals continue to express male sexual behavior long after castration (Kinsey, et al., 1948). The effect of androgen on male sexual behavior is most easily documented in normal adults who have experienced a decline in androgens and subsequent testosterone replacement therapy (Bagatell, et al., 1994). Although, androgens are important for male sexual behavior, measurements of serum hormones alone often are not adequate to predict sexual behavior.

In humans, erectile "potency" is sometimes considered a measure of sexuality. In adult men with low levels of gonadal hormones (hypogonadism), nocturnal erections are generally infrequent and it was originally assumed that such individuals were "impotent." However, even infants and young children with low levels of androgens are capable of erections, indicating that erections can occur in the absence of high levels of testicular activity. There are case reports of sexual activity in men who have been castrated for years or even decades (Kinsey, et al., 1948). In addition, laboratory studies have shown that even profoundly hypogonadal men are capable of responding with erections to highly erotic stimuli, such as pornography. Erectile responses under these conditions were remarkably similar in normal and hypogonadal men, and in some cases hypogonadal men may even maintain erections for a longer period than gonadally normal men (Kwan, et al., 1982). Such findings with children and hypogonadal men indicate that androgen levels alone do not explain erectile potency.

In support of a role for androgens in sexual behavior is the finding that sexual interest or motivation is low in prepubertal boys and in men with various forms of hypogonadism (Burris, et al., 1992). Androgen treatments are typically associated with increased interest in sexual activities, as measured by self-report, as well as increases in nocturnal erections. However, increases in sexual behavior as a result of androgen treatments are less reliable, probably in part because men with a history of sexual inactivity may lack social skills or opportunities for sexual behavior. Social and historical variables, possibly also experienced as changes in other hormones, are critical determinants of masculine sexuality.

6. AGGRESSION

6.1 The Role of the Testes

Sex differences in the expression of aggression are well documented and aggression may increase around the time of puberty. In addition, castration long has been used to manipulate tameness or docility in domestic animals (Bronson and Desjardins, 1971; Archer,

1991). These observations have lead to the hypothesis that testicular secretions play a causal role in aggression.

In general, in animals castration is associated with reductions in aggression and androgen replacement with increased aggression. However, social and sexual experience can have important effects on the expression of aggression. For example, a history of winning or social dominance, may be more important than hormonal status in determining the outcome of aggressive encounters. The effects of social rank are particularly apparent in primates (Bouissou, 1983).

Most of the research dealing with androgens and aggression in humans is correlational and retrospective, and of course complicated by issues associated with studying human aggression (Albert, et al., 1993). In general, the strongest relationships between endogenous hormones and aggressive behaviors have been reported in studies of adolescents (Susman, et al., 1987; Inoff-Germain, et al., 1988; Paikoff and Brooks-Gunn, 1990). Only a small number of studies have shown correlations between androgen levels in adulthood and measures of aggression (Archer, 1991; Alberts, et al., 1993). In one study prisoners with a history of adolescent violence tended to have comparatively high levels of testosterone when studied compared to nonviolent offenders (Kreuz and Rose, 1972). In the latter work, as well as several other studies on aggression, correlations were not observed between adult testosterone levels and various measures of aggressivity. However, in more recent studies, free testosterone levels taken from saliva have been related to a variety of aggressive traits in normal men (Dabbs, et al., 1996). Free testosterone in cerebrospinal fluid also has been correlated with aggression in violent criminals (Virkkunen, et al. 1994).

It also has been suggested that "the hormones of individuals may shape the culture of groups" and that the effects of testosterone may be more apparent when studied in groups, rather than in individuals (Dabbs, et al., 1996). For example, salivary testosterone levels in fraternity men have been correlated with "rambunctious" or unruly behavior characteristic of the fraternity. Groups with lower average levels of salivary testosterone (12.3 ng/Dl) tended to be more academically successful, more likely be characterized as friendly and more likely to smile when compared to groups with higher testosterone (14.3 ng/Dl). Although these group differences were statistically significant, it is not known whether individuals respond differently to small differences in hormone levels (Alberts, et al., 1993).

6.2. Anabolic-Androgenic Steroids

The comparatively recent, but widespread illegal use of synthetic anabolic-androgenic steroids (AAS) has generated interest in the behavioral effects of androgens. Anecdotal reports are common of increased aggression during AAS use, sometimes called "roid rage." Some, but not all, steroid users report episodes of violence and aggression (Pope and Katz, 1994). However, prospective and controlled studies of the behavioral effects of AAS are rare.

Studies attempting to demonstrate effects of androgens on mood have been somewhat more successful. Although, positive effects of androgens on mood have not been universally reported, most studies do report mood improvements when hypogonadal men are treated with androgens (Bahrke, et al., 1990). For example, in a double-blind study in hypogonadal men (who are incapable of producing androgens) low levels of androgen were associated with feelings of anger and depression; both anger and depression were reduced during androgen replacement therapy. A weaker androgen (DHEA) also has been associated with mood-enhancement in both men and women.

6.3. Hormones and Female Aggression

Hormonal influences on aggression in women have been studied only infrequently. In adolescents studied within the context of their family, higher levels of estradiol and a weak androgen (androstenedione) have been correlated with the incidence of aggressive behaviors, such as angry responses toward parental authority (Inoff-Germain, et al., 1988). The effects of estrogen treatments in hypogonadal girls also supports the hypothesis that exposure to estrogen (at least in adolescence) may increase some forms of aggressive behavior. In contrast DHEA is negatively associated with reports of aggressive behavior in girls (Susman, et al., 1987; Paikoff and Brooks-Gunn, 1990).

7. PARENTAL BEHAVIOR

7.1. Hormones of Pregnancy and Parturition

In mammals, maternal behavior is typically observed at approximately the time of birth, and may be demonstrated prior to parturition. The hormonal events of pregnancy, including exposure to high levels of estrogen and progesterone, and a subsequent dramatic prepartum decline in progesterone and increases in oxytocin and prolactin are associated with both maternal responses and lactation (Pedersen and Prange, 1979; Pedersen, 1996; Bridges, 1990; Rosenblatt, 1995). However, hypotheses regarding hormonal causes of parental behavior are confounded by the fact that apparently normal parental behavior is observed in virgin females, even after removal of the ovary and uterus, as well as in males of many species. Thus, the experience of pregnancy and birth are not essential for parental behavior.

7.2. Hormonal Treatments and Parental Behavior

Various hormonal "cocktails" that approximately mimic the hormonal events of pregnancy and birth can facilitate the onset of maternal behavior in virgin female rats. Estrogen treatment generally hastens the onset of maternal behavior. Progesterone treatment followed by a cessation of progesterone also contributes to a the onset of pup-directed behaviors. Treatments with prolactin (Bridges, 1990) are associated with maternal behavior in rats, as well as with paternal behavior in California mice (Gubernick, et al., 1994). In addition, oxytocin, which is normally released during parturition or by vaginal-cervical stimulation, has been implicated in the onset of maternal behavior in sheep (Keverne, et al., 1983) and rats (Pedersen and Prange, 1979). In addition, in rats recent evidence suggests that oxytocin is involved in the maintenance of maternal behavior (Pedersen, 1996).

7.3. Nonhuman Primate Parental Behavior

In nonhuman primates, including humans, research which directly examines the hormonal regulation of parental care is virtually nonexistent (Rosenblatt, 1995; Corter and Fleming, 1995). However, behavioral observations of pregnant rhesus monkeys have indicated that primiparous (first-time) mothers may not show reliable increases in parental care (Gibber, 1986). In contrast, in squirrel monkeys, responsiveness to infants does increase during pregnancy (Rosenblum, 1972). Variation in the degree to which maternal behavior is hormonally regulated probably reflects the social conditions under which parental behavior would be shown in nature. For example, it has been suggested that in nomadic primates, such as baboons, dramatic behavioral changes associated with parenting would be disadvantageous. Animals living under these conditions must exhibit appropriate infant care behaviors, while continuing to maintain continuous contact with a troop (Altmann, 1987).

A few studies in New World primates have examined hormonal correlates of parenting responsiveness. In red-bellied tamarins a positive correlation was detected between "good mothering" and urinary estradiol measured prior to delivery (Pryce, et al., 1988). In addition, in common marmosets operant responses (bar-pressing) in response to crying infants correlated with the presence of estrogen and progesterone in late pregnancy, and increased in females that were treated with a combination of estrogen and progesterone that was similar to that experienced during pregnancy and parturition (Pryce, et al., 1993). Male parental care in marmosets also has been associated with increases in prolactin (Dixson and George, 1982).

7.4. Human Parental Behavior

In humans, studies of hormone and behavior interactions affecting maternal behavior have been limited to correlations between endogenous hormones and a variety of behavioral events normally associated with childbirth. Individual differences, of largely unknown origins, have a major impact on human parental behavior. In addition, child rearing experience is a powerful positive determinant of maternal responsivity, and the social environment, including the presence of supportive companions, can affect subsequent mother-infant interactions (Sosa, et al., 1980).

Women in their first pregnancy report that they experience increased maternal feelings and responsivity around the twentieth week of gestation, at approximately the time fetal movements are first detectable (Corter and Fleming, 1995). The adrenal hormone, cortisol, has been associated with maternal behavior (Fleming, et al., 1987; Corter and Fleming, 1995). High levels of cortisol on days 2 or 3 postpartum were correlated with positive maternal behaviors and attitudes. Cortisol levels also were positively correlated with positive responses to odors from infants, suggesting a role for cortisol in sensory processes. Although attempts to correlate levels of other steroid hormones, including estrogen and progesterone, and maternal responsivity have been unsuccessful, the paucity of data precludes conclusions regarding the role of sex steroids in human maternal behavior.

7.5 Oxytocin and Mother-Infant Interactions

Because oxytocin is a uniquely mammalian hormone with a critical role in both birth and lactation, it was an obvious candidate for involvement in maternal behavior. In fact oxytocin has been termed the hormone of "mother love" (Klopfer, 1971; Newton, 1973). As mentioned above, animal research offers support for this hypothesis (Pedersen, 1996). However, evidence implicating oxytocin in human behavior only recently has begun to accumulate, and direct evidence for a behavioral effect of oxytocin in humans remains to be demonstrated.

Various autonomic, physiological and behavioral effects of oxytocin could contribute to the development and expression of maternal behavior (Uvnas-Moberg, 1996). Parasympathetic (vagal) activity, is closely associated with both lactation and maternal behavior, while the more activated states associated with sympathetic-adrenal activity are antithecal to many components of maternal function. Oxytocin plays an important role in vagal activity and is associated with energy conservation, which is necessary for milk production. Concurrently oxytocin pulses play an essential role in the delivery of milk. Lactating women are less physiologically and autonomically reactive to stress than nonlactating women, including those who have given birth and elect to bottle feed their infants (Altemus, et al., 1995). Oxytocin injections and/or lactation have many effects (reviewed Carter and Altemus, 1996) and can inhibit the activity of the sympathoadrenal axis, reducing cortisol production and blood pressure.

Oxytocin production during breast-feeding is correlated with personality traits and behaviors generally associated with parental behavior. For example, basal levels of oxytocin are related to "calmness," while pulsatile patterns of oxytocin production are associated with a desire to "please, give and interact socially" (Uvnas-Moberg, et al., 1990). In one study of Swedish women, those who had acute caesarean sections had fewer oxytocin pulses during the postpartum period and also were less likely than vaginally-delivered women to describe themselves as exhibiting a calm personality and high levels of sociality (Nissen, et al., 1996).

8. STEROIDS AND COGNITIVE FUNCTIONS

In addition to the above behaviors, sex steroids also have been implicated in many other components of mammalian life including learning and memory, and especially verbal fluency and spatial learning. In animals, spatial orientation generally is more critical for males, who presumably need these skills to locate mates. Significant gender differences in spatial abilities have been reported in humans (Hampson, 1995), and there are recent reports from the study of people electing hormone therapies prior to sex-change operations that androgen treatments can improve testable spatial abilities in women. In contrast, verbal communication may be especially important to females, and verbal skills actually declined during testosterone treatment. Conversely, when men are given estrogens cognitive testing indicates a loss in spatial skills and a concurrent improvement in verbal abilities (Van Goozen, et al., 1995).

Alzheimer disease, which involves memory loss and other forms of cognitive dysfunction, is two to three times more common in women than in men after the age of 65 (Sherwin, 1996). In addition, women who have used estrogen during the postmenopausal women are less likely to develop Alzheimer disease. The amount and duration of estrogen

Table 3. Hormones and cognitive function

	Estrogen	Androgens[1]	Oxytocin	Vasopressin
Verbal memory	INC or nc [2,3]	DEC [3]		
Spatial memory or abilities	DEC [3]	INC [3]		
Attention demanding tasks	INC	INC [3]	DEC	INC

[1]Testosterone or DHEA
[2]nc = no change; INC = increase; DEC = decrease
[3]Sherwin, 1996. Obstet. Gynecol. 87:20S; Roca, et al., 1996. Infertil. Reprod. Med. Clin. N. Amer. 7:341; Hampson, E. 1995. J. Psychiat. Neurosci. 20: 397; Van Goozen, et al., 1995, Psychoneuroendocrinology 20:343.

Table 4. Steroid hormones and cns dysfunction

	Estrogen	Progesterone	Androgens[4]
Depression	DEC [1]	INC [1]	DEC ?
Schizophrenia	DEC [2]		INC [3]
Alzheimer's	DEC [1]		

[1]Sherwin, 1996; Roca, et al., 1996; Rubinow and Schmidt, 1996.
[2]Seeman, 1996.
[3]Pubertal onset in males; Hafner, et al., 1993.
[4]Testosterone or DHEA

treatment also has been associated with a lower risk of this disorder. In addition, in uncontrolled studies elderly women who have been diagnosed with Alzheimer disease showed improvements in both memory and mood following estrogen treatments. A number of animal studies have indicated that estrogen can have dramatic effects on neural growth and regeneration (Luine and Harding, 1994), lending support to the hypothesis that estrogen is protective against age-related cognitive dysfunction.

9. STEROIDS AND EMOTIONAL DISORDERS

A variety of kinds of evidence have linked emotional behavior to hormones. Two conditions, the menstrual cycle and menopause, have been the focus of a great deal of research on human behavior. In addition, gender differences in the prevalence of mental illnesses have been used as indirect evidence for possible hormonal effects on emotional disorders. For example, depression is more common in women than in men. In contrast, a pubertal onset of schizophrenia is more common in males than females (Hafner, et al., 1993), although the lifetime occurrence of schizophrenia is approximately equal in men and women (Seeman, 1996). Effects of hormones on emotional lability in men are described above in the context of aggression.

9.1. The Menstrual Cycle

Women often report fluctuations in mood states associated with the menstrual cycle, leading to the description of a complex of emotional and physical symptoms described collectively as the "premenstrual syndrome" or "PMS." Cyclic mood fluctuations, including anxiety and depression, are commonly considered components of PMS. Because cyclic changes in gonadal hormones characterize the menstrual cycle, it has been assumed that steroid hormones were involved in PMS. In spite of many attempts, a clear relationship between PMS and gonadal hormones has not been established. After an exhaustive review of the literature, Block and associates (1996) have concluded that there is "no consistent or convincing evidence that PMS is related to a simple deficiency or excess of gonadal steroids or gonadotropins or to an abnormal secretory pattern of these hormones." However, there are anecdotal reports and one recent study which indicate that progesterone, administered as vaginal suppositories, has therapeutic effects for PMS (Baker, et al., 1995). In addition, a clinical treatment that blocks ovarian function and menstrual cyclicity ("Lupron") is effective in about 70% of PMS cases. There also are recent reports of improvements in PMS in about 60% of women receiving low doses of a selective serotonin reuptake inhibitor ("prozac"), which probably increases serotoninergic activity. Exercise and dietary manipulations also may be effective in some women (Johnson, et al., 1995). Thus, effective therapies for PMS do exist. However, the nature of the physiological causes of mood changes associated with the menstrual cycle are not related in any simple way to gonadal hormones, and remain to be described.

9.2. Menopause

The midlife cessation of ovarian function, known as the menopause or perimenopause, has long been associated with emotional changes in some, but not all, women. However, life experiences have a major impact on the expression of depression and women vary considerably in their behavioral responses during the perimenopausal period. It is estimated that less than 10% of women experience serious emotional symptoms (Schmidt, et al., in press). Follicles, which are the major source of estrogen, are no longer present in the postmenopausal ovary. However, postmenopausal women vary widely in their production of endogenous hormones, and estrogens continue to be synthesized postmenopausally in both ovarian and adrenal tissues. Adipose tissue is a site of estrogen biosynthesis, so "fatness" can have a significant effect in maintaining endogenous estrogen levels following menopause.

In spite of these issues, clinical consensus has begun to emerge regarding the role of hormones in behaviors associated with the menopause and aging (Sherwin, 1996; Roca, et al., 1996; Rubinow and Schmidt, in press; Schmidt, et al., in press). In at least some women who have low levels of endogenous estrogen, treatment with this hormone may ameliorate depressive-like symptoms. Other hormones, including androgens (testosterone and DHEA), have been suggested to prevent depressive-life symptoms during the menopause, and are gaining clinical acceptance. However, both of these hormones may have masculinizing side-effects and can be converted to estrogens; at present research is insufficient to allow strong conclusions regarding their effectiveness.

The onset of schizophrenic symptoms also may be associated with the menopause, and it has been suggest that estrogen protects women from schizophrenia (Hafner, et al., 1993; Seeman, 1996). Schizophrenia may be exacerbated by the menopause and during other periods when estrogens are low, such as the postpartum period.

Table 5. Androgens and human behavior

Correlations in adulthood:
Positive (but not essential)
+ Sexual behavior, esp. interest
+ Irritability & anti-sociality
+ Aggression, esp. impulsive
+ Spatial ability
+ Psychoses
+ Cardiovascular disease
+ Muscle size & strength
Negative
– Recovery from CNS damage ?
– Verbal fluency

reviewed Archer, 1991; Alberts, et al., 1993; Van Goozen, et al., 1995; Dabbs, et al., 1996; Rubinow and Schmidt, 1996.

10. STEROIDS AND BEHAVIOR

10.1. Androgens and Human Behavior

As described above, steroid hormones can affect many aspects of human behavior. However, correlations between specific steroid hormones and a given behavior often are weak. Many factors, including genetics, developmental experiences, cognitive states, and context variables determine human behavior, and may of course contribute to behavioral expression.

Among these factors, social organizations, including inter- and intrasexual relationships, provide a matrix within which other behaviors occur. The behavioral effects of steroid hormones, especially in humans, have for the most part not been considered in these contexts.

For example, animal studies have revealed that the effects of steroid hormones may vary among polygynous versus monogamous mammals (Carter, et al., 1995; Wingfield, et al., 1996). Androgens are responsible for the greater muscle mass and larger body size seen in polygynous males. [A relative lack of physical sexual dimorphism is a defining trait of monogamy (Kleiman, 1977)]. Polygynous males may need to compete repeatedly for females and also tend to produce large quantities of sperm (which may be necessary for successful sperm competition) (Dixson, 1996). Male animals, particularly in polygynous species, are usually more territorially aggressive and better able to defend themselves from physical aggression than are females. However, as summarized by Wingfield and associates (1996), significant costs are associated with exposure to prolonged high levels of testosterone. Among these are energetic costs, related to the anabolic effects of androgens, and indirect energetic costs associated with high levels of activity. Aggressive males also are vulnerable to predation and wounding.

Positive social behaviors, including gregariousness and the formation of social bonds, are less frequently studied than reproductive and agonistic behaviors, but are critical for primate reproduction. Agonistic behavioral traits associated with androgen exposure may be in conflict with social behaviors needed for courtship, the formation of social bonds and subsequent parental behavior. Higher levels of male sociality are probably nec-

Table 6. Estrogen and human behavior

Correlations
Positive
+ Sexual behavior
+ Maternal behavior
+ Positive social behaviors
+ Verbal fluency
+ Good health (CNS, cardiovascular)
Negative
− Spatial learning
− Schizophrenic symptoms
− Aggression ?
− Uterine cancer
− Breast cancer ?

(reviewed Hafner, et al., 1993; Sherwin, 1996; Roca, et al., 1996; Hampson, 1995; Van Goozen, et al., 1995.

essary for monogamy and cooperative breeding. The traits of monogamous mammals, including male-female pair bonding, male parental care, incest avoidance, and easily suppressed male sexual behavior (Kleiman, 1977), may represent the effects of comparatively low level of androgenic activity during both sexual development and adulthood. Evidence supporting this hypothesis comes from research with rodents including monogamous prairie voles (reviewed Carter, et al., 1995; Carter and Roberts, 1996).

The apparent absence in humans of strong correlations between androgens and behaviors could in part be a function of the high sociality of humans. Behavioral effects of androgens which prevent successful social interactions, including androgen-stimulated aggression, might interfere with traits necessary for reproduction, including successful pair bonding and parental care. Humans engage to varying degrees in the latter behaviors, and may be an example of a species in which the behavioral costs and benefits of androgens are in conflict.

10.2. Estrogen and Human Behavior

Estrogen is an exceptionally potent hormone, with effects in both males and females. At least some of the behavioral effects of testosterone probably require conversion of this hormone to an estrogen (Meisel and Sachs, 1994). Estrogens are necessary for the normal growth and development of the CNS in both sexes (Luine and Harding, 1994). Women generally live longer than men. In part this is because estrogen is cardioprotective. In addition, estrogen is apparently protective against schizophrenia and depression. Estrogen also may reduce the risk for other cognitive dysfunctions including loss of verbal fluency and memory disorders including Alzheimer disease.

Estrogen is critical for mammalian reproduction, with regulatory roles in sexual and maternal behavior. In general, estrogen is associated with increased sociality. Estrogen can regulate a variety of neural functions associated with social interactions. In addition, estrogens, or androgens converted to estrogens, can influence the actions of other hormones, such as oxytocin and vasopressin, which may in turn affect behavioral states and reactions to social stimuli.

11. PEPTIDES, STEROIDS AND BEHAVIOR

The effects of both oxytocin and vasopressin are particularly responsive to reproductive steroids (Tables 7 and 8). Oxytocin and vasopressin are best known as neurohypophyseal peptides, released by the posterior pituitary. However, both peptides also are released within the nervous system. Receptors for oxytocin and vasopressin are distributed in areas of the CNS that have been implicated in reproduction, emotion and autonomic functions.

Because neuropeptides are active in very small amounts within the CNS they are more difficult to manipulate and study than steroid hormones. In addition, at present there is no simple method for administering neuropeptides within the CNS. (In animals, it is necessary to inject chemicals or their antagonists directly into the CNS, which is not feasible in humans.) Thus, most research on the behavioral effects of either oxytocin or vasopressin in humans is correlational, and subsequently difficult to interpret. However, animal studies also support the hypotheses described here (Carter, 1992; Uvnas-Moberg, 1996).

Table 7. Oxytocin and human behavior

Correlations:
Positive
+ PARASYMPATHETIC NERVOUS SYSTEM ACTIVITY
+ Sexual behavior (released during orgasm in both sexes)
+ Birth (released by vaginal cervical stimulation)
+ Lactation (released by breast stimulation & prolactin)
+ Exposure to an infant (increased number of pulses)
+ Maternal behavior
+ Massage
+ Energy conservation, digestive efficiency, weight gain
+ Calm personality, high sociality
+ HIGHER IN WOMEN (or women more sensitive to effects)
Negative
− SYMPATHETIC NERVOUS SYSTEM ACTIVITY
− Stress reactivity (decreased by OT treatment)
− Memory (amnesic)
− Autism (OT low in some types)

(reviewed Carter, 1992; Uvnas-Moberg, 1996; Carter and Altemus, 1996)

In female mammals, many aspects of oxytocin's effects are influenced by estrogen, which is capable of regulating the expression of oxytocin's receptors. Very little is known regarding oxytocin receptors in primates, but in other species oxytocin receptors tend to show species-specific patterns of expression within the CNS (Insel and Shapiro, 1992; Insel, et al., 1996). Although oxytocin is a hormone that traditionally is associated with female reproductive functions, both sexes produce and can respond to oxytocin (Carter, 1992). Uvnas-Moberg (1996) has proposed that oxytocin regulates both the energetic and behavioral needs of female reproduction (Table 7). Oxytocin enhances parasympathetic activity, regulates the digestive system, reduces sympathoadrenal activity and is analgesic. Women who are lactating and thus producing comparatively large amounts of oxytocin, are more digestively efficient and are calmer and less emotionally reactive to stress (Uvnas-Moberg, 1996; Altemus, et al., 1995; Carter and Altemus, 1996). Vasopressin, particularly within the areas of brain that have been implicated in reproduction, is sexually-dimorphic and androgen-dependent (Bamshad, et al., 1993). In animals, vasopressin also been implicated in territoriality (Ferris and Delville,

Table 8. Vasopressin and human behavior

Correlations
Positive
+ SYMPATHETIC NERVOUS SYSTEM ACTIVITY
+ ANTIDIURETIC HORMONE (ADH), Water retention
+ Blood pressure increase
+ Stress reactivity and release of adrenal hormones
+ Sexual behavior (released during arousal in males)
+ Attention and certain kinds of learning (in males)
+ Addiction
+ Obsessiveness or "jealousy"
+ HIGHER LEVELS IN MEN (Androgen dependent)

(reviewed van Wimersma Greidanus and van Ree, 1990; Dantzer and Bluthe, 1992; Altemus, et al., 1992; Bruins, et al., 1995.)

1994), mate guarding and a variety of other defensive behaviors, which may protect either the individual or the family (Carter, et al., 1995). Vasopressin also has been implicated in obsessiveness in humans (Altemus, et al., 1992).

Oxytocin and vasopressin are released centrally and peripherally under a variety of social conditions (reviewed Carter, 1992; Uvnas-Moberg, 1996). Oxytocin and vasopressin differ in only two amino acids and are so structurally similar that these molecules can occupy each other's receptors, acting (perhaps dose-dependently) as antagonists and/or agonists. In addition, the release and/or actions of both hormones are steroid-dependent. It is possible that oxytocin and vasopressin, in conjunction with other neurotransmitters and hormones, mediate some of the behavioral effects, especially in social contexts, that have been attributed to estrogen and testosterone respectively. Interactions among oxytocin and vasopressin may provide a part of the physiological substrate that allows animals, including humans, to make transitions from one behavioral state to another. Positive social behaviors, such as those needed to live within a family and reproduce may be influenced by oxytocin, while more defensive behaviors, including territoriality and some forms of aggression, might be facilitated by vasopressin. Evidence for this hypothesis is currently being collected in our work with prairie voles; in this monogamous and communal mammal, the effects of adult gonadal hormones on social behaviors appear comparatively unimportant, while both oxytocin and vasopressin have been implicated in several reproductive and social behaviors (Carter, et al., 1995). We are currently using the vole model in which peptides are particularly obvious, to examine the hypothesis that oxytocin may act as an endogenous "anti-vasopressin," permitting the high levels of sociality that are necessary for mammalian reproduction. In turn, the effects of vasopressin may permit the responses needed for self-defense in the face of environmental challenges.

12. MECHANISMS OF PEPTIDE-STEROID INTERACTIONS

Oxytocin and vasopressin possess unique properties including the ability to influence neural systems that regulate selected behavioral and autonomic functions and the capacity to interact with each other's receptors. The effects of oxytocin and vasopressin may be both rapid, and long-lasting. Their functions are regulated by steroids. For example, the concurrent presence of steroid and peptide receptors in a particular neural system offers one mechanism through which steroid hormones might specify these functions. Other neurochemicals such as the endogenous opiates, the catecholamines and serotonin regulate the release and actions of oxytocin and vasopressin. Catecholamines, including dopamine and norepinephrine, may be necessary to activate or reward various behavioral processes such as male sexual behavior and aggression (Meisel and Sachs, 1994). Serotoninergic drugs, such as "prozac," have effects on emotional states and may inhibit sexual responses. It is likely that peptides, including oxytocin and vasopressin are involved in these behaviors through interactions with the serotoninergic, opioid and catecholaminergic systems.

The effects of peptides on the autonomic nervous system may be especially important to our understanding of hormonal influences on human behavior. Many specific components of human behavior, such the motor patterns involved in sexual behavior, aggression or parenting are regulated at least in part by cognitive processes. However, the decision to engage in a given behavior and mood states are strongly determined by visceral processes or states, which may in turn "motivate" the occurrence of specific behaviors.

Oxytocin is important in parasympathetic function (Uvnas-Moberg, 1996), and particularly the control of subdiaphragmatic vagal activity (Porges, 1996). Specific brainstem nuclei that provide the cells of origin for the different vagal pathways are differentially influenced by oxytocin and vasopressin. Oxytocin and vasopressin may interact to influence organs regulated by vagal efferent pathways originating in the dorsal motor nucleus of the vagus (DMNX) located in the brainstem (Uvnas-Moberg, 1996). Sensory feedback to the CNS from the visceral organs travels through vagal afferents to the nucleus tractus solitarius (NTS), providing a mechanism through which oxytocin may influence visceral experiences and bodily states. Thus, the influence of oxytocin and/or vasopressin on visceral states "feeds back" to the brain via vagal afferents and, given the perceived change in visceral state, may influence the probability that a particular behavior will occur. In addition, vagal efferent pathways, which originate (ventral to the DMNX) in the nucleus ambiguus, provide the primary innervation of supradiaphragmatic organs (e.g., heart, lungs). This component of the vagus, termed the "smart" vagus, coordinates sucking, swallowing, vocalizations with breathing. Moreover, this vagal component is neuroanatomically linked with the brainstem nuclei that regulate and monitor facial muscles and the muscles of mastication to form a ventral vagal complex (Porges, 1996). The ventral vagal complex has vasopressin, but not oxytocin, receptors. Hormones may act directly or indirectly through this system to integrate attention, emotional states and social communication with metabolic demands.

13. SUMMARY

In laboratory rats reproductive hormones, especially androgens and estrogens, have marked effects on behaviors associated with reproduction (Pfaff, et al., 1994; Meisel and Sachs, 1994). The effects of androgens are particularly strong in polygynous animals, and less obvious in monogamous or communal species (Carter, et al., 1995; Wingfield, et al., 1996). However, the physiological and behavioral costs associated with exposure to either androgens or estrogens may have minimized the effectiveness of these hormones in social mammals.

It has proven difficult to identify specific effects of the sex steroids on human behavior. However, recent studies have implicated neuropeptides, including oxytocin and vasopressin, in the social control of behavior (Carter, et al., 1995; Uvnas-Moberg, 1996). Humans are highly social and rely heavily on social stimuli for the regulation of reproductive processes. These peptides, working against a steroid background, may provide substrates for the rapid behavioral changes and emotional states that are necessary to the complexity of human behaviors.

ACKNOWLEDGMENTS

I am grateful to Dr. Peter Schmidt for his help in obtaining references relevant to human behavioral endocrinology, and to Dr. Stephen Porges for suggestions regarding mechanisms through which the autonomic nervous system can influence behavior. Research from my laboratory described here was sponsored by NIH, NIMH and NSF.

REFERENCES

Albert, D. J., Walsh, M. L., and Jonik, R. H. 1993. Aggression in humans: What is its biological foundation? Neuroscience and Biobehavioral Reviews 17:405–425.

Altemus, et al., Pigott, T., Kalogeras, K., 1992. Abnormalities in the regulation of vasopressin and corticotropin releasing factor secretion in obsessive-compulsive disorder. Archive of General Psychiatry 49:9–20.

Altemus, M, Deuster, P. A., Galliven, E., Carter, C. S., and Gold, P. W. 1995. Suppression of hypothalamic-pituitary-adrenal axis responses to stress in lactating women. Journal of Clinical Endocrinology and Metabolism 80:2954–2959.

Altmann, J. 1987. Life span aspects of reproduction and parental care in anthropoid primates. In J. B. Lancaster, J. Altmann,, A. S. Rossi, & L. R. Sherrod (Eds.). Parenting across the life span. New York: Aldine deGruyter. pp. 15–29.

Archer, J. 1991. The influence of testosterone on human aggression. British Journal of Psychology 82:1–28.

Baker, E. R., Best, R. G., Manfredi, R. L., Demers, L. M., and Wolf, G. C. 1995. Efficacy of progesterone vaginal suppositories in alleviation of nervous symptoms in patients with premenstrual syndrome. Journal of Assisted Reproduction and Genetics 12:205–209.

Bagatell, C. J., Heiman, J. R., Rivier, J. E., and Bremner, W. J. 1994. Effects of endogenous testosterone and estradiol on sexual behavior in normal young men. Journal of Clinical Endocrinology and Metabolism 78:711–718.

Bamshad, M., Novak, M. A., and De Vries, G. J. 1993. Species and sex differences in vasopressin innervation of sexually naive and parental prairie voles, *Microtus ochrogaster* and meadow voles, *Microtus pennsylvanicus*. Journal of Neuroendocrinology 5: 247–255.

Becker, J., Breedlove, S. M., and Crews, D. 1992. Behavioral Endocrinology. Cambridge, MA, Bradford Books, MIT Press.

Block, N., Schmidt, P. J., and Rubinow, D. R. 1996. Clinical aspects of premenstrual syndrome. Infertility and Reproductive Medicine Clinics of North America 7:315–329.

Bouissou, M. 1983. Androgens, aggressive behaviour and social relationships in higher mammals. Hormone Research 18:43–61.

Bronson, F. H., and Desjardins, C. 1971. Steroid hormones and aggressive behavior in mammals. In: Eleftheriou, B.E., and Scott, J. P. (eds) Physiology of Aggression and Defeat. New York: Plenum Press, pp.

Bronson, F. H., and Heideman, P. D. 1994. Seasonal regulation of reproduction in mammals. In: The Physiology of Reproduction. Second Edition. Ed. E. Knobil and J. D. Neill. New York: Raven Press, pp. 541–583.

Brown, R. E. 1994. An introduction to neuroendocrinology. Cambridge University Press: Cambridge.

Brown-Sequard, C. E. 1889. The effects produced on man by subcutaneous injections of a liquid obtained from the testicles of animals. Lancet 2:105–107.

Bruins, J., Hijman, R., and Van Ree, J. M. 1995. Effect of acute and chronic treatment with desglycinamide-[arg^8]vasopressin in young male and female volunteers. Peptides 16:179–186.

Bridges, R. S. 1990. Endocrine regulation of parental behavior in rodents. In: Mammalian Parenting. N. A. Karsnegor and R. S. Bridges. New York: Oxford University Press, pp. 93–117.

Burris, A. S., Banks, S. M., Carter, C. S., Davidson, J. M., and Sherins, R. J. 1992. A long-term, prospective study of the physiologic and behavioral effects of hormone replacement in untreated hypogonadal men. Journal of Andrology 13:297–304.

Carter, C. S. 1992. Oxytocin and sexual behavior. Neuroscience and Biobehavioral Reviews 16:131–144.

Carter, C. S. 1995. Physiological substrates of monogamy: The prairie vole model. Neuroscience and Biobehavioral Reviews 19:303–314.

Carter, C. S., and Roberts, R. L. 1996. The psychobiological basis of cooperative breeding in rodents. In: Solomon, N., and French, J. New York: Cambridge Press, pp. 231–266.

Carter, C. S., and Altemus, M. 1996 Integrative functions of lactational hormones in social behavior and stress management. Ann. N. Y. Acad. Sci. in press.

Corter, C. M., and Fleming, A. S. 1995. Psychobiology of maternal behavior in human beings. In M. H. Bornstein. Handbook of Parenting, Vol. 2: Biology and Ecology of Parenting. Mahwah, New Jersey: Lawrence Erlbaum Assoc., pp. 87–116.

Dabbs, J. M., Jr., Hargrove, M. F., and Heusel, C. 1996. Testosterone differences among college fraternities: Well-behaved versus rambunctious. Personality and Individual Differences 20: 157–161.

Dantzer, R., and Bluthe, R.-M. 1992. Vasopressin involvement in antipyresis, social communication, and social recognition: A synthesis. Critical Reviews in Neurobiology 16:243–255.

Dennerstein, L, Gotts, G., Brown, J. B., Morse, C. A., Farley, T. M. M., and Pinol, A. 1994. The relationship between the menstrual cycle and female sexual interest in women with PMS complaints and volunteers. Psychoneuroendocrinology 19:293–304.

Dixson, A. F. 1996. Evolutionary perspectives on primate mating systems and behavior. Ann. N.Y. Acad. Sci, in press.

Dixson, A. F., and George, L. 1982. Prolactin and parental behaviour in a male New World monkey. Nature 299:551–553.

Ferris, C. F., and Delville, Y. 1994. Vasopressin and serotonin interactions in the control of agonistic behavior. Psychoneuroendocrinology 19:593–602.

Fleming, A. S., Steiner, M., and Anderson, V. 1987. Hormonal and attitudinal correlates of maternal behavior during the early postpartum period. Journal of Reproductive and Infant Psychology 5:193–205.

Gibber, J. R. 1986. Infant-directed behavior of rhesus monkeys during their first pregnancy and parturition. Folia Primatologica 46: 118–124.

Gubernick, D. J., Schneider, K. A., and Jeannotte, L. A. 1994. Individual differences in the mechanisms underlying the onset and maintenance of paternal behavior and the inhibition of infanticide in the monogamous biparental California mouse, *Peromyscus californicus*. Behavioral Ecology and Sociobiology 34:225–231.

Hafner, H., Riecher-Rossler, A., An der Heiden, W., Maurer, K., Fatkenheuer, and Loffler, W. 1993. Generating and testing a causal explanation of the gender difference in age at first onset of schizophrenia. Psychological Medicine 23:925–940.

Hampson, E. 1995. Spatial cognition in humans: Possible modulation by androgens and estrogens. Journal of Psychiatry Neuroscience 20:397–404.

Inoff-Germain, G., Arnold, G. S., Nottelmann, E. D., Susman, E. J., Cutler, G. B., Jr., and Chrousos, G. P. 1988. Relations between hormone levels and observational measures of aggressive behavior of young adolescents in family interactions. Developmental Psychology 24:129–139.

Insel, T. R., and Shapiro, L. 1992. Oxytocin receptor distribution reflects social organization in monogamous and polygamous voles. Proceedings of the National Academy of Sciences 89:5981–5985.

Insel, T. R., Young, L., and Wang, Z. 1996. Molecular aspects of monogamy. Annals of the New York Academy of Sciences, in press.

Johnson, W. G., Carr-Nangle, R. E., and Bergeron, K. C. 1995. Marconutrient intake, eating habits and exercise as moderators of menstrual distress in healthy women. Psychosomatic Medicine 57:324–330.

Keverne, E. B., Levy, R., Poindron, P., and Lindsay, D. R. 1983. Vaginal stimulation: An important determinant of maternal bonding in sheep. Science 219:81–83.

Kinsey, A. C., Pomeroy, W. B., and Martin, C. E. 1948. Sexual Behavior in the Human Male. Philadelphia: W. B. Saunders

Kleiman, D. 1977. Monogamy in Mammals. Quarterly Review of Biology **52**:39–69.

Klopfer, P. H. 1971. Mother love: What turns it on? American Scientist 59:404–407.

Kraemer, G. W. 1992. A psychobiological theory of attachment. Behavioral and Brain Sciences 15:493–511.

Kravitz, H. M., Haywood, T. W., Kelly, J., Wahlstrom, C., Liles, S., and Cavanaugh, J. L, Jr. 1995. Medroxyprogesterone treatment for paraphiliacs. Bulletin of the American Academy of Psychiatry Law 23:19–33.

Kreuz, L. E., and Rose, R. M. 1972. Assessment of aggressive behavior and plasma testosterone in a young criminal population. Psychosomatic Medicine 34:321–332.

Kwan, M., Greenleaf, W. J., Mann, J., Crapo, L., and Davidson, J. M. 1983. The nature of androgen action on male sexuality: a combined laboratory self-report study on hypogonadal men. Journal of Clinical Endocrinology and Metabolism 57:557–562.

Luine, V. N., and Harding, C. F. (Eds,) 1994. Hormonal restructuring of the adult brain. Annals of the New York Academy of Sciences, Volume 743.

McCoy, N. L. The menopause and sexuality. 1992. In: The Menopause and Hormonal Replacement Therapy ed. R. Sitruk-Ware and W.H. Utian. New York: Marcel Dekker, pp. 73–100.

Meisel, R. L., and Sachs, B. D. 1994. The physiology of male sexual behavior. In: The Physiology of Reproduction. Second Edition. Ed. E. Knobil and J. D. Neill. New York: Raven Press, pp. 3–105.

Michael, R.P., and Zumpe, D. 1993. A review of hormonal factors influencing the sexual and aggressive behavior of macaques. American Journal of Primatology 30:213–241.

Nelson, R. J. 1995. Behavioral Endocrinology. Sunderland, MA: Sinauer.

Newton, N. 1973. Interrelationships between sexual responsiveness, birth, and breast feeding. In: Contemporary Sexual Behavior: Critical Issues in the 1970s. J. Zubin and J. Money. Baltimore: Johns Hopkins University Press. pp. 77–98.

Numan, M. Maternal behavior. In: The Physiology of Reproduction. Second Edition. Ed. E. Knobil and J. D. Neill. New York: Raven Press, pp. 221–302.

Nissen, E., Uvnas-Moberg, K., Svensson, K., Stock, S, Widstrom, A. M., and Winberg, J. 1996. Different patterns of oxytocin, prolactin but not cortisol release during breast feeding in women delivered by caesarean section or by the vaginal route. Early Human Development, in press.

Paikoff, R. L, and Brooks-Gunn, J. 1990. Associations between pubertal hormones and behavioral and affective expression. In Psychoneuroendocrinology C. S. Holmes, Ed. New York: Springer-Verlag. pp. 205–226.

Pedersen, C. A. 1996. Oxytocin control of maternal behavior: Regulation by sex steroids and offspring stimuli. Annals of the New York Academy of Sciences, in press.

Pedersen, C. A. and Prange, A. J, Jr. 1979. Induction of maternal behavior in virgin rats after intracerebroventricular administration of oxytocin. Proceedings of the National Academy of Sciences 76:6661–6665.

Pfaff, D. W., Schwartz-Giblin, S, McCarthy, M. M., and Kow, L.-M. 1994. Cellular and molecular mechanisms of female reproductive behaviors. In: The Physiology of Reproduction. Second Edition. Ed. E. Knobil and J. D. Neill. New York: Raven Press, pp. 107–220.

Pope, H. G., and Katz, D. L. 1994. Psychiatric and medical effects of anabolic-androgenic steroid use. Archives of General Psychiatry 51:375–382.

Porges, S. W. 1996. Emotion: An evolutionary by-product of the neural regulation of the autonomic nervous system. Annals of the New York Academy of Sciences, in press.

Pryce, C. R., Dobeli, M., & Martin, R. D. 1993. Effects of sex steroids on maternal motivation in the common marmoset (Callithrix jacchus): Development and application of an operant system with maternal reinforcement. Journal of Comparative Psychology 107:99–115.

Pryce, C. R., Abbott, D. H., Hodges, J. K., and Martin, R. D. 1988. Maternal behavior is related to prepartum urinary estradiol levels in red-bellied tamarin monkeys. Physiology and Behavior 44:717–726.

Robel, P., and Baulieu, E.-E. 1995. Dehydroepiandrosterone (DHEA) is a neuroactive neurosteroid. Annals of the New York Academy of Sciences 774:82–110.

Roca, C. A., Schmidt, P. J. and Rubinow, D. R. 1996. Clinical aspects of climacteric mood disorders. Infertility and Reproductive Medicine Clinics of North America 7:341–353.

Rosenblatt, J. S. 1995. Hormonal basis of parenting in mammals. In M. H. Bornstein. Handbook of Parenting, Vol. 2: Biology and Ecology of Parenting. Mahwah, New Jersey: Lawrence Erlbaum Assoc., pp. 3–25.

Rosenblum, L. A. 1972. Sex and age differences in response to infant squirrel monkeys. Brain, Behavior and Evolution 5:30–40.

Rubinow, D. R., and Schmidt, P. J. 1996. Androgens, brain and behavior. American Journal of Psychiatry, in press.

Seeman, M. V. 1996. The role of estrogen in schizophrenia. Journal of Psychiatry Neuroscience 21:123–127.

Sherwin, B.B. 1996. Hormones, mood, and cognitive functioning in postmenopausal women. Obstetric and Gynecology 87:20S-26S.

Sosa, R., Kennell, J., Klaus, M., Robertson, S., and Urrutia, J. 1980. The effect of a supportive companion on perinatal problems, length of labor, and mother-infant interaction. New England Journal of Medicine 303:597–600.

Su, T.-P., Pagliaro, M., Schmidt, P. J., Pickar, D., Wolkowitz, O., and Rubinow, D. R. 1993. Neuropsychiatric effects of anabolic steroids in male normal volunteers. Journal of the American Medical Association 269:2760–2764.

Susman, E. J., Inoff-Germain, G., Nottelmann, E. D., Cutler, C. B., Jr., and Chrousos, G. P. 1987. Hormones, emotional dispositions, and aggressive attributes in young adolescents. Child Development 58:114–1134.

Uvnas-Moberg, K. 1996. Physiological and endocrine effects of social contact. Annals of the New York Academy of Sciences, in press.

Uvnas-Moberg, K., Windstrom, A. M., Nissen, E., and Bjorvell, H. 1990. Personality traits in women 4 days post partum and their correlation with plasma levels of oxytocin and prolactin. Journal of Psychosomatic Obstetrics and Gynaecology 11:261–272.

Van Goozen, S. H. M., Cohen-Kettenis, P. T., Gooren, L. J. G., Frijda, N. H., and Van de Poll, N. E. 1994. Activating effects of androgens on cognitive performance: Causal evidence in a group of female-to-male transsexuals. Neuropsychologia 32:1153–1157.

Van Goozen, S. H. M., Cohen-Kettenis, P. T., Gooren, L. J. G., Frijda, N. H., and Van de Poll, N. E. 1995. Gender differences in behaviour: Activating effects of cross-sex hormones. Psychoneuroendocrinology 20: 343–363.

van Wimersma Greidanus, T. B., and van Ree, J. M. 1990. Behavioral effects of vasopressin. Current Topics in Neuroendocrinology 10:61–79

Virkkunen, M., Kallio, E., Rawlings, R., Tokola, R., Poland, R. E., Guidotti, A., Nemeroff, C., Bissette, G., Kalogeras, K., Karonen, S.L., and Linnoila, M. 1994. Personality profiles and state aggressiveness in Finnish Alcoholic, violent offenders, fire setters, and healthy volunteers. Archives of General Psychiatry 51:28–31.

Wallen, K. 1990. Desire and ability: Hormones and the regulation of female sexual behavior. Neuroscience and Biobehavioral Reviews 14:233–241.

Wallen, K., and Tannenbaum, P. L. 1996. Hormonal modulation of sexual behavior and affiliation in rhesus monkeys. Annals of the New York Academy of Sciences in press.

Witt, D. M. 1996. Regulatory mechanisms of oxytocin-mediated sociosexual behavior. Annals of the New York Academy of Sciences, in press.

Wingfield, J. C., Jacobs, J., and Hillgarth, N. 1996. Ecological constraints and the evolution of hormone-behavior interrelationships. Annals New York Academy of Sciences, in press.

9

COPULATION, MASTURBATION, AND INFIDELITY

State-of-the-Art

R. Robin Baker

School of Biological Sciences
3.239 Stopford Building
University of Manchester
M13 9PT, United Kingdom

ABSTRACT

When a woman copulates with two or more different men within five days, the sperm from those men compete for the 'prize' of fertilising any egg she may produce. This 'sperm competition' is probably both a lottery and a race, but more than anything it could also be a war, with sperm of different morphologies playing different roles. The risk of sperm competition has been argued to shape the evolution of almost every aspect of human sexuality: from testis size to penis shape; from the 'wet sheet' phenomenon to masturbation and the female orgasm. Most male behaviour can be seen as an attempt either to prevent sperm competition or to win any competition that occurs if he fails. Most female behaviour can be seen - as she 'shops around' for resources and genes - as a continual attempt to optimise any advantageous opportunities for promoting sperm competition. Here, I summarise and test some hypotheses as to how the risk of sperm competition has shaped male and female sexuality. In particular, I evaluate suggestions that men with larger testes, men of greater bilateral symmetry, and bisexual men are morphs adapted to greater involvement in sperm competition. The timing of copulations which could lead to sperm competition varies with the risk of conception in different ways in different phases of a woman's reproductive ontogeny. So too does the woman's retention of sperm as determined by her orgasm pattern. Direct evidence is presented that men with larger testes and more symmetrical men are more likely to become involved in sperm competition. Both also inseminate women with more sperm during copulation and shed more sperm during masturbation. Bisexual men, however, ejaculate fewer sperm. Heterosexual men inseminate established partners with more sperm when the risk of sperm competition is high but inseminate extra-pair women with fewer sperm. Low sperm numbers during extra-pair copulations and in the ejaculates of bisexual men are achieved via masturbation and may

New Aspects of Human Ethology, edited by Schmitt et al.
Plenum Press, New York, 1997

be strategies for success in sperm competition. If ejaculate competitiveness is a trade-off between sperm numbers and sperm age, smaller but younger ejaculates may be a better compromise when the male has only a limited opportunity to inseminate a particular woman, whereas larger, albeit older, ejaculates may be a better compromise when a male has more frequent opportunity to inseminate.

1. INTRODUCTION

When a woman copulates with two or more different men within a few days, the sperm from those men compete for the 'prize' of fertilising any egg she may produce. This 'sperm competition' (Parker 1970) is both a lottery (Parker 1982) and a race (Gomendio and Roldan 1991), but more than anything it could also be a war (Baker and Bellis 1988, 1995; Baker 1996), with sperm of different morphologies playing different roles. Fewer than 1% of sperm are programmed to seek and fertilise an egg (at any one time) (Lee 1988), and it has been hypothesised that the remainder are programmed for a variety of offensive and defensive activities (Baker and Bellis 1988, 1989a,1995).

Following Smith (1984) I have argued that the risk of sperm warfare has helped to shape the evolution of just about every aspect of human sexuality - from testis size to penis shape, and from the 'wet sheet' phenomenon to masturbation and the female orgasm (Baker and Bellis 1995; Baker 1996).

Smith (1984) suggested that sperm competition would occur whenever a woman had sex with two different men within 7–9 days, this being the maximum active life of human sperm once inside the cervix. He was probably correct (Baker and Bellis 1995), but to be conservative, I have in the past used a 5-day criterion, using the fertile rather than the active life of sperm (Baker and Bellis 1995). Other authors (e.g. Gomendio and Roldan 1993) would prefer to be even more conservative and use a criterion of only 2–3 days but, as shown previously (Baker and Bellis 1995), even this actually makes very little difference to estimates of the level of sperm competition. If a woman is going to have sex with two different men within the space of 7 days, she almost always does so within 2–3 days, perhaps precisely so as to promote the most active sperm competition (Bellis and Baker 1990). Estimates for the UK suggest that about 4% of children are the result of sperm competition, each being conceived while their mother contains sperm in her reproductive tract from two or more different men (Baker and Bellis 1995).

Over the past few years, research at Manchester has led to a relatively detailed hypothesis of the way that sperm competition may have shaped the sexual strategies of both men and women (Baker and Bellis 1995; Baker 1996). The various suggestions are outlined at different points in this paper, alongside results from new experimental investigations.

The main theme of this rather multi-faceted paper is that the importance of sperm competition to human sexuality varies not only from person to person but also from one stage of sexual ontogeny to another.

2. SOURCES OF DATA

2.1. The Manchester Study of Whole Ejaculates and Flowbacks

The main materials and methods used in the Manchester studies of whole ejaculates and flowbacks have been described in detail elsewhere (Baker and Bellis 1993a,b, 1995).

Only a brief résumé is needed here, concentrating on the elements of protocol relevant to this paper.

Subjects were recruited through staff and students at the University of Manchester. The majority were aged 19–22 years and were undergraduates or their friends. The identities of the majority were unknown to the principal investigators and for the remainder, confidentiality was assured throughout. The subjects collected ejaculates during their normal sexual activity. They were provided with condoms, a beaker, a jar of fixative (2% glutaraldehyde in phosphate buffer; Pursel and Johnson 1974), a questionnaire and instructions on how to process the ejaculates after collection. They then handed the completed questionnaire and jar of fixed ejaculate to an intermediary who delivered them to the laboratory.

The total number of sperm in each ejaculate was estimated using an improved Neuebauer Haemacytometer (Belsey et al. 1987). All sperm counts were carried out 'blind', the worker knowing neither the identity of the subject nor the contents of the questionnaire that accompanied the ejaculate.

Most of the personal details collected need no explanation (e.g. height; weight; number of cigarettes smoked per day). Men were also asked to measure the length and width of their left testis to the nearest millimetre, using plastic callipers. From these measures, testis volume was estimated in cm^3, using the equation for an ovoid.

Between 1988 and 1996, four main types of samples were collected: copulatory, flowback, anal intercourse, and masturbatory. In all, 83 males and 58 females have generated 700 samples for examination.

Copulatory samples were collected either within established partnerships (> 1 month) (designated as In-Pair Copulations or IPCs) or outside of an established partnership (= Extra-Pair Copulations or EPCs). In this study, the following situations qualified as EPCs:

1. Copulation by either the male or female from an established partnership with somebody other than the established partner *between* copulations with the established partner;
2. The first copulation with a new sexual partner;
3. Any copulation with a female who has at least one other current sexual partner.

All other copulations were designated IPCs.

Flowback is the material that flows back out of the vagina 15–120 minutes after insemination. Females collected flowbacks in a beaker, then fixed and thereafter processed the material as for whole ejaculates. The detailed instructions given to females for the collection of flowbacks have been published in full elsewhere (Baker and Bellis 1995). All flowbacks so far collected have been IPCs.

2.2. The Company Questionnaire Study: A Nation-Wide (UK) Survey of Female Sexual Behaviour

In March 1989, in collaboration with Mark Bellis and Company Magazine, I carried out a nation-wide questionnaire survey of female sexual behaviour. The questions asked (Bellis et al. 1989), the characteristics of the nearly 4000 respondents (Baker et al. 1989), and many of the analyses (Bellis and Baker 1990, Baker and Bellis 1993b, 1995) have been published elsewhere and need not be described further here.

2.3. The Calahonda Field Study: Mate Choice in a Competitive Situation

For a fortnight in July every year since 1986, I have run a Field Course for Biology undergraduate students on the south coast of Spain. The location is Calahonda, about 100 km from both Malaga to the West and Almeria to the East. The total number of people (staff, postgraduate students, and undergraduates) in any one year has varied from as few as 30 to as many as 70. In 1995 and 1996, the two years of detailed study described below, the group stayed in Calahonda for 16 days. In 1995 it consisted of 53 undergraduates, 8 postgraduates and 3 senior staff; in 1996 of 57 undergraduates, 8 postgraduates and 4 senior staff.

Most people who attend the Calahonda course and who have an established partner leave that partner behind in the UK for the duration of the field course. However, in 1995 there were two, and in 1996 four, established couples in the group.

The modal (about 80%) age for the group that travels to Spain each summer is 20 years. In most years, the sex ratio among undergraduates is about 45% males. Given the structure of the group, the nature and climate of the location, and the fact that the group is engaged in academic pursuits for only about 12 hours each day, it would be surprising if in most years there were not a degree of sexual activity.

A unique feature of the Calahonda situation is the almost total lack of privacy. For example, the group sleeps communally under the stars on the flat roof of the villa at which the field course is based. Three 'proper' bedrooms are primarily for people who are ill and, for safety reasons, cannot be locked. Sexual activity takes place mainly at night, either on the beach, on the path to the villa, in the open bedrooms, or occasionally in the bathrooms or shower cubicles. This unique combination of circumstances means that most sexual liaisons are noticed by *somebody* - and word quickly spreads. By the end of the field course, who has interacted sexually with whom is public knowledge.

In 1995 (LC) and 1996 (JH, CL & JW), students carried out projects as part of their Field Course training that, although not initiated for the purpose, allowed a unique analysis to be carried out when retrospectively linked to the public knowledge over sexual liaisons. This analysis was of the characteristics of people who, in this brief and competitive situation, did and did not copulate (or nearly copulate).

Most of the sexual liaisons analysed in this paper involved copulation (as either actually witnessed by other field course participants or freely confirmed by the couple to their friends). In fact, four of the EPC ejaculates in the whole-ejaculate study were collected at Calahonda. Naturally, however, some of the intense interactions seen by other people may not actually have involved penetration. To be included in the analysis, however, a couple need only to have been seen (by at least two other field course participants) engaged in intense sexual activity in a situation in which copulation was possible.

The copulation data in this study therefore fall rather uneasily into several different categories. Unusually for studies of human sexual behaviour, some of the copulations were actually, though inadvertently, witnessed by one or other of the investigators. Others were reported second-hand to the investigators and yet others were surmised (as described above) from either first-hand or second-hand observations. The data are thus better than just gossip, but obviously not as good as the systematic observations that are sometimes possible for other species. The data should also be better than are possible from retrospective self-reporting. However, the way in which the data were collected should obviously be borne in mind in evaluating the conclusions reached in the sections that follow.

The student projects were aimed at evaluating the physical factors that rendered some people more attractive than others, as measured by questionnaire. To this end, LC &

JH measured fluctuating asymmetry (see next section), height, weight, circumference of waist and hips, and the colour of hair, eyes and skin. Attractiveness was measured via anonymous questionnaire, all course participants being asked to list who among the group, if any, they would consider as either (a) a short-term, purely sexual, partner and/or (b) a long-term partner. Mean return rate over the two years was 65 percent. The majority of non-returns seemed to be due to the non-respondent finding nobody attractive in the group. In addition, in 1996, course participants were asked (by CL) whether they had an established relationship, the length of that relationship, and (females only) whether they were using oral contraceptives and the stage of their menstrual cycle.

In the analyses that follow, a general measure of attractiveness is obtained from the number of short- and long-term votes each person obtained summed over the completed questionnaires (assuming non-returns = zero votes). To allow comparison across years, the total number of votes were then standardised to a population of the opposite sex of fifty.

2.4. Measuring Bilateral Symmetry (= Fluctuating Asymmetry)

The term 'fluctuating asymmetry' has a precise meaning. It refers to any bilateral feature, the mean of the signed size differences (left-right) of which, measured across a population, is not significantly different from zero (Van Valen 1962). As such, it is distinguished from those features for which asymmetry is usual (e.g. testis size; testis height; biceps size). However, as anybody who has ever read papers on fluctuating asymmetry will know, the way the term is normally used and measured leads to horrendous difficulties of description and interpretation. In this paper, I make a stand for sanity and both measure and discuss the phenomenon in terms of bilateral symmetry, not asymmetry.

Of the half dozen or so standard measures of symmetry that have been used by other authors e.g. (Thornhill *et al.* 1995), I chose to use just four: index finger lengths; ear lengths; wrist widths and ankle widths. Measurements of each person were taken to the nearest millimetre either with callipers or with dividers. Symmetry was then calculated by the formula :

$$100 - ((IFD*100/MIF) + (ELD*100/MEL) + (WWD*100/MWW) + (AWD*100/MAW)/4)$$

IFD	absolute difference between lengths of left and right index fingers
MIF	mean length of left and right index fingers
ELD	absolute difference between lengths of left and right ears
MEL	mean length of left and right ears
WWD	absolute difference between widths of left and right wrists
MWW	mean width of left and right wrists
AWD	absolute difference between widths of left and right ankles
MAW	mean width of left and right ankles.

In the Calahonda study, the measurements in 1995 were all taken blind by one person (LC). In 1996 they were measured by JH. In the Manchester study of whole ejaculates, however, the priority need for anonymity combined with the collection of data over several years meant that no one person could take the measurements. I opted instead, therefore, for a conservative protocol involving each subject being measured by a different person, usually a friend or partner, who communicated the measurements on a questionnaire via intermediaries. As such, it was not possible to check the accuracy of the measurements. I should stress, therefore, that there may be more 'noise' in these symmetry data than in those of most researchers. However, as there cannot possibly be any correlation between the direction of any error of measurement and the counts of sperm number and morphology, the result of this noise is to make it *less* likely for symmetry to emerge as a significant factor, not more likely. It is any *absence* of influence of symmetry in our studies, therefore, that should be viewed with caution, more than its presence.

2.5. Data Analysis

Most of the aspects of sexuality studied here (e.g. sperm numbers; attractiveness) are influenced by many variables which cannot be controlled experimentally. Instead, they are controlled statistically, using the following procedure.

Multiple Regression analysis was performed by entering all potentially relevant variables, then deleting each least significant variable until all remaining variables had a P-value less than 0.05.

In the analysis of sperm numbers, this procedure was an intermediate step for the calculation of residuals rather than an end in itself. The entire data base was used (including multiple samples from individuals) and missing values were replaced by means. Under such circumstances, the 'P'-value represents an arbitrary threshold, not formal significance, because of pseudo-replication. The final step in evaluating single variables was to exclude the target variable from the relevant multiple regression equation, then analyse the residuals by an appropriate parametric or non-parametric test with respect to that target variable. At this step, missing values were excluded and pseudo-replication was avoided by using only either the first or non-weighted mean measure for each person, as appropriate. P-values are then legitimate measures of probability.

In other analyses, such as the analysis of attractiveness, multiple regression analysis did not involve pseudo-replication and could be used as a test of significance in its own right.

All analyses are two-tailed unless shown otherwise.

3. COPULATION, MASTURBATION AND INFIDELITY: SPERM COMPETITION AND FEMALE SEXUAL STRATEGY

One of the themes of this paper is that the importance of sperm competition varies at different stages in both male and female sexual ontogeny. This section is concerned with the role of sperm competition in female sexual strategy.

3.1. Female Sexual Strategy

Female sexual strategy in humans and other monogamous animals with bi-parental care can be described as a 'shopping around' for genes and resources (Baker 1996). In this strategy, copulation (both In-Pair and Extra-Pair) is both a means of collecting genes and a means of attracting resources (support, protection and security) via the offering and confusing of paternity. The promotion of sperm competition (by mating with different males within five days) is a mechanism for selecting the genes of males (Bellis and Baker 1990; Baker 1996).

The first phase (Phase I) in a woman's reproductive ontogeny usually begins during adolescence and is a process of meeting a succession of males, first to assess their potential as partners and secondly to attract them to provide company, support, protection and other resources. Part of this phase is the pairing with one male while remaining vigilant for a male who is better, either in terms of resources offered, genes, or both. Seeking or allowing copulation may be part of this strategy and sperm competition may often occur (17% of females generate sperm competition in their first 50 lifetime copulations; Company survey; Baker and Bellis 1995).

The second phase (Phase II) begins when the female (ideally) has found and attracted a male who is an acceptable compromise in terms of the genes and resources offered. Reproduction eventually begins. The woman's second child is the most likely of all her children to have been fathered by her partner, the first sometimes having been fathered by one of her previous male partners (towards the end of Phase I) (Schacht and Gershowitz 1963; Baker and Bellis 1995). In the middle of Phase II, fidelity is at its greatest and the risk of sperm competition at its lowest.

The third phase (Phase III) is one in which, from the springboard of her established relationship and existing children, the female renews her shopping around for genes and resources better than those provided by her current partner. For infidelity to occur, the female needs to meet someone with genes and/or resources sufficiently better than her partner's to outweigh the costs of infidelity (Baker 1996). The chances of infidelity and sperm competition slowly increase as the woman ages and each child after the second is less and less likely to be fathered by her partner (Schacht and Gershowitz 1963; Baker and Bellis 1995). Increasingly, the evidence on paternal discrepancy suggests that women paired to men in higher socio-economic groupings (SEG), with more to lose and less to gain, are the least likely to enter this third phase (High SEG, paternal discrepancy about 1%; middle about 5–14%; low SEG up to 20–30%; see Baker and Bellis 1995; Baker 1996).

One of the major female weapons in this shopping around is a loss of œstrus, or 'sexual crypsis' (Hrdy 1981). Indeed, it now seems that it was the evolution of sexual crypsis in a variety of primate lineages, including humans, that pre-adapted some of those lineages to evolve a reproductive strategy of monogamy plus infidelity (Sillén-Tullberg and Møller 1993). By hiding the fertile phase of their menstrual cycle, females make it difficult for a male to guard them at their most fertile time. They thus gain greater freedom to shop around for genes via infidelity and to gain resources from several or a succession of males by confusing paternity (Hrdy 1981).

The fertile period in humans is from five days before ovulation to up to a day after, with the maximum chance of conception of about 1 in 3 coming with copulations two days before ovulation (Barrett and Marshall 1969). However, many features of the human menstrual cycle have evolved which hide this phase and peak very successfully, not only from surrounding males but also from the female herself. Among these features are: a pre-disposition to seek or allow copulation throughout the cycle; lack of visual or olfactory clues that ovulation is imminent; very variable time interval from menstruation to ovulation; a high incidence of anovulatory cycles (60% in 20 year olds, Döring 1969; Baker, unpublished data); and erratic mood swings throughout the cycle (see review in Baker and Bellis 1995).

The crypsis is further enhanced by there being a responsive element to ovulation. From about day 5 of the menstrual cycle, ovulation seems to be 'on hold' while the woman meets and assesses males, including perhaps collecting sperm (Clark and Zarrow 1971; Jöchle 1975; Baker 1996). This phase may last anything from 2–21 days. Depending on events during this phase ovulation may or may not occur. It seems particularly likely to occur a couple of days after the female has a brief opportunity to collect sperm from an attractive male.

Such crypsis and responsiveness makes life very difficult, not only for prospective mates but also for the experimenter. A crude rule of thumb, which still ignores the prob-

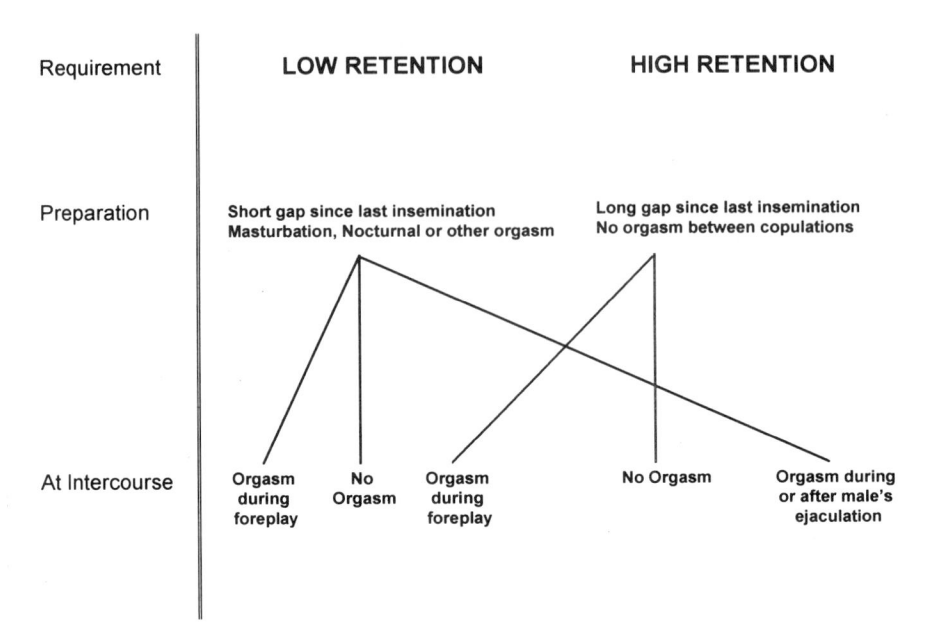

Requirement	LOW RETENTION	HIGH RETENTION
Preparation	Short gap since last insemination Masturbation, Nocturnal or other orgasm	Long gap since last insemination No orgasm between copulations
At Intercourse	Orgasm during foreplay / No Orgasm / Orgasm during foreplay	No Orgasm / Orgasm during or after male's ejaculation

Figure 1. Summary of the orgasm patterns that lead to high and low levels of sperm retention (constructed from data in Baker and Bellis 1993b).

lem of anovulatory cycles and ovulatory responses to copulation, is to assume that copulations by females taking oral contraceptives or not taking oral contraceptives but in the second half of their menstrual cycle, are non-fertile and part of shopping around for resources. In contrast, copulations by females not taking oral contraceptives and in the first half of their menstrual cycle are much more likely to be fertile and hence part of shopping around for genes.

Added to the crypsis provided by a woman's menstrual cycle is a crypsis from being able to hide how many sperm she retains during copulation. Some time after being inseminated, female mammals eject part or all of the ejaculate (Morton and Glover 1974; Ginsberg and Huck 1989; Baker and Bellis 1993b). This ejected portion is the flowback. Women eject an average of around 50% of sperm within 15–120 minutes of insemination (Baker and Bellis 1993b). The proportion ejected, however, is variable and depends on the occurrence, frequency and timing of orgasms in the days before and at the time of the copulation concerned. The rather complex pattern is summarised in Fig. 1. At its most simple, we can say that at any given copulation, females show a high or low retention of sperm but that the mechanisms they use to influence retention are cryptic to their partner.

3.2. Sperm Competition and Female Behaviour from Phase I to Phase III of Reproductive Ontogeny

All three of the groups of women studied for this paper contained individuals who claimed to have sexual partners, yet the behaviour associated with infidelity to these partners was different. Most of the differences can be related to the phase of reproductive ontogeny that the different groups of women were in. Two of the groups studied for this paper were in Phase I (Calahonda; Manchester whole-ejaculate); the third was just about in Phase II (Nation-wide Company survey), with some moving on to Phase III.

In the Calahonda and Manchester whole-ejaculate studies, the average age of women with sexual partners was 21 and 22 years, respectively. None had children at the time of study and relatively few (perhaps <5%) were paired to the man who would be their partner when they first reproduced. In contrast, the average age of the women with partners in the Company survey was 24 years (N = 2914). Many were with the man who was or would be their partner when they first reproduced (mean number of children = 0.26 ± 0.01; range = 0–6; N = 2559), and some were being unfaithful to that partner (Baker *et al.* 1989; Baker and Bellis 1995).

Although 73% (24/33) of the women in the Calahonda study (1996) had partners, the average length of existing relationships was only 22 ± 4 months. The only factor that influenced the chances of a woman having a partnership was age (more older women having partners), none of the measures of attractiveness (i.e. votes, symmetry or waist:hip ratio) having a significant association with length of relationship (unpublished data).

Over the two years of study at Calahonda, 37% (22/59) of the women who arrived in Spain unaccompanied by a partner formed a sexual liaison during the two weeks of their stay. Of these, 2% had three sexual partners during the two weeks, 12% had two, and 24% had one. Having an established partner back in the UK significantly reduced the chances of a sexual liaison in Spain. Even so, 21% were unfaithful to their distant partner, compared with 63% (5/8) of women with no partner in the UK who formed a liaison in Spain ($\chi^2 = 4.3$; P = 0.037).

Multiple regression analysis revealed only two factors which influenced the number of sexual partners a female had during her two weeks in Spain. As predicted (Thornhill and Gangestad 1994), bilateral symmetry was a significant factor though it was secondary to length of relationship ($F_{2,27} = 4$; P = 0.03; $t_{relationship} = 2.3$; P = 0.031; $t_{symmetry} = 1.9$; $P_{1\text{-tailed}} = 0.032$). Whether or not the woman was using oral contraceptives had no significant effect.

The involvement of symmetry in the probability of a woman mating while in Spain seemed to be due to its contribution to her attractiveness (Grammer and Thornhill 1994). In fact, in the Calahonda study symmetry was the only factor of those measured that was significantly associated with female attractiveness ($F_{1,62} = 9.3$; P = 0.003). As expected (Singh 1993), females with a more gynoid waist:hip ratio tended to be favoured also, but the multiple regression coefficient was not significant (t = 0.9; $P_{1\text{-tailed}} = 0.19$).

The highest mean number of sexual partners while in Spain was shown by females taking oral contraceptives (0.63, N = 8) followed by those not taking contraceptives but in the second half of their menstrual cycle (0.55; N = 11) and finally those in the first half of their menstrual cycle (0.50; N = 12). The differences, however, were not significant (ANOVA: $F_{2,28} = 0.06$; P = 0.94). Differences in risk of conception for the three groups were in any case probably minimal, given that the menstrual cycles of well over half of even the women not taking oral contraceptives would be anovulatory (Baker, unpublished data).

This hint of a negative association with risk of conception for women in Phase I shown in Calahonda is shown much more strongly by women in the Manchester whole-ejaculate study. Here, the proportion of copulations that were EPCs decreased significantly as the risk of conception increased (Fig 2a).

The indication was, therefore, that for women in Phase I of their reproductive ontogeny, EPCs (and any sperm competition that might incidentally result) were part of a shopping around for resources not genes. In contrast, the proportion of copulations that were EPCs by the Phase II or III females in the Company survey *increased* as the probability of conception increased (Fig 2a). The indication was, therefore, that EPCs and any sperm competition that resulted were part of this population's shopping around for genes.

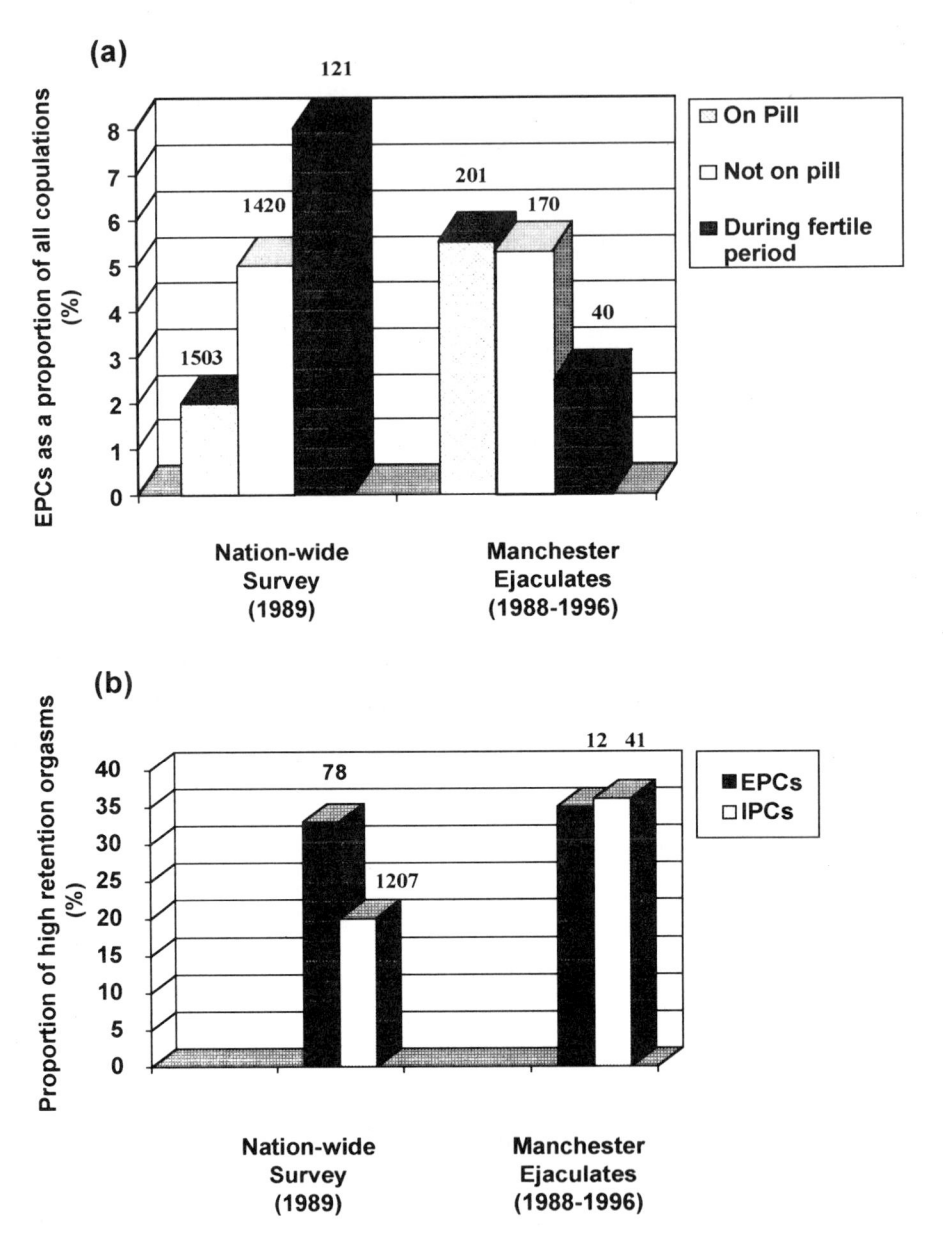

Figure 2. Summary of changes in female strategy for EPCs with stage of reproductive ontogeny. The women in the UK nation-wide survey were on average three years older than those in the Manchester whole-ejaculate study and were beginning to reproduce. Women in the nation-wide survey were (a) more likely to be unfaithful to their partner when the risk of conception was high and (b) showed an orgasm pattern/sperm retention pattern that favoured the EPC male rather than the partner. Women in the whole-ejaculate study were (a) less likely to be unfaithful when the risk of conception was high and (b) made no distinction between EPC males and current partner with respect to sperm retention. Nation-wide data from Baker and Bellis (1995).

One further difference illustrated in Fig. 2 supports this interpretation - the two groups of women differ in their orgasm pattern and hence (Fig. 1) in the relative proportion of sperm they were likely to retain from IPCs and EPCs. The Phase II-III women in the Company survey showed a higher proportion of high-retention orgasms when copulating with an extra-pair male than when copulating with their partner (Fig. 2b). No such difference was seen for the Phase I women in the whole-ejaculate study who seemed not to discriminate between a current partner and an extra-pair male in terms of sperm retention. The proportion of high-retention orgasms for these Phase I females was similar to the level with EPC males for Phase II-III females and higher than the level with IPC males.

The changes shown in Fig. 2 are consistent with a gradual shift during Phase II from using copulation and infidelity to shop around for resources during Phase I to using them to shop around for genes during Phase III.

4. NUMBER OF SPERM INSEMINATED DURING COPULATION

If women are using infidelity, and hence sperm competition, in different ways at different stages in their lives, we might also expect men to adjust their ejaculates in different ways at different stages in their lives. However, before we can examine the ways that men adapt to sperm competition at their different ontogenetic stages, it is necessary first to know what other factors influence ejaculate size.

Previous studies of the number of sperm men ejaculate have had medical origins and have almost universally involved samples collected via masturbation. The multiple regression equations that are the basis of the single variable analyses presented in this paper are given in Table 1. The variables of greatest interest to ethologists are discussed in later sec-

Table 1. Multiple regression equations for the number of sperm inseminated during copulation or ejaculated during masturbation. Also shown are the variables entered but subsequently removed from the equations because P>0.05. The equations are based on the total data sets with missing values for independent variables replaced by means as explained in Materials and Methods. Both equations are a very significant fit to the data (P<0.001) and all variables have P<0.05. However, because of pseudo-replication, the regression equations are a tool rather than ends in themselves

Independent variables	Copulation	Masturbation
Time since last ejaculation (h)	2.4	2.0
Height (cm)	4.5	5.1
Weight (Kg)		
Height/weight ratio		
Waist circumference (cm)		−10.5
Hip circumference (cm)		6.3
Waist/hip ratio		
Volume of left testis (cm^3)	4.6	1.9
Symmetry (finger	ear	wrist
Age (y)	−12.9	
Year of birth	−17.6	−6.4
Alcohol (units/day)	23.1	12.9
Cigarettes (/day)		
Heterosexual (0) or Bisexual (1)		−87.0
Time with partner since last copulation (or last 10 days) (%)	−0.6	−0.8
r^2	0.69	0.37

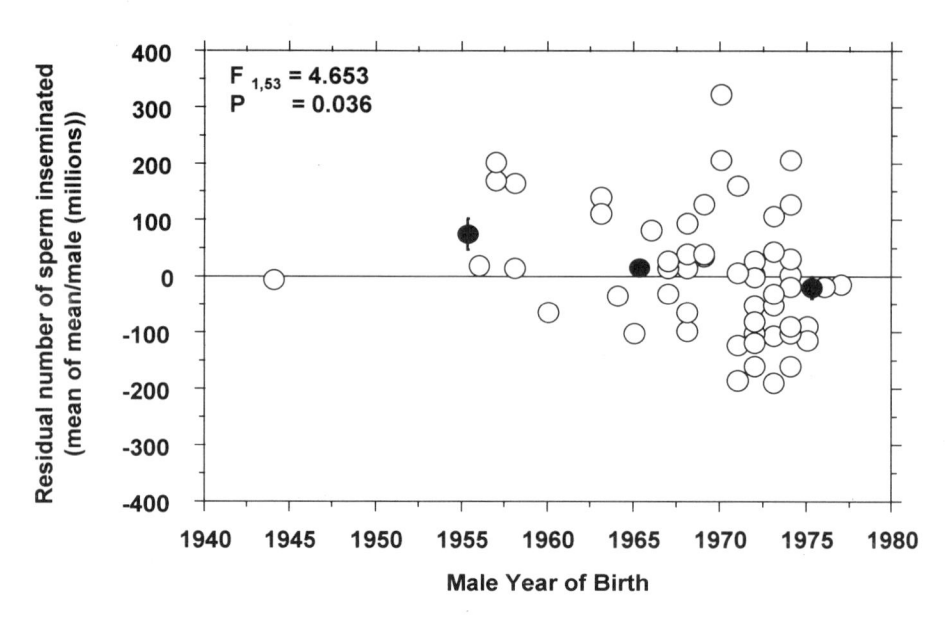

Figure 3. Number of sperm inseminated during copulation declines with year of birth from the 1950s to the 1970s. Residual sperm numbers were calculated for each of 232 ejaculates using the multiple regression equation shown in Table 1, modified by removal of 'year of birth'. Mean residual was then calculated for each of 55 males and plotted against year of birth (open circles). Solid black dots show the means ± SE of the means/male split by decade of birth. (Baker, Bellis and Bainbridge, unpublished data.)

tions. Many of the remainder are consistent with earlier work by other authors using only masturbatory ejaculates.

Hours since last ejaculation (Jouannet *et al.* 1981) and testis size (Kim and Lee 1982; Harvey and May 1989) both show the expected positive relationships (Table 1). Chronic alcoholism has a major negative influence on semen quality (Gomathi *et al.* 1993), but small amounts of alcohol are associated with more sperm in masturbatory ejaculates (Gerhard *et al.* 1992), as in Table 1. Autopsy studies have shown that the number of sperm manufactured per day by human testes declines by about 5 million for every year of a man's life after the age of about 25 y (Neaves *et al.* 1984). The decline per insemination of about 13 million for each year of age found in the Manchester study matches this autopsy work well, being the decline expected if the interval between ejaculations were 2–3 days.

Multiple regression analyses suggest that, as shown for masturbatory ejaculates produced by men in, for example, Scotland (Irvine *et al.* 1996) and Paris (Auger *et al.* 1995) - but not, for example, Toulouse (Bujan *et al.* 1996) - the Manchester subjects show a decline in sperm number in both copulatory and masturbatory ejaculates with year of birth. Those born in the 1970s inseminate fewer sperm than those born in the 1960s or 1950s. This indication is supported by more detailed analysis, which represents the first demonstration of a decline in the number of sperm in copulatory ejaculates (Fig. 3).

5. SPERM COMPETITION AND SPERM NUMBERS DURING IN-PAIR COPULATION

A major feature of human sexual ontogeny is that a man eventually settles into a (relatively) long-term relationship with a woman, most often as a prelude to reproduction.

During this relationship, the man will routinely inseminate his partner at more or less frequent intervals, depending on many factors. During this routine in-pair sexual activity, the risk of sperm competition will vary from one copulation to another, again depending on many factors.

Sperm Competition Theory has long predicted that when the risk of sperm competition is higher, males should inseminate more sperm (Parker 1982; Parker 1990). This prediction has now been supported by many different studies of many different animals, and the principle has been found to apply across species, within species and from insemination to insemination by the same male (Harvey and Harcourt 1984; Bellis *et al.* 1990; Gage and Baker 1991).

One of the first successes in the study of human sperm competition was the demonstration that the number of sperm in a man's ejaculate during In-Pair Copulation (IPC) with an established partner was a function of the risk of that partner already containing sperm from another man (Baker and Bellis 1989b). Later, it was shown that individual males made this adjustment from one IPC to the next (Baker and Bellis 1993a, 1995).

The parameter used in those studies to measure the risk of sperm competition was the percentage of time the couple had spent together since their last copulation (or in the last ten days, whichever was the shorter). This parameter has since been verified as a good index of the risk of sperm competition (Baker and Bellis 1995).

In these previous tests of Sperm Competition Theory, sample sizes prevented controlling for any (Baker and Bellis 1989b) or more than one (hours since last IPC; Baker and Bellis 1993a) variable other than percent time together. The possibility that the observed association with sperm number was an artefact of some other factor was therefore always a matter for concern (e.g. perhaps males who drank more alcohol, and thus inseminated more sperm, also spent less time with their partners). The current data set allows all those factors listed in Table 1 to be controlled, at least statistically. Yet the prediction generated by Sperm Competition Theory is still supported (Fig. 4). The possibility of artefact will, of course, always remain, but the more factors that are controlled, the lower the risk. For the moment, therefore, the data suggest that, for any given time since his last ejaculation, the less time a man has spent with his partner, the more sperm he will inseminate at their next copulation.

6. COPULATION AND INFIDELITY: SPERM COMPETITION SPECIALISTS AMONG MEN?

Once within a long-term relationship, therefore, men adjust their ejaculate according to the risk of sperm competition, increasing sperm number when the risk of sperm competition increases. As their stage of reproductive ontogeny, the security of their long-term relationship, and hence the risk of sperm competition all change with time, so too will the frequency with which they inseminate large ejaculates.

Stage of reproductive ontogeny, however, is not the only factor to influence the security and nature of a man's long-term relationship. Another factor will be the behaviour of the man himself, and in this respect men vary. Differences in men's behaviour can very often influence the extent to which their sperm are exposed to competition which in turn should influence the characteristics of their ejaculates.

Within any phylogenetic group of animals, the females of some lineages are more polyandrous than the females of other lineages. Males of the more polyandrous lineages will have been exposed to a greater level of sperm competition during their evolution than

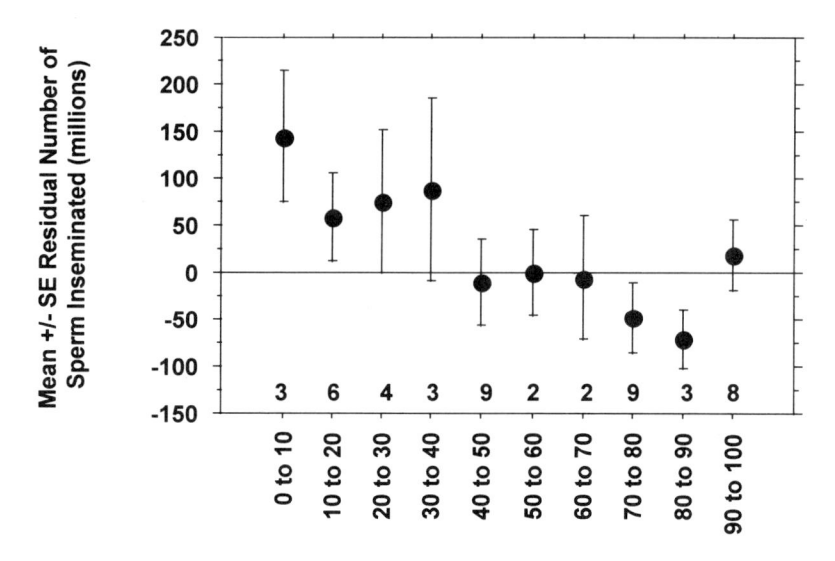

Figure 4. The number of sperm inseminated during In-Pair Copulation (IPC) increases with the risk that the female contains sperm from another man. Risk of sperm competition is measured as the percent of time the couple have spent together since their last IPC (or the last 10 days, whichever is the shorter). Lower percent times together are associated with higher risks of sperm competition. Residual sperm numbers calculated from the parameters listed in Table 1 modified by exclusion of percent time together. All of the parameters listed are therefore statistically controlled (including hours since last ejaculation). Pseudo-replication of data is avoided by including only the first IPC inseminate per couple. Number of couples per data point as shown, $F_{1,47}$ =5.7, P=0.022.

the males of other lineages. Analyses across species of groups as diverse as butterflies (Svärd and Wiklund 1989), birds (Møller 1988b) and primates (Harvey and Harcourt 1984; Møller 1988a) have shown that the males of the more polyandrous lineages have certain characteristics that seem to be adaptations to the increased level of sperm competition to which they are exposed. Larger testes relative to body size and more sperm per ejaculate are just two of these characteristics.

Such analyses use mean values for testis size, sperm numbers and other parameters for the different species. Yet within species, differences between individuals can often be as extreme as differences in means across species. Moreover, these inter-individual differences are often heritable.

On the basis of such observations, I suggested (Baker and Bellis 1995) that heritable intra-specific variation in relevant characteristics could reflect a balanced polymorphism (Hartl and Clark 1989; Cook 1991) with respect to sperm competition. The corollary was that some males were actually genetically programmed to pursue a lifetime strategy which involved exposing their sperm to competition more than others. Men with large testes were suggested to be one such morph; bisexual males another. Here, I suggest further that more-symmetrical men will expose their sperm to competition more than less-symmetrical men.

6.1. Men with Large Testes

The range of testes size in adult men is considerable (Diamond 1986). In our own data, for example, the volume of the left testis ranges from 7 to 52 cm^3. Testis size is heritable (probably via genes on the Y-chromosome, Short 1979), a fact reflected by consistent racial and population differences (Kim and Lee 1982; Diamond 1986; Rushton and Bogaert 1987).

I hypothesised (Baker and Bellis 1995; Baker 1996) that the within-population frequency distribution of testes size reflects a balanced polymorphism. In this hypothesis, men with larger testes were programmed to spend less time with partners, ejaculate often, and expose their sperm more often to sperm competition. Men with smaller testes were programmed to invest more in mate guarding and only infrequently to expose their sperm to competition. Between the two extremes, the intermediate men were programmed to pursue a more mixed strategy.

In support of this hypothesis, it was shown that: (1) the percentage of time men spent with their partners was a negative function of testis size; (2) men with larger testes had a greater sperm production rate; and (3) men with larger testes were more likely to become involved in sperm competition (as judged by a panel of 20 people) (Baker and Bellis 1995). This evidence has since been criticised on the grounds that: (a) the measurement of sperm production rate was invalid; (b) absolute, rather than relative, testis size was used; and (c) the panel ranking was of whether men with larger testes *would*, rather than *did*, become more involved in sperm competition (Birkhead 1995; Barrett 1996).

The current data set now allows these criticisms to be answered. First, the number of sperm inseminated during copulation is higher for males with larger testes (Fig. 5). Secondly, the testis size of the men who engaged in EPCs during the course of the whole-ejaculate study was significantly larger than the testis size of men who produced only IPC ejaculates during the study (Mean Volume of Left Testis: EPC males, 37.9 ± 5, N = 11;

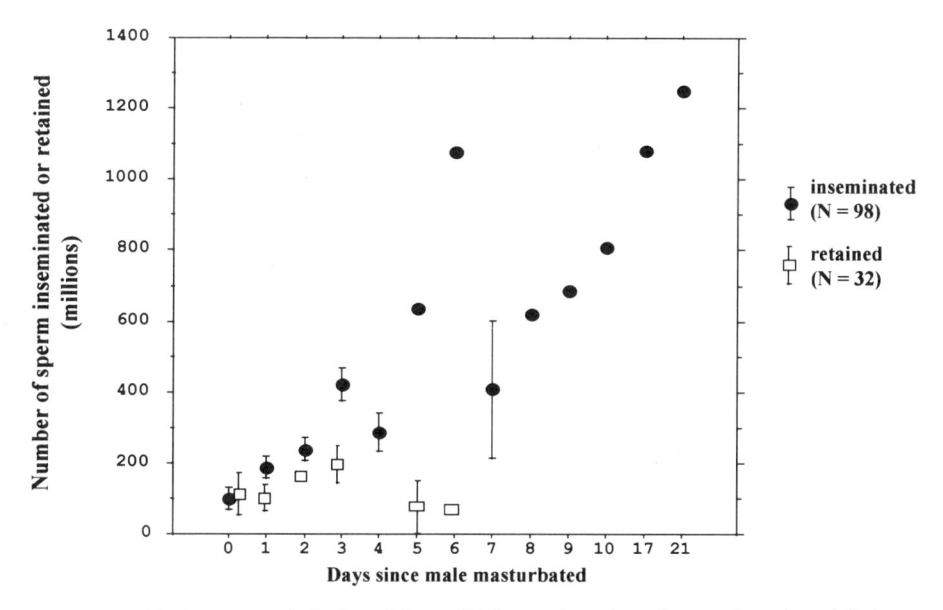

Figure 5. Relationship between male body weight, testis size, and number of sperm inseminated during copulation for 51 couples when controlled for the factors listed in Table 1. (a) There is no significant relationship between body weight and testis size ($F_{1,57}$ = 1.06; P = 0.31). (b) Men with larger testes inseminate more sperm during copulation (R^2 = 0.10; $F_{1,49}$ = 5.6, P = 0.022). Each data point and the analysis are for the first inseminate collected per couple. If relative testis size (volume as a proportion of body weight) is used, the relationship is still significant ($F_{1,49}$ = 5.7, P = 0.021).

IPC males, 25 ± 2.1, $N = 32$. $t = 2.77$; $P = 0.004$). This remained true if relative testis size was used ($t = 2.45$; $P = 0.018$). The hypothesis continues to be supported, therefore, by the current data and amended analyses.

It is in fact a moot point whether relative testis size is a more appropriate measure than actual testis size. Although the measure is superior across species, it is less clear within a species. What matters when two men inseminate the same woman is the number of sperm they inseminate, not how many they inseminate relative to their body size. The use of relative testis size is further devalued in the Manchester data by the lack of association between testis size and body size (Fig. 5a). Although the hypothesis is supported whichever measure is used, actual testis size is probably the more valid measure of the two.

For the moment, therefore, the hypothesis that men with larger testes are a morph genetically programmed for a greater involvement in sperm competition continues to be supported.

6.2. Bisexual Males

In Britain, about 6% of men engage in homosexual activities at some time in their lives, most first doing so before the age of 20 y (Johnson *et al.* 1992). Of these, around 80% also have heterosexual experience at some time during their lives and thus qualify as bisexual, rather than exclusively homosexual (even though they may have phases of exclusive homosexuality at various times in their lives). Similar levels of bisexuality are found in other industrialised countries (e.g. France, ACSF 1992), but much higher levels may be found in other societies (Ford and Beach 1952; Davenport 1965).

Familial, twin and linkage studies have now established beyond reasonable doubt that sexual orientation has a genetic basis (Bailey and Pillard 1991; Bailey and Bell 1993; Hamer *et al.* 1993). On the grounds that such a genetic complex could only be maintained in populations at observed levels if it conveys a reproductive advantage under some circumstances, I have hypothesised that bisexuality is an evolved alternative to the majority reproductive strategy of exclusive heterosexuality (Baker and Bellis 1995; Baker 1996). The suggestion is that, when rare, bisexuals have a reproductive advantage over heterosexuals but that spread of the bisexual genetic complex through industrial populations is limited by an associated enhanced risk of sexually transmitted disease. Theoretically, the level of the genetic complex for bisexuality in countries such as Britain should represent the equilibrium point of a balanced polymorphism, 6% thus being the level at which both bisexuals and heterosexuals have the same mean reproductive success.

Bisexuals begin their sexual explorations at an earlier age, have more sexual partners (both male and female), and are at a greater risk to disease than heterosexuals (Kinsey *et al.* 1948; Johnson *et al.* 1994). Whereas bisexuals probably have fewer children than heterosexuals within long-term relationships, they probably have them younger (like lesbian bisexuals), and have more via short-term sexual liaisons and sperm competition (see review in Baker and Bellis 1995).

The prediction was, therefore, that bisexuals should produce ejaculates that are more competitive and more 'prepared' for sperm competition than heterosexuals. On the basis of Sperm Competition Theory, it was predicted that bisexuals should inseminate women with ejaculates containing more sperm than heterosexuals.

Five of the 83 (6%) males in the whole ejaculate study were involved in homosexual activities and all five had previously also had sexual relationships with women and so could be considered bisexual. These levels are not significantly different from the national average (Johnson *et al.* 1992).

For the time being, direct testing of the prediction that bisexuals should inseminate women with more sperm than heterosexuals is not possible. Although 39 masturbatory ejaculates have been collected from five bisexual males we have only two 'copulatory' ejaculates (from two different males), and both of those are from anal intercourse with another male. We do not yet have vaginal inseminations from bisexual males.

The indication so far, however, is that bisexuals produce fewer sperm per ejaculate than heterosexuals, not more. If we allow pseudo-replication and use the total data set, multiple regression analysis gives a significant difference ($t = 3.0$; $P = 0.003$) between the masturbatory ejaculates of bisexuals and heterosexuals, bisexuals ejaculating an average of 87 million fewer sperm per ejaculate (Table 1). More rigorous analysis of the first ejaculate in each category/male gives a difference in the means for masturbatory ejaculates of 143 million sperm (heterosexuals: 203 ± 30 million, $N = 65$; bisexuals: 60 ± 21, $N = 5$). When controlled for either just hours since last ejaculation or for all the factors listed in Table 1, the difference reduces to 105 million. For 'copulatory' ejaculates, the difference is 177 million (heterosexuals: 199 ± 28 million, $N = 64$; bisexuals: 22 ± 14, $N = 2$). When controlled for either just hours since last ejaculation or for all the factors listed in Table 1, the difference reduces to 100 million and 81 million, respectively. However, with only five bisexuals contributing masturbatory ejaculates, and two 'copulatory', none of these more rigorous analyses yield significance (lowest P-value 0.19, t-test).

It would, of course, be no surprise to find that men ejaculate relatively few sperm during anal intercourse with other men. There is no obvious reason, however, why bisexual men should ejaculate fewer sperm during masturbation. The other putative sperm competition specialists, men with large testes and more-symmetrical males, ejaculate more sperm during copulation (Figs. 5 and 6) as well as during masturbation (controlling for all factors in Table 1, $t_{testis\ size} = 1.8$; $P_{1\text{-tailed}} = 0.035$; $N = 56$, $t_{symmetry} = 2.2$; $P_{1\text{-tailed}} = 0.016$).

At first sight, therefore, there is no support from a study of bisexuals' ejaculates for the hypothesis that bisexuality is a syndrome adapted to sperm competition. The situation is discussed further below.

6.3. Symmetrical Males

Whereas some features of male anatomy are genetically programmed to be asymmetrical (e.g. right and left testes, Diamond 1986) and others are asymmetric through differential usage (e.g. biceps circumference), others are genetically programmed to be symmetrical (Van Valen 1962). The features measured in our studies fall into this latter category and over our study population as a whole the signed difference between each of these features on the right and left sides of the body is zero (unpublished data). Asymmetry of these features, therefore, although not particularly important per se, reflects a perturbation in the developmental process that may be due either to basic genetic sub-optimality and/or to reduced genetic resistance to disease organisms that disrupt development. Although such asymmetry seems trivial in its own right, its measure provides the experimenter (and, presumably, potential mates and rivals) with a phenotypic window into an individual's genetic fitness.

An initial widespread scepticism over fluctuating asymmetry has had to be re-appraised in the light of many demonstrations (e.g. for insects, birds, and humans) that symmetry is associated with various facets of sexual behaviour. Almost universally, for a whole range of animals, it has been found that males who are more symmetrical are faster, stronger, fitter and more attractive to females (Møller and Pomiankowski 1993; Manning and Ockenden 1994). Humans appear to be no exception (Gangestad et al. 1994; Grammer and Thornhill 1994; Thornhill and Gangestad 1994). Moreover, symmetry (as fluctuating asymmetry) has a mean heritability of 0.27 and an overall statistically significant effect size of 0.15 (Møller and Thornhill, in press).

Previous studies suggest indirectly that more-symmetrical men are likely to expose their sperm to competition more than less symmetrical men. This is because symmetrical men tend to have more sexual partners (Thornhill and Gangestad 1994), more extra-pair sexual relations and to have sex with women after shorter periods of courtship (Gangestad and Thornhill, in press).

Much of this can be attributed to the fact that more-symmetrical men are judged to be more attractive than other men (Gangestad et al. 1994; Grammer and Thornhill 1994; Thornhill and Gangestad 1994). In which case, their greater involvement in sperm competition is in part an incidental by-product of the fact that women may be more eager to collect their genes (see section 3.1). However, this may not be the whole explanation because more-symmetrical men appear to 'flirt' with women other than their partner more than do other men (Gangestad and Thornhill, in press). The possibility emerges, therefore, that symmetry is part of a syndrome that includes greater pre-adaptation to sperm competition. In which case, we should again expect that more symmetrical men should inseminate more sperm during copulation.

Previous studies of male attractiveness have usually involved asking women to assess male attractiveness from photographs, and mating success has been measured using retrospective self-reporting. The Calahonda and whole-ejaculate studies allow the conclusions from these previous studies by other authors to be checked using much more direct measures.

Most (82%; 99/121) of the men in the Calahonda and whole-ejaculate studies were aged 19–25 (mode 50% at 19–20 years) and were in the transitional phase between adolescent exploration and opportunism (Phase I) and the gaining of exclusive access to an individual female (Phase II). Only 34% (23/67) of the men aged 19–25 years claimed to have an established partnership. The mean duration of those partnerships was 10 ± 3 months (and over all males aged 18–25 years, 4 ± 1.5 months, scoring those of males without partners as 0 weeks). Only 22% (16/74) had such exclusive access that they were spending most nights with their partner, thus spending over 50% of their time with her between copulations.

In the studies reported here, the more-symmetrical men: (1) were voted to be more attractive (Calahonda study); (2) obtained more sexual partners during a two-week period in a competitive situation (Calahonda study); (3) were more likely to be involved in an established sexual partnership of longer duration (Calahonda study); (4) had more sexual access to their partner (i.e. spent a greater proportion of their time with them) (whole-ejaculate study); and (5) inseminated more sperm at each copulation (whole-ejaculate study) (Fig. 6).

This preliminary study, therefore, supports the hypothesis that symmetrical males are a genetic morph with a predisposition for greater involvement in sperm competition. They appear, however, to be distinct from the 'large-testes' morph, for in the whole-ejaculate study there was no significant relationship between symmetry and testis size ($r = -0.31$; $P = 0.061$; $N = 37$ men). In fact, the relationship was nearly significantly negative, more-symmetrical males having smaller testes, not larger.

One of the strongest associations with male symmetry in our studies was with the proportion of high uptake orgasms experienced by the female member of a couple. A rela-

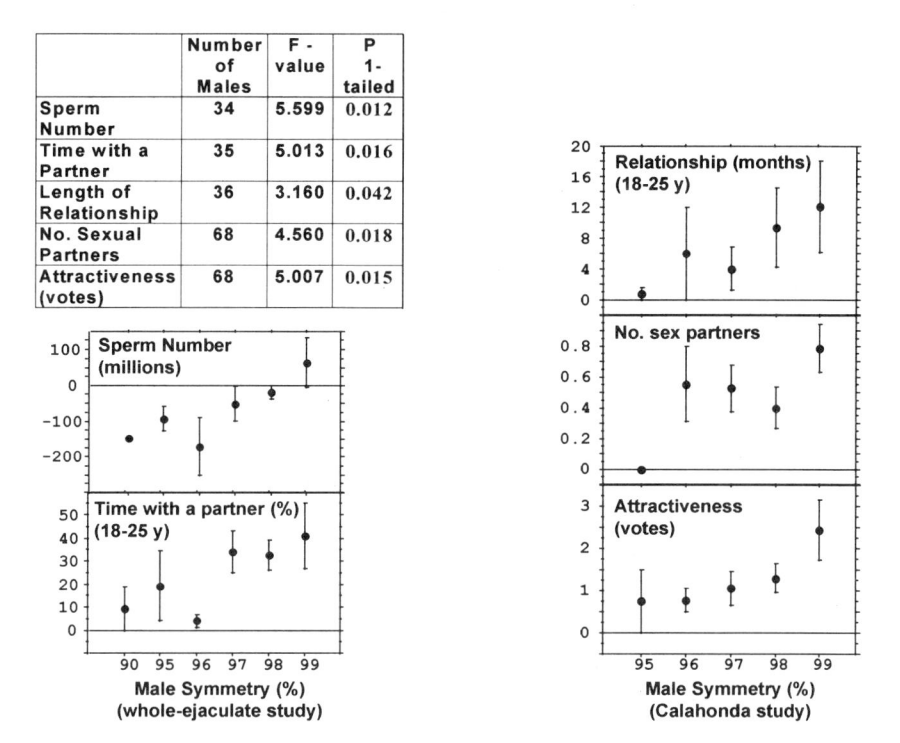

	Number of Males	F-value	P 1-tailed
Sperm Number	34	5.599	0.012
Time with a Partner	35	5.013	0.016
Length of Relationship	36	3.160	0.042
No. Sexual Partners	68	4.560	0.018
Attractiveness (votes)	68	5.007	0.015

Figure 6. Regression analyses of the relationship between male bilateral symmetry and five measures of male sexual performance in the Manchester whole-ejaculate and Calahonda field studies.

tionship between these two measures for couples in the transition from Phase I to Phase II in Albuquerque was also found by Thornhill et al. (1995). However, whereas Thornhill et al. found that females showed a higher proportion of high uptake orgasms when paired with more-symmetrical males, in our study exactly the opposite was found. Two factors influenced the proportion of high retention orgasms in the Manchester whole-ejaculate study: male symmetry and the mean percentage time the couple spent together between copulations ($F_{2,19} = 5.4$; $P = 0.014$; $t_{symmetry} = 3.1$; $P = 0.006$; $t_{time\ together} = 2.6$; $P = 0.017$). The sign of the association for males with larger testes was also negative, but not significant.

One possibility from the multiple regression analysis is that females paired to males who inseminate more sperm have a lower proportion of high uptake orgasms. However, the difference between the Manchester and Albuquerque findings remains, for the moment, unexplained.

7. SPERM COMPETITION AND SPERM NUMBERS DURING EXTRA-PAIR COPULATION (EPC)

Thus far in this paper, I have established that once men reach the stage in their sexual ontogeny when they have a long-term partner, they meet the expectation of Sperm Competition Theory and inseminate their partner with more sperm when the risk of sperm competition is higher. Some men, particularly those with large testes and/or with greater symmetry, seem more likely to behave in a way that will involve them in sperm competi-

Table 2. A test of the prediction by Sperm Competition Theory that a male should inseminate more sperm during Extra-Pair Copulations than during copulations with an established partner. The prediction is not supported. Data from Baker & Bainbridge (unpublished)

	Extra-pair copulations (N=15)	In pair copulations (N=49)	t-value	P (one-tailed)
Actual numbers of sperm (millions)	101 ± 24	229 ± 34	−2.0	0.98
Residual numbers of sperm (millions) (Controlled for hrs since ejaculation)	−71 ± 16	−11 ± 22	−1.5	0.93
Residual numbers of sperm (millions) (Controlled for all factors, Table 1)	−44 ± 21	18 ± 17	−1.9	0.97

tion. Accordingly, they also seem to manufacture and inseminate more sperm during IPC. The next important question is whether men also inseminate more sperm during EPC when the risk of sperm competition is again very high.

The current data set contains 20 ejaculates (from 11 different men with 13 different women) which were collected under circumstances other than within a long-term partnership. There were three main situations (see Material and Methods). All were outside of a partnership (and thus qualify as EPCs) and all were situations in which the man would have had every reason, consciously and/or subconsciously, to assume that the woman already contained sperm from another man. Sperm competition theory, therefore, would predict that the male should inseminate an above average number of sperm (Parker 1990). However, neither actual sperm numbers, numbers controlled for hours since last ejaculation, nor numbers controlled for all of the factors in Table 1, supported the prediction (Table 2). On the contrary, males seemed to inseminate fewer sperm into women who were not their established partner.

This apparent departure from expectation does not seem to be due to the men who provided the EPC ejaculates being low sperm producers. First, they had testes larger than other males (Section 4.1). Secondly, they did not produce significantly fewer sperm than other males during IPCs or masturbation (Table 3). Moreover, when the number of sperm EPC males produced in EPCs, IPCs and masturbation were compared, these males themselves ejaculated different numbers of sperm in each situation, least being ejaculated during EPCs (EPC 82 ± 20 million; IPC 121 ± 28; masturbation 224 ± 35; N = 20, 10 and 42 respectively; ANOVA: $F_{2,69}$ = 4.3; P = 0.016; analysis of all ejaculates). It was not therefore that the men who contributed EPC ejaculates were men who ejaculated fewer sperm.

Table 3. Number of sperm ejaculated during IPC and masturbation by men who did and did not contribute EPC ejaculates: The indication is that it is EPC ejaculates that are different, not the men in the Manchester whole-ejaculate study who contributed them. Analysis of first sample of each type per couple. Data from Baker & Bainbridge (unpublished)

Men who contributed EPCs			Men who did not contribute EPCs				
Type of ejaculate	N (pairs)	Sperm (millions)	Type of ejaculate	N (pairs)	Sperm (millions)	t-value	P-value
EPC	15	101 ± 24	IPC	44	238 ± 38	−2.1	0.043
IPC	5	150 ± 49	IPC	44	238 ± 38	0.8	0.44
Masturbatory	8	147 ± 53	Masturbatory	62	199 ± 31	0.6	0.57

Rather, it was that they ejaculated fewer sperm when inseminating a woman who was not a partner.

8. MASTURBATION AND INFIDELITY

It seems, therefore, that men respond to an elevated risk of sperm competition during IPCs by inseminating their long-term partners with more sperm but to an elevated risk during EPCs by inseminating their short-term partners with fewer sperm. Why - and is it in some way due to a link between masturbation and sperm competition?

When a variable with the values EPC = 1, IPC = 0 is entered into a multiple regression analysis of the number of sperm in the first copulatory sample produced by each couple, it has a regression coefficient significantly different from zero. As soon as hours since last ejaculation is entered into the equation, however, the EPC variable loses significance, suggesting that the reduced number of sperm in EPC ejaculates may be linked to the fact that EPCs may often be preceded by a more recent ejaculation than other ejaculations.

Analysis of the first sample of each type produced by the male of each pair shows that there is a significant heterogeneity in the time to last ejaculation for the different ejaculates (ANOVA: $F_{2,144}$ = 3.33; P = 0.038) with EPCs following more quickly after a recent ejaculate than IPCs and masturbations (EPCs 25 ± 6 hrs after the previous ejaculation; IPCs 53 ± 7; masturbations 41 ± 3; N = 15, 60 and 72 respectively). Moreover, the recent ejaculation was more likely to be a masturbation (60%, 12/20 before EPC; 33%, 116/350 before IPC; χ^2 = 6.031; P = 0.014).

It is not, however, that the males who contributed EPC ejaculates simply had a higher masturbation rate. Although the sample size for IPCs from these men is small and the heterogeneity not significant (P = 0.24), the men who contributed EPCs show a pattern of abstinence times similar to that by the total contributors (EPCs 22 ± 5 h; IPCs 39 ± 12 h; Masturbations 46 ± 9 h; N = 20, 10 and 44 respectively). Moreover, the masturbation rate for EPC contributors (every 46 h) is about the same as for men without partners (every 42 ± 4 h; N = 27). Bisexuals, however, maintain a higher rate (every 28 ± 7 h; N = 5) with an abstinence interval about the same as that which precedes EPCs.

Whether the short abstinence interval before EPCs reflects the latter being unexpected events roughly half-way between masturbations, or whether the males were anticipating the EPCs by masturbating a day in advance as hypothesised (Baker 1996) cannot yet be determined. Either way, the fact remains that in these EPCs, males inseminated an ejaculate that contained relatively few sperm because about a day earlier the male had ejaculated, most often via masturbation. The interesting possibility is that the masturbatory activity could be an adaptation to the fact that the next copulation, when it comes, is likely to be with a woman with a high chance of containing sperm from another male. An insemination preceded by masturbation may thus be more competitive in some way than one not so preceded. And one obvious way that this may be so is because the average age of the albeit fewer sperm inseminated will be younger.

Other authors have proposed that masturbation occurs because sperm have a limited shelf life (Levin 1975). Here, however, the hypothesis is that masturbation produces a more competitive EPC ejaculate. If so, it could imply that, because of their high ejaculation rate, bisexuals are permanently prepared for EPCs whereas heterosexuals anticipate each event opportunistically. In which case it remains possible that, as hypothesised, bisexuals are a morph adapted to an above average involvement in sperm competition (section 5.2).

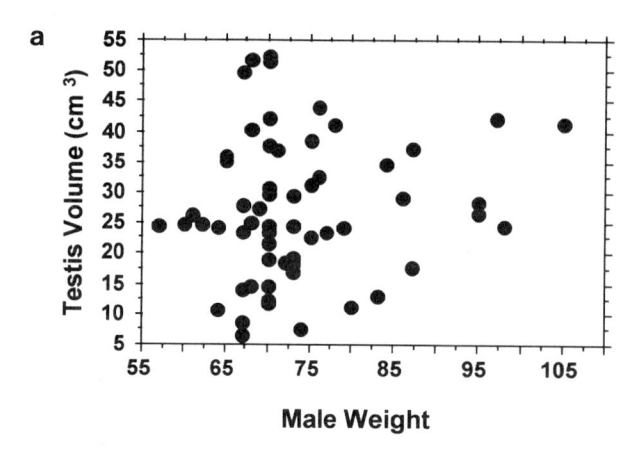

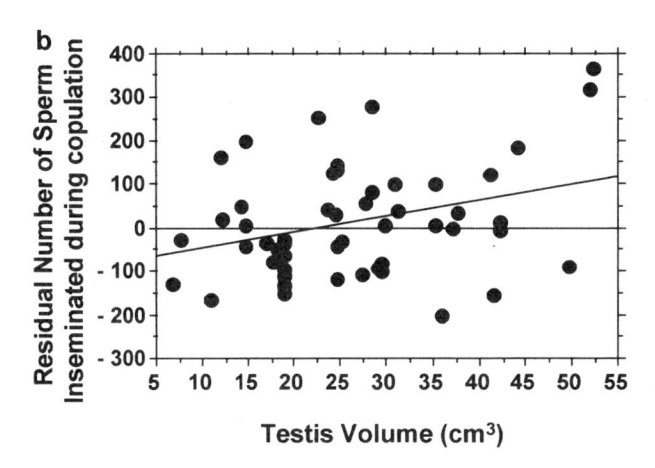

Figure 7. Influence of the time interval between a male masturbating and then inseminating a female on the number of sperm inseminated and the number of sperm retained. Although longer abstinence times (masturbation to copulation) increases the number of sperm ejaculated into the woman, the number of sperm she retains does not increase (Baker and Bellis 1993a,b). If the competitiveness of an ejaculate is a trade-off between the age and number of sperm, ejaculates preceded by masturbation a day beforehand may well be the most competitive for an EPC male (Baker 1996).

Men inseminate fewer sperm if they have masturbated beforehand, the amount depending on how recently the masturbation occurred (Baker and Bellis 1993a). The flowback study shows, however, that the number retained by the female does not decrease (Fig. 7). Inseminates preceded by masturbation are retained in higher proportions and similar numbers as inseminates not so preceded. Yet the sperm inseminated are younger. They should therefore live longer inside the female and, whether the sperm competition proceeds as a lottery, a race or as warfare, could be at a competitive advantage.

There remains the question of why EPC males should inseminate smaller, younger ejaculates for sperm competition whereas IPC males inseminate larger, older ejaculates. The main difference between IPC and EPC males is their frequency of access to a female. IPC males can afford to inseminate older ejaculates because they will have sexual access to the female again before the sperm from the previous inseminate die. EPC males, however, often have only a one-off chance at inseminating the woman and, under such circumstances it may be that having fewer sperm live longer in the female is a greater advantage than having more sperm die sooner.

If this is the explanation, men with large testes and more-symmetrical men would gain a double advantage. Because they seem to manufacture more sperm all together (ejaculating more sperm during both copulation and masturbation), they can inseminate more younger sperm after masturbating beforehand and should thus be even more competitive in the EPC role.

9. CONCLUDING COMMENTS

As Sperm Competition Theory is applied to more and more organisms, its early promise as a paradigm for the understanding of sexual behaviour has been maintained. I suggest that human sexuality is no exception. As I hope this paper continues to show, there are few aspects of human copulation, masturbation and infidelity that have not been shaped during evolution, at least in part, by the threat of sperm competition.

This does not mean that the earliest predictions of Sperm Competition Theory have all been supported. In particular, it now seems that there are many more elements to a competitive ejaculate than simply sperm number, with sperm size, sperm morphology and now sperm age being just three that have so far been implicated. It also seems, from this paper, that the most strategic ejaculate may well be different for the IPC male and the EPC male.

The ways in which the selective pressures associated with sperm competition manifest themselves are being found to be more varied than originally thought. The work reported here highlights three ways in which the situation is more complex.

First, the role of the female is increasingly being seen to be important. No longer can the female reproductive tract be thought of merely as the arena in which males play out their sperm competition games. In humans, the female has a major role in deciding when sperm competition will occur, which males she will select to compete, and which she will give an initial advantage. The order in which she collects different men's sperm, the time interval between doing so, and how many sperm from each male she will allow to enter the arena for competition, would all seem to be crucial factors in determining the outcome of any given competition. Only after the female has had her influence might ejaculate competitiveness be an important factor.

Secondly, the pressures of sperm competition seem to have produced different types of males with different strategies for avoidance, involvement and success in sperm competition. In this paper, I have drawn attention to three: men with large testes; bisexual men; and symmetrical men. All of these types are already known to have some genetic determination. This does not mean, even though to emphasise my point I have referred to them as 'morphs', that the genetics of the different situations are simple. All three types are clearly part of a continuous distribution, the genetics of which need much more detailed analysis than at present is possible. No matter what the precise nature of the inheritance of the different strategies, however, the prediction that all will be maintained in the population as a balanced polymorphism provides a theoretical starting point for further investigation. At the very least, the prediction that the extremes of the strategies, those most involved in sperm competition, should have the population mean for reproductive success but a higher variance could be tested. Except, of course, that it is very difficult to measure the reproductive success of males without more global DNA fingerprinting of children than at present is ethically possible.

Thirdly, this paper has highlighted the possibility that both men and women respond to the potential of sperm competition in different ways at different stages of their sexual ontogeny. This is interesting in its own right but perhaps is most important as a warning that future studies of human sexuality need to be much more explicit, and perhaps much more selective, about the ontogenetic stage of their subjects. Mixing subjects at different stages of mate selection and reproduction may well confuse interpretation. The equally strong, but diametrically opposed relationship between female orgasm pattern and symmetry of the male partner found in the Manchester and Albuquerque studies may well be explained by differences in ontogenetic stage between the two groups. However, comparison is difficult because both groups were made up of couples at a variety of ontogenetic stages.

Unfortunately, the most important questions are going to be the most difficult to answer. In eight years, I have managed to obtain only 20 EPC ejaculates, one IPC ejaculate from a man whose partner was being unfaithful, and no vaginal inseminates from bisexual males or flowbacks from EPCs. Ethical constraints prevent systematic collection of such ejaculates. Yet without such ejaculates it will be impossible to answer directly the main questions about how men and women manipulate sperm and ejaculates when engaging in, or promoting, sperm competition.

In contrast to this ever-nearer dead-end for the direct investigation of ejaculates, the possibilities for investigating the way that testis size and symmetry influence male (and, for symmetry, female) reproductive strategy at different stages of their lives seem to be widening. There is still much more to discover about human copulation, masturbation and infidelity than is discussed in this interim description of the state-of-the-art.

10. ACKNOWLEDGEMENTS

I thank all of those people who provided the samples and information on which this paper is based and Louise Cracknell (1995) and Jessica Hedley, Chloë Leland and Jessica Whitmore (1996) for their work in the Calahonda studies. As ever, I am particularly grateful to Mark Bellis (1988–94) and Chris Bainbridge (1994-present) for their invaluable collaboration.

11. REFERENCES

ACSF. (1992). AIDS and sexual behaviour in France. *Nature, London* **360**, 407–409.
Auger, J., Kunstmann, J. M., Czyglik, F. and Jouannet, P. (1995). Decline in semen quality among fertile men in Paris during the past 20 years. *New England Journal of Medicine* **332**, 327–8.
Bailey, J. M. and Bell, A. P. (1993). Familiality of male and female homosexuality. *Behavioral Genetics* **23**, 313–322.
Bailey, J. M. and Pillard, R. C. (1991). A genetic study of male sexual orientation. *Archives of General Psychiatry* **48**, 1089–1096.
Baker, R. R. (1996). *Sperm Wars: Infidelity, Sexual Conflict and other Bedroom Battles*, pp. 364. London: Fourth Estate.
Baker, R. R. and Bellis, M. A. (1988). "Kamikaze" sperm in mammals? *Animal Behaviour* **36**, 936–939.
Baker, R. R. and Bellis, M. A. (1989a). Elaboration of the kamikaze sperm hypothesis: a reply to Harcourt. *Animal Behaviour* **37**, 865–867.
Baker, R. R. and Bellis, M. A. (1989b). Number of sperm in human ejaculates varies in accordance with sperm competition theory. *Animal Behaviour* **37**, 867–869.
Baker, R. R. and Bellis, M. A. (1993a). Human sperm competition: Ejaculate adjustment by males and the function of masturbation. *Animal Behaviour* **46**, 861–885.
Baker, R. R. and Bellis, M. A. (1993b). Human sperm competition: Ejaculate manipulation by females and a function for the female orgasm. *Animal Behaviour* **46**, 887–909.
Baker, R. R. and Bellis, M. A. (1995). *Human Sperm Competition: Copulation, Masturbation and Infidelity*, pp. 353. London: Chapman and Hall.
Baker, R. R., Bellis, M. A. and Hudson, G. (1989). The orgasm: you've redefined it. *Company*, 60–62.
Barrett, J. C. and Marshall, J. (1969). The risk of conception on different days of the menstrual cycle. *Population Studies* **23**, 455–461.

Barrett, L. (1996). Believe it or not. *BioEssays* **18**, 338–339.

Bellis, M. A. and Baker, R. R. (1990). Do females promote sperm competition?: Data for humans. *Animal Behaviour* **40**, 997–999.

Bellis, M. A., Baker, R. R. and Gage, M. J. G. (1990). Variation in rat ejaculates is consistent with the kamikaze sperm hypothesis. *Journal of Mammalogy* **71**, 479–480.

Bellis, M. A., Baker, R. R., Hudson, G., Oram, E. R. and Cook, V. (1989). The orgasm: your chance to redefine it. *Company* , 90–92.

Belsey, M. A., Eliasson, R., Gallegos, A. J., Moghissi, K. S., Paulsen, C. A. and Prasad, M. R. N. (1987). *WHO laboratory manual for examination of human semen and semen-cervical mucus interaction*. Cambridge: Cambridge University Press.

Birkhead, T. R. (1995). Book Review. *Animal Behaviour* **50**, 1141–1142.

Bujan, L., Mansat, A., Pontonnier, F. and Mieusset, R. (1996). Time series analysis of sperm concentration in fertile men in Toulouse, France between 1977 and 1993. *British Medical Journal* **312**, 471–472.

Clark, J. H. and Zarrow, M. X. (1971). Influence of copulation on time of ovulation in women. *American Journal of Obstetrics and Gynecology* **109**, 1083–1085.

Cook, L. M. (1991). *Genetic & Ecological Diversity: the sport of nature*. London: Chapman & Hall.

Davenport, W. (1965). Sexual patterns and their regulation in a society of the southwest pacific. In: *Sex and Behaviour* (Ed. Beach, F. A.), pp. 164–207. New York: John Wiley & Sons.

Diamond, J. M. (1986). Variation in human testis size. *Nature, London* **320**, 488–489.

Döring, G. K. (1969). The incidence of anovular cycles in women. *Journal of Reproduction and Fertility, Supplement* **6**, 77–81.

Ford, C. S. and Beach, F. A. (1952). *Patterns of Sexual Behaviour*. London: Eyre & Spottiswoode.

Gage, M. J. G. and Baker, R. R. (1991). Ejaculate size varies with socio-sexual situation in an insect. *Ecological Entomology* **16**, 331–337.

Gangestad, S. W. and Thornhill, R. (In Press). Human sexual selection and developmental stability. In *Evolutionary Social Psychology* (ed. J. A. Simpson and D. T. Kenrick). Hillsdale, New Jersey: Lawrence Erlbaum.

Gangestad, S. W., Thornhill, R. and Yeo, R. A. (1994). Facial attractiveness, developmental stability, and fluctuating asymmetry. *Ethology and Sociobiology* **15**, 73–85.

Gerhard, I., Lenard, K., Eggert-Kruse, W. and al., e. (1992). Clinical data which influence semen parameters in infertile men. *Human Reproduction* **7**, 830–837.

Ginsberg, J. R. and Huck, U. W. (1989). Sperm competition in mammals. *Trends Ecol. Evol.* **4**, 74–79.

Gomathi, C., Balasubramanian, K., Vijaya, B. N. and al., e. (1993). Effect of chronic alcoholism on semen studies on lipid profiles. *International Journal of Andrology* **16**, 175–181.

Gomendio, M. and Roldan, E. R. S. (1991). Sperm competition influences sperm size in mammals. *Proceedings of the Royal Society of London* **243**, 181–185.

Gomendio, M. and Roldan, E. R. S. (1993). Mechanisms of sperm competition: Linking physiology and behavioural ecology. *Trends Ecol. Evol.* **8**, 95–100.

Grammer, K. and Thornhill, R. (1994). Human (*Homo sapiens*) facial attractiveness and sexual selection: The role of symmetry and averageness. *Journal of Comparative Psychology* **108**, 233–242.

Hamer, D. H., Hu, S., Magnuson, V. L., Hu, N. and Pattatucci, A. M. L. (1993). A linkage between DNA markers on the X chromosome and male sexual orientation. *Science* **261**, 321–327.

Hartl, D. L. and Clark, A. G. (1989). *Principles of population genetics. 2nd edition*. Sunderland, Mass: Sinauer Associates.

Harvey, P. H. and Harcourt, A. H. (1984). Sperm competition, testis size, and breeding systems in primates. In *Sperm Competition and the Evolution of Animal Mating Systems* (ed. R. L. Smith), pp. 589–600. London: Academic Press.

Harvey, P. H. and May, R. M. (1989). Out for the sperm count. *Nature, London* **337**, 508–509.

Hrdy, S. B. (1981). *The Woman that never Evolved*. Cambridge: Harvard University Press.

Irvine, S., Cawood, E., Richardson, D., MacDonald, E. and Aitken, J. (1996). Evidence of deteriorating semen quality in the United Kingdom: birth cohort study in 577 men in Scotland over 11 years. *British Medical Journal* **312**, 467–471.

Jöchle, W. (1975). Current research in coitus-induced ovulation: A review. *Journal of Reproduction and Fertility* **Suppl. 22**, 165–207.

Johnson, A. M., Wadsworth, J., Wellings, K., Bradshaw, S. and Field, J. (1992). Sexual lifestyles and HIV risk. *Nature, London* **360**, 410–412.

Johnson, A. M., Wadsworth, J., Wellings, K., Bradshaw, S. and Field, J. (1994). *Sexual Attitudes and Lifestyles*. London: Blackwell.

Jouannet, P., Czyglik, F., David, G., Mayaux, M. J., Spira, A., Moscato, M. L. and al., e. (1981). Study of a group of 484 fertile men. I: Distribution of semen characteristics. *International Journal of Andrology* **4**, 440–449.

Kim, D. H. and Lee, H. Y. (1982). Clinical investigations of testicular size. *Journal of the Korean Medical Association* **25**, 135–144.
Kinsey, A. C., Pomeroy, W. B. and Martin, C. E. (1948). *Sexual behaviour in the human male*. Philadelphia: W. B. Saunders.
Lee, S. (1988). Sperm preparation for assisted conception. *Conceive* **12**, 4–6.
Levin, R. J. (1975). Masturbation and nocturnal emissions - possible mechanisms for minimising teratozoospermie and hyperspermie in man. *Medical Hypotheses* **1**, 130–131.
Manning, J. T. and Ockenden, L. (1994). Fluctuating asymmetry in racehorses. *Nature, London* **370**, 185–186.
Møller, A. P. (1988a). Ejaculate quality, testis size and sperm competition in primates. *Journal of Human Evolution* **17**, 479–488.
Møller, A. P. (1988b). Testes size, ejaculate quality and sperm competition in birds. *Biological Journal of the Linnean Society* **33**, 273–283.
Møller, A. P. and Pomiankowski, A. (1993). Fluctuating asymmetry and sexual selection. *Genetica* **89**, 267–279.
Møller, A. P. and Thornhill, R. (In Press). On the heritability of developmental stability: a review. *Journal of Evolutionary Biology*.
Morton, D. B. and Glover, T. D. (1974). Sperm transport in the female rabbit: the role of the cervix. *Journal of Reproduction and Fertility* **38**, 131–138.
Neaves, W. B., Johnson, L., Porter, J. C., Parker, C. R. and Petty, C. S. (1984). Leydig cell numbers, daily sperm production, and serum gonadotropin levels in aging men. *Journal of Clinical Endocrinology and Metabolism* **59**, 756–763.
Parker, G. A. (1970). Sperm competition and its evolutionary consequences in the insects. *Biological Reviews* **45**, 525–567.
Parker, G. A. (1982). Why are there so many tiny sperm? Sperm competition and the maintenance of two sexes. *Journal of theoretical Biology* **96**, 281–294.
Parker, G. A. (1990). Sperm competition games: sneaks and extra-pair copulations. *Proceedings of the Royal Society of London* **242**, 127–133.
Pursel, V. G. and Johnson, L. A. (1974). Glutaraldehyde fixation of boar spermatozoa for acrosome evaluation. *Theorio.* **1**, 63–69.
Rushton, J. P. and Bogaert, A. F. (1987). Race differences in sexual behavior: Testing an evolutionary hypothesis. *J. Res. Person.* **21**, 529–551.
Schacht, L, E. and Gershowitz, H. (1963). Frequency of extra-marital children as determined by blood groups. In: *Proc. Second Internat. Cong. Human Gent.*, (Ed. Gedda, L.), pp. 894-897, G. Rome: Mendel.
Short, R. V. (1979). Sexual selection and its component parts, somatic and genital selection, as illustrated by man and the great apes. *Advances in the Study of Behavior* **9**, 131–158.
Sillén-Tullberg, B. and Møller, A. P. (1993). The relationship between concealed ovulation and mating systems in anthropoid primates: a phylogenetic analysis. *American Naturalist* **141**, 1–25.
Singh, D. (1993). Body shape and women's attractiveness: the critical role of waist-to-hip ratio. *Human Nature* **4**, 297–321.
Smith, R. L. (1984). Human sperm competition. In *Sperm Competition and the Evolution of Animal Mating Systems* (ed. R. L. Smith), pp. 601–660. London: Academic Press.
Svärd, L. and Wiklund, C. (1989). Mass production rate of ejaculates in relation to monandry-polyandry in butterflies. *Behavioral Ecology and Sociobiology* **24**, 395–402.
Thornhill, R. and Gangestad, S. W. (1994). Fluctuating asymmetry and human sexual behaviour. *Psychological Science* **5**, 297–302.
Thornhill, R., Gangestad, S. W. and Comer, R. (1995). Human female orgasm and mate fluctuating asymmetry. *Animal Behaviour* **50**, 1601–1615.
Van Valen, L. (1962). A study of fluctuating asymmetry. *Evolution* **16**, 125–142.

ABSTRACTS OF THE 13TH BIENNIAL CONFERENCE OF THE INTERNATIONAL SOCIETY FOR HUMAN ETHOLOGY (ISHE), VIENNA 5–10 AUGUST, 1996

FATHERHOOD CERTAINTY, PATERNAL INVESTMENT AND CHILDREN'S COMMUNICATIVE COMPETENCE

Grazia Attili and Patrizia Vermigli

Universita 'dell' Aquila, Dipartimento di Culture Comparate, Palazzo Camponeschi, L'Aquila, Italy; fax 06-824737; e mail: Vermigliat kant.irmkant.rm.cnr.it, tel:06-86090374

According to evolutionary theory, fathers should invest in their children just in case they are certain about their paternity. In our species, fatherhood certainty can be gained by fathers through a clear physical resemblance with them. We can expect that when this happens, they should invest more in their offsprings. 36 eight-year-old children involved in free-play interactions in their home with fathers and mothers were observed and videotaped. Children's social success in school was assessed by sociometrics. Fathers were asked individually whether they felt that their children were resembling more them than mothers. Furthermore 12 judges were asked to judge the degree of similarity between children and parents. The quality of the father-child relationship was assessed in terms of frequency of helping to play, encouraging, guiding, teaching, controlling, criticizing. Fathers whose resemblance to their children was acknowledged both by judges and by them were more positive and warm to their offspring. The positive quality of the father-child relationship was linked to children's communivative competence in the peer-group.

ENVIRONMENTAL AESTHETICS IN URBAN SPACES

Klaus Atzwanger,[1,2] Katrin Schäfer,[1] Christa Sütterlin,[2] and Kirsten Kruck[2]

[1]Ludwig-Boltzmann-I for Urban Ethology, Althanstr 14, A-1090 Vienna; e-mail: A8111GCA@vm.univie.ac.at
[2]Research Center for Human Ethology in the Max-Planck-Society, D-82346 Andechs

Evolutionary approaches to environmental aesthetics hypothesize that humans prefer places where exploration is easy and which indicate the availability of resources necessary for survival, e.g. water and food, social contacts, and high prospect refuge quality. Thus, in urban environments, the quality of public areas could influence human behaviour, depending on whether phylogenetic requirements are met. The structure of public areas may induce general well-being and consequently the probability of encounters. The latter is of high importance since game theory predicts the amount of potential future interactions to be responsible for the probability of cooperation. To determine the impact of public areas on behaviour, we observed interactions, collected questionnaire data and developed an inventory to quantify the structural features of a place. Our results strongly support our hypothesis. The amount of green space and essential ressources as well as the presence of prospect and refuge features correlated with user´s behaviour and mood: The better the quality of an area the more interactions were observed and the more satisfaction was reported. Graduation in urban environmental features covaries with differences in subjective evaluation and behavioural data. The design of public areas enriching personal contacts may be a means to fight anonymity and rising criminality in cities.

WALKING SPEED AND DEPRESSION: ARE SLOW PEDESTRIANS SAD?

Klaus Atzwanger and Alain Schmitt

Ludwig-Boltzmann-Inst. for Urban Ethology, c/o Human Biology, Althanstr 14, 1090 Vienna, Austria, e-mail: A8111GCA@vm.univie.ac.at

It is known since the early days of psychiatry (e.g. Kraepelin) that a depressed mood goes along with psychomotor slowdown (see DSM-III). This has been quantitatively demonstrated by sophisticated and rather complex (Fisch et al. 1983, J. Abnormal Psychol. 92 :307) and by quite easy to handle ethological methods (Sloman et al. 1987, Can. J. Psychiatry 32 :190). The latter found that low mood individuals show a smaller push off force during normal gait and thus have a „heavier step" than those in a good mood. The biologist´s claim is that individuals signal their inner state to others by body motion. Our model (Schmitt & Atzwanger 1995, Ethol. & Sociobiol. 16 :451) suggests that besides factors like culture, sex, age, body height, men´s socioeconomic status and women´s attractiveness, mood correlates with walking speed: The better the mood, the more dynamic the gait and the higher the walking speed. We measured walking speed and mood (Beck Depression Inventory) of randomly selected pedestrians (N= 279). Pedestrians who were depressed walked slower than those who were in good mood. This result was stable when factors like sex, age, bodyheight and status were controlled for (simple factorial ANOVA). We discuss whether the measurement of walking speed can be developed into a simple and reliable tool for investigating the nonverbal behaviour associated with affective disorders.

REAL-TIME PATTERNS IN CHILDREN'S INTERACTIVE PROBLEM SOLVING

Jeanne Beaudichon [1] and Magnus S. Magnusson [2]

[1]Laboratoire de Psychologie, University of Paris V.; lapsydee@msh-paris.fr, Fax: (33)140462993
[2]Human Behavior Laboratory, University of Iceland. email: msm@rhi.hi.is, Fax: (354)5625219

The aim of this ongoing research is to get new insight into the structure of children's interactions during problem solving. Such insight could also be an important key to a better understanding of the underlying cognitive processes. The 32 child-child dyads of children aged between 8 and 9 (mean 8.10) were given a puzzle solving task. The two two-level experimental variables are a) complexity of the task, and b), symmetry in the instruction of the partners (either equal but partial task information is given to each partner, or one is taught the task and then asked to teach the other, i.e. tutoring). The interactions were videotaped with a timer in the image and coded using Beaudichon's categories. The Theme program was then used to search for T-patterns. Even if the categories involve more interpretation than is usual in ethological studies, highly regular repeated patterns were found varying in complexity and content with the experimental conditions. Some detected patterns and other results are presented.

THE DARK SIDE OF LOVE: ABOUT BREAKING UP ROMANTIC RELATIONSHIPS

Michael Bechinie and Karl Grammer

Ludwig-Boltzmann-Institute for Urban Ethology, c/o Institut für Humanbiologie, Althanstraße 14, A-1090 Vienna, Austria; e-mail: m.bechinie@magnet.at

Most studies on romantic relationships concentrate on the "sunny side" of the topic. In this project lovesickness is defined as an emotional complex that comes over a person being left by his/her partner. It is argued that lovesickness is a depressionlike state and therefore it should affect a person's behaviour. 11 men and 11 women with above defined lovesickness have undergone a semistructured interview. The subjects' instant emotional state was recorded with a list of adjectives. Portrait photographs were taken and in addition the subjects made images of their expartners available. The interviews were videotaped unobtrusively. All pictures were rated by 19 men and 15 women using a semantic differential test. A defined scene (30sec) of every video was digitized and analyzed with "Automatic Movie Analysis". The results of this analysis were then related to questionnaire and rating data.

EARLY CRYING—DEVELOPMENTAL CONSTANT OR CIVILIZATORY ARTIFACT?

Joachim Bensel

Forschungsgruppe Verhaltensbiologie des Menschen (FVM), Albertstr. 21a, D-79104 Freiburg, Germany, Fax ++49-761-39579, email: bensel@ruf.uni-freiburg.de

Crying in the first 3 months of the infant's life takes a particular course. It begins shortly after birth, reaches its climax in the 2nd month and afterwards gradually subsides. Noteworthy is its decrease at the time of increase in sensomotoric and social competence. Is this a developmental phenomenon? The transcultural comparison reveals, however, considerably reduced crying periods, no excessive crying, no apparent 6-week-crying peak in less sophisticated societies. In western industrial countries, comparable trends are also demonstrated if more carrying around and breast feeding than usual occurs. Is increased (excessive) crying a civilisation artifact by reason of the distance of our life today to the environment of evolutionary adaptedness? Some data, gained with the help of three month diaries and interviews on 103 infants, demonstrate the variance of infant crying and its interweavement with biological rhythm and activity patterns. First indications for the explanation of the nature-nurture question ensue.

THE COMPLEXITY CRITERION IN LINGUISTICS

Bernard H. Bichakjian

University of Nijmegen, Department of French, P.O. Box 9103, Nijmegen, The Netherlands, Phone: + 31 24 361-2203, Fax: + 31 24 361-5939, E-mail: Bichakjian@let.kun.nl

Evolution is not an issue in biology and evolutionary ethology is (again) gaining acceptance. In mainstream linguistics, evolution is taboo, and linguists are using the complexity argument to deny its existence. It is claimed that because all natural languages are complex systems, their features cannot be ranked on a developmental scale. Therefore, languages have not evolved; language evolution does not exist (Bickerton, Pinker, and Nichols). The fatuity of this line of reasoning is patent because it would prove that biological evolution has not occurred either, since sharks, crocodiles, and apes are all complex animals. This paper argues that complexity must be used discriminately—the Irish elk's antlers and the human cortex do not belong to the same type of complexity; one is a huge encumbrance without commensurate yield, the other an enormous as—set for a tiny burden. Likewise the complexity of linguistic features can be a huge encumbrance without matching return or an economical instrument with broad functional possibilities. This discriminating analysis reveals that languages have evolved by trading their less profitable features for more advantageous alternatives. Language evolution exists, and the guiding principles are the same in biology, behavior, and linguistics.

THE EVOLUTIONARY PSYCHOLOGY OF POLITICAL COMMUNICATION

Daniela Lenti Boero

Corso di Laurea in Psicologia, Facoltà di Magistero, Via Saffi 15, 61029-Urbino, Italy. fax: 0722-327916

The wide introduction of television and its utilization in political talk show, debates and social information has radically changed political communication within modern democracies in recent times. Though blamed by many parts, TV, a sophisticated technological means, paradoxically allows in modern societies a very archaic way of leader choice: the direct visual gathering of information due to the possibility of seeing them speaking and acting, though in an artificial set. However, due to acoustic and written media interactions, TV is not the only means of social information. In this preliminary work the evolution of different ways of social and political knowledge exchange from prehistory through archaic societies to present times is examined and debated. Due to the very fast political changes in Italy, this country in recent years was a very interesting place in order to observe "natural experiments" in social communication and leader choice and to record their impact.

REDUCED PERCEPTION OF FACIAL EXPRESSIONS PREDICTS PERSISTENCE OF DEPRESSION

Netty Bouhuys, Erwin Geerts, and Peter-Paul Mersch

Dept. of Psychiatry, Academic Hospital Groningen, P.O. Box 30.001, 9700 RB Groningen, The Netherlands, Fax: +31-50-3696727; e-mail: A.L.Bouhuys@med.rug.nl

Background: It was hypothesized that deficits in the decoding of facial emotional expressions may play a role in the persistence of depression. Method: In a prospective longitudinal study, 33 depressed out-patients judged schematic faces with respect to the emotions they express (fear, happiness, anger, sadness, disgust, surprise, rejection and invitation) at admission (T_0), and 6 and 30 weeks later. Severity of depression was assessed at these three moments (Beck Depression Inventory). Results: Those patients who perceived less sadness, rejection or anger in faces at T_0 were less likely to show a favourable course of depression after 6 weeks (sadness, anger) or after 30 weeks (sadness, rejection, anger). These relationships could not be ascribed to initial levels of depression, age or gender. The perception of sadness and rejection did not change over time, and therefore may have trait-like qualities. Conclusion: Depression is more persistent in the subgroup that is hypo-sensitive to (negative) facial signals. This hypo-sensitivity may affect interpersonal processes (see Geerts et al. and Hale et al., this volume).

BEHAVIOURAL MODIFICATION IN DEPRESSED CHILDREN: HEAD AND TRUNK MOVEMENTS BEFORE AND AFTER TREATMENT DURING CLINICAL INTERVIEWS

Gérard Brand and Jean-Louis Millot

Laboratoire de Psychophysiologie, U.F.R des Sciences et Techniques, Route de Gray, 25000 Besançon, FRANCE, Ph: (33) 81.66.64.34; Fax: (33) 81.66.64.31

The study of behavioral activity in depressed adults by ethological methods has produced several publications. On the other hand, there is currently no published research on behavioral modifications in child depression. The clinical observations concerning body movements in depressed children are quite similar to those found in adults: the patients in acute stage appear to be less mobile. This work uses the "Berner System" to quantify the head and trunk mobility of children (aged 7 to 13 years and hospitalized for a major depressive episode according to the diagnostic criteria of the DSM III-R) during a standardized interview recorded on video. The analysis compares the results obtained on the day of admission (before treatment), on the day of discharge (after clinical improvement), and for a control sample. The frequencies of sagittal movements of head and trunk differed between both phases of illness. This was not the case for rotational and lateral movements. This result can be considered to be of predictive value for the evolution of child depression. Note however that the "head turned downwards"-position was not characteristic of the depressed state.

IS THE ACUTE NEUROLEPTIC-INDUCED AKATHISIA A DISPLACEMENT ACTIVITY?

Martin Brüne

Zentrum für Psychiatrie und Psychotherapie, Universitätsklinik, Alexandrinenstr. 1-3, D-44791 Bochum

The term "akathisia" was coined by Haskovec in 1901. Since the introduction of neuroleptics it is observed in up to 75% of the patients on these drugs. The syndrome of akathisia (Sachdev & Loneragan, The present status of akathisia, J Nerv Ment Dis 1991, 179: 381) consists of an urge to move and objective signs of repetitive movement patterns. The most accepted pathophysiological hypothesis is a postsynaptic blockade in the mesocortical dopaminergic pathways. Method: The literature is reviewed from an ethological perspective using Grant´s (Man 1969, 4: 425–36) check list of behavioural elements and with respect to possible neurophysiological explanations. Results: The repetitive movements of akathisia resemble behavioural elements which are typical for conflict situations or flight. Akathisia is sometimes associated with dysphoria and anxiety states as well as with other movement patterns which are regarded as displacement activities. Discussion: Neuroleptic-induced akathisia is subjectively often distressing. The exact pathophysiological mechanism is unclear. In animal models neuroleptics induce catalepsy and block flight behaviour. In humans akathisia might be a reaction of blocked escape during psychotic states with a high motivation.

HOMO-SPECIFIC SYMBOL-SYNTACTIC PAIGNIOTIC-KOLYMBETIC ETHOLOGY

Michael Bujatti-Narbeshuber

Museum of Natural History, Dept of Anthropology, Burgring 7, 1014 Vienna, Austria, Fax +43 1 523 52 54

The serotonin based neurochemistry of rest and fulfilment socioemotions (Bujatti & Riederer 1976, J. Neural Transm. 39:257) is used to characterize the homo specific socialization pattern and homo specific behavioral plasticity. This novel hagiosophic rank order overrides but temporarily regresses to adrenergic fight or flight based primate hedonic-agonic social rank order described by Chance & Jolly (1970) which manifests itself as structure of attention. Fundamental to homo specific hagiosophic rank order it is a novel structure of attention on the inner transcendental locus of control. According to the Littoral Double Niche Transition (DNT-) theory of hominine evolution (Bujatti-Narbeshuber 1985) this critical evolutionary advantage of associated species specific behavioral plasticity for problem solvingand socialization consists in the evolutionary stabilization of mammalian play-dream behavior together with an—all behavior calming—hierarchically superior ethology of vertebrate diving co-evolving in a coastal terraquatic double-niche setting to teleonomic creative intelligence. Allowing for authentic play now even under emergency, this oxygen and glucose conservation program of the diving response reduces neural ion pumping to a state of rest and fulfilment of pure self-awareness without specific information processing (transcendence, self-conscience) whose maintenance becomes the teleonomy determining factor of this syntactic-sound symbol initiated creative intelligence ethology.

GENDER DIFFERENCES IN SOCIAL STRATEGIES OF RUSSIAN URBAN PRIMARY SCHOOLCHILDREN

Marina L. Butovskaya[1] and Alexander G. Kozintsev[2]

[1]Inst. of Ethnol. & Anthropol., Leninsky prosp. 32-A, Moscow 117334, Russia, fax:7-095-9380600; e-mail: recon@iea.msk.su
[2]Museum of Anthropology and Ethnography, Saint-Petersburg 199034, Russia, e-mail: sasha@kozintsev.spb.su

Our aim is to examine the gender differences in social strategies and individual practices of Russian urban primary schoolchildren. Focal sampling method was used and 15 5-min videos were collected for each child. Fifty-six characteristics of social behaviour (aggression, help, avoidance, demonstrations, friendly contacts, smiling, laughing etc.) were assessed by principal components (PC) analysis. PC1 was interpreted as a measure of a child's general involvement in social life. PC2 and PC3 show strong dependence on gender. Each of them describes an independent strategy used by girls: PC2—to look at more children and laugh together with others; PC3- to perform more demonstrations and avoidance. The ratio of aggression was higher when boys were active aggressors. A higher friendly motivation of girls interactions was found. Girls were evidently inclined to choose friends among children with similar social patterns, while for boys, such a criterium was not sufficient.

ORGANIZATION OF DYADIC INTERACTIONSIN INFANTS

C. Casagrande,[1] Magnus S. Magnusson,[2] and Hubert Montagner

[1]Lab. of Psychophysiol., Univ. Franche-Comté, route de Gray, 25030 Besancon Cédex, France, fax 81666431
[2]Human Behavior Lab., Univ. of Iceland, Skölabaer, Sudurgata 26, IS-101 Reykjavik, Iceland, MSM@RHI.HI.IS

This study introduces a new approach to interactive abilities of 4/5 months-old infants within the framework of an experimental situation without parental "guidance". Communicative exchanges were coded using 374 items classified in 24 behavioural categories. A micro-analytic and real-time approach and data analyses using a special purpose sofware package (THEME) allow to detect recurrent "patterns" and to identify a syntactic and hierarchical behavioural organization. It also permits to identify sequentially organized communicative bouts (both intra- and interindividual interactive patterns). The main results show that 4–5 months-old infants present a high level of competence in the coordination of individual behavioural acts and of complex, sustained, structured behavioural sequences in interaction processes. Furthermore, if each child appears to have a personal behavioural dynamic, the infants especially present a quantitative and qualitative communicative adjustment through a process of mutual "entrainment". This approach shows a surprising ability of infants to coordinate their own acts and to organize the interaction in a "symetric situation" and thus may be valuable to detect early disturbances in handicapped or "at risk" children.

THE FCST: A MEASURE OF FEMALE-FEMALE COMPETITION STRESS

Charles Crawford and Sally Walters

Department of Psychology, Simon Fraser University, Burnaby, B.C., Canada, V5A 1S6, E-mail: Charles_Crawford@sfu.ca, Phone/FAX 604-291-3660/3427

Surbey (1987), Voland & Voland (1989), and Crawford (1989) have developed an explanation of anorexia nervosa based on the idea that weight control was an adaptation for adjusting ancestral girl's reproductive effort in response to environmental stresses, and that exaggeration, in the current environment, of stresses producing adaptive ancestral weight loss leads to pathological dieting. We are developing a paper and pencil measure to study the role of female-female competition as a trigger of ancestral reproduction suppression mechanisms. In three studies done on undergraduates we developed a 29 item test. In the fourth study, the test, as well as measures of body dissatisfaction, the Body Mass Index, and the Drive for Thinness Scale from the Eating Disorders Inventory (Garner, Olmstead & Polivy 1983), was administered to 316 junior high school girls. A factor analysis produced two factors: Female-Female Competition Stress and Body Image Dissatisfaction. The coefficient alpha for the test is 0.89.

ATTRACTIVENESS: SOME CUES WEIGH INFINITELY MORE THAN OTHERS

Jean Czerlinski and Daniel G. Goldstein

Center for Adaptive Behavior and Cognition, Max Planck Institute for Psychological Research, Leopoldstr. 24, 80802 Munich, Germany, Phone: 89/38 602 238, Fax: 89/38 602 252; email: czerlinski@mpipf-muenchen.mpg.de

How do an organism's features work together to make it attractive? The literature suggests many variables which affect attractiveness including age, hip-to-waist ratio, and facial symmetry. However, there are few examples of studies on how these numerous cues create an overall impression. Common sense suggests attractiveness is a weighted combination of cues, implying that features are compensatory; for example, a host of nice features should make up for any defective one. In contrast, we propose a simple model which uses a non-compensatory rank ordering of cues wherein some features matter infinitely more than others. This model considers and processes less information than weighted-combination models. In computer simulations, our Take The Best algorithm did as well or better at predicting attractiveness ratings than a weighted linear model of three predictors. Since our algorithm is also simpler, it may be a better model of how organisms deal with multiple features in judging attractiveness.

EVOLUTION, BRAIN, AND ARTS

Georgi Dimitrov

Kliment Ohridsky University of Sofia, 1000, Sofia, Bulgaria, 15 Ruski Bd. , Telex: 23296 Suko R BG

The functional asymmetry between the brain hemispheres plays an important role in the behaviour of man. It manifests itself in the arts in the opposing pairs semantic & aesthetic, left & right, male & female, etc. The semantic, e.g., with its symbolic character is explicable & logically determined, while the aesthetic depends on inner feelings and is therefore inexplicable. Surrealism e.g. emphasizes the semantic, while abstract painting emphasizes the aesthetic. For a number of neuroethologists these preferences are brain determined. The functional asymmetry is even more clearly manifested in the relation between left and right in the arts. Investigating 50 000 images from ancient times up to present, Hufschmidt (1980, Arch. Psychiat. Nerverkr. 229:17) asserts that right-handed people paint the human profile turned to the left. His neurophysiological arguments are based on the role of the right brain hemisphere in image processing. According to Spennemann (1984, Neuropsychol. 22:613) left orientation has its roots in the stove age, but it firmly establishes itself during the age of the Greek Epos. E.g. Levy (1984, Neuropsychol 14:431) Schwartz and Hewitt (1970, Percept. & Motor Skills. 30:991) are of the opinion that the aesthetic preferences of different people, cultures and time periods are greatly influenced by biological development. Leroi-Gourhan shows that the raised left hand expresses the female. These and other investigations offer sufficient proof to assume that this duality is innate.

THE MILGRAM EXPERIMENTS AND EVOLUTIONARY PSYCHOLOGY

Charles Elworthy

Free University of Berlin, and the European Academy, Schloss Wartin, D-16306 Wartin, Germany; Fax: +49.33331/65218; E-Mail: Elworthy@T-Online.de

Stanley Milgram's 1960's research on conformity and obedience to authority continues to provide an important foundation for social psychology. The results of Milgram's experiments are generally treated by text books in considerable detail, but the explanatory hypotheses he advanced for the observed behaviour are seldom discussed. Milgram created, however, a sophisticated theoretical framework whose key elements and methodology can be seen as precursors of modern evolutionary psychology. My paper reviews relevant recent developments in evolution and cognition, and the role of the individual/kin-selection—group-selection controversy, which has led to the relative neglect of phenomena which appear to presuppose group selection. It then considers Milgram's research results, and assesses his hypotheses and methodology in the context of an appropriately developed evolutionary psychology. Finally it considers the general implications of Milgram's work for our understanding of evolution and cognition, and the influence of the social environment on behaviour.

CO-EVOLUTION OF COGNITIVE FUNCTIONS AND NATURAL LANGUAGE

Gertraud Fenk-Oczlon and August Fenk

[1]Institute of Linguistics and Computerlinguistics
[2]Department of Cognitive Research
University of Klagenfurt, Universitätsstraße 67, A-9020 Klagenfurt; e-mail: gertraud.fenk@uni-klu.ac.at

Language can be either considered as a subsystem of the cognitive apparatus or at least as a system which develops together—in „co-evolution" (e.g. Holenstein 1990)—with other cognitive functions. The following results were found when starting from a more specific assumption relating the immediate memory span (cf. Pöppel 1985, Schleidt 1992) to speech segmentation:

a. the more syllables per clause, the fewer phonemes per syll. (r= -0,77)
b. the more syllables per word, the fewer phonemes per syllable (r= -0,45)
c. the more syllables per clause, the more syllables per word (r= +0,38)
d. the more words per clause, the fewer syllables per word (r= -0,69).

This set of mutually dependent cross-linguistic correlations found in 29 typologically different languages (Fenk & Fenk-Oczlon 1993) clearly indicates that there are time-related constraints which are effective in typological differentiation in the sense of set-points in a self-regulating system (Fenk-Oczlon & Fenk 1995). A relevant language-related memory limit is described in Baddeley's (e.g. 1986) articulatory loop model saying that one can recall as many items as one could pronounce in 2 (1,5–3) seconds. This span corresponds to about 7 plus minus 2 syllables (Fenk-Oczlon & Fenk 1985): it covers about 5 very complex and long syllables or 10 very simple and short syllables.

AVERAGENESS AND SYMMETRY: THE ASSESSMENT OF FEMALE BEAUTY

Martin Fieder, Karl Grammer, and Gudrun Ronzal

Ludwig-Boltzmann-Inst. for Urban Ethology, Althanstrasse 14, A-1090 Vienna; e-mail: A8402796@UNET.univie.ac.at

The host-parasite paradigm predicts that the evaluation of human attractiveness should be connected with assessments of parasite resistance and developmental stability. **Two** possible features which could reflect this are symmetry and averageness of faces and bodies. Yet there are problems with the determination of symmetry and averageness. Measurement errors are sometimes higher than the observed differences. In this experiment 100 female faces and bodies of two cultures (Americans and Japanese) where rated by independent male raters for attractiveness. Faces where then analysed for symmetry and averageness with the help of digital image analysis. Bodies were analysed only for symmetry. Symmetry dominates averageness with familiar stimuli (Caucasian Americans). Non-familiar stimuli (Japanese) are judged according to their deviation from the averageness of the familiar stimuli. Moreover symmetry of front and back view of bodies and face correlate with each other and averageness. These relations between symmetry, averageness and beauty judgements suggest the existence of prototype templates for the recognition of beauty.

DIGITAL IMAGE ANALYSIS AND THE STRUCTURE OF HUMAN BEHAVIOUR

Valentina Filova,[1] Martin Fieder,[2] and Karl Grammer[2]

[1]IBL-Sistemi, Zelezna cesta 18, 61000 Ljubljana, Slovenia
[2]Ludwig-Boltzmann-Institute for Urban Ethology Althanstrasse 14 A-1090 Vienna/Austria

In recent years it has been shown that traditional ethological approaches to the analysis of behaviour may fail for the assessment of time-structures and quality of continuous non-verbal behaviour. This is the case because of reliability problems in the determination of beginnings and ends of behaviour categories and the impossibility to reliably describe behaviour. The workshop outlines the problems of such an analysis and demonstrate possible solutions and their limitations. First, a simple 2D solution on the basis of difference pictures and edge-detectors is presented which is able to analyse body movements in terms of motion energy detection and changes in body contours. The resulting data stream then is translated to time-structure, emphasis, expressiveness and speed of digitised video data and speech automatically. The practical part is a demonstration of such a system implemented on a PowerPC. Second, three different approaches to 3D tracking of the human body are discussed and a new model-based approach to pose recovery and tracking of body movements which is currently under development is outlined. The possible applications of this approach are automatic recognition of individuals through gait analysis, monitoring of therapy success, the determination of synchronisation between individuals and the analysis of the relation between speech and body movements.

HORMONES AND FEMALE SELF-PRESENTATION

Bettina Fischmann,[1] Karl Grammer,[1] and John Dittami[2]

[1]Ludwig-Boltzmann-Institute for Urban Ethology, Institute for Human Biology;
[2]Abteilung für Ethologie, Institute for Zoology, Univ. of Vienna

Objective: Can sexual advertising be related to hormonal conditions and circumstances? Method: Prospective field study, conducted at ten dancing-evenings including a total of 380 female probands. Data collection by questionnaire, video-tape and saliva collection, furthermore computerised assessment of bare skin and tightness of clothing and laboratory measurements of estradiol, progesterone, testosterone and cortisol. Standard statistical procedures were applied. Results: No general association between measured and determined variables were observed. Positive correlations between estradiol and clothing were found for young women, aged 15 to 19 years, as well as for females with low cortisol levels and with partner absent. Naked skin and tight clothing are commonly considered to be „sexy". Mood and expectations for flirtation/sexual encounter were highest with those below 19 years and those taking oral contraceptives without partner, and this expectation correlated with tightness of clothing. Conclusion: Selfpresentation seems to be very deliberate. Women know about the effects of their appearance and this is operationalized with growing age. In an individual context, self-presenting behaviour may be under hormonal influence.

COMPETITION AND SOCIAL BONDING AMONG FEMALE GORILLAS (GORILLA G. GORILLA) AND BONOBOS (PAN PANISCUS)

Cornelia Franz[1] and Iris Weiche[2]

[1]Karl-Franzens-Universitaet, Institut fuer Zoologie II, Universitaetsplatz 2, A-8010 Graz, Austria, Fax: 0043/316/381-255, e-mail: Franzc@kfunigraz.ac.at
[2]Morgenstelle 28, D-72076 Tübingen, e-mail: I.Weiche@t-online.de

Socioecological factors are expected to determine spatial, hierarchical and affiliative relationships between females. In gorillas and bonobos, as in most human societies, females transfer between groups and feeding competition between females appears to be relaxed. According to theory, weak female dominance hierarchies and infrequent female-female interactions are expected. This hypothesis is tested by quantitative comparison of data from 100 observation hours of one gorilla and one bonobo group (n=7 each) of the Stuttgart Zoo, Germany. Both species exhibited linear dominance hierarchies with female dominance in bonobos and male dominance in gorillas. Bonobo females had significantly higher rates of aggression, affiliation and association with other females than expected, while gorilla females more frequently associated with the male than with other females (G-Tests, $p<0.001$). Results indicate high levels of competition within females of both species. However, bonobo society resembles the female-bonded pattern while gorillas exhibit a non-female-bonded social structure.

HUNTING OR FOLLOWING EACH OTHER : ORIGINS OF MIMETIC SKILLS?

Anton Fürlinger

Isbarygasse 13, A-1140 Vienna, Austria

Merlin Donald (*Origins of the modern mind,* 1991) described a phase of mimetic skills in infrahominid primate development. Taking evolutionary psychology seriously and looking back along the line of vertebrates that has led towards man, several steps in „mimicking" behaviors can be shown. Hunting another animal involves, on the side of the hunter, a sensory part (watching the behavior of the prey) and a motor part (following the path or trajectory of the prey). Simultaneously, on the side of the prey, behavior of the predator is watched (sensory part), direction and speed of his movements will influence flight movements (motor part). Given in general that, starting with the first fish ancestors, most of all vital object-related behavior was on mobile, animated objects (prey and predator animals, competing animals, partners, other conspecifics), hunting was among the first action tasks in vertebrate phylogeny. Let us look closer into interactive "following" behavior: motor patterns of the one feed directly into sensory patterns of the other, thereby steering the motor patterns of the latter, which will in turn feed into sensory patterns of the first. The remaining environment "behaves" in much more predictable ways, physical structures giving contrast and support. The coupled, better crossed sensory-motor system between partners of interaction may not only have been the first mimicking-type behavior, but represent one of the minimum-essential nervous functions. Brains always develop(ed) together!

NONVERBAL INTERPERSONAL ATTUNEMENT PREDICTS COURSE OF DEPRESSION

Erwin Geerts and Netty Bouhuys

Dept. of Psychiatry, Academic Hospital, PO Box 30.001, 9700 RB Groningen, The Netherlands, Fax: +31-50-3696727

Background: Satisfaction with interpersonal relationships may play a causal role in the course of depression. In non-depressed subjects interpersonal satisfaction is related to equal levels of interpersonal behaviour. Such equal levels are reached via mutual attunement. Support seeking by patients and support giving by the social environment are the primary focus in interpersonal theories of depression. We investigated the pre-treatment attunement between nonverbal elements of depressed patients' support seeking behaviour and of interviewers' support giving behaviour in relation to subsequent improvement. Method: Two patient groups were studied (major depression [n=31]; seasonal affective disorder [n=56]). We assessed the attunement as the absolute difference between the patients' support seeking behaviour and the interviewers' support giving behaviour.

Results: In both groups, the time-course of the attunement predicted subsequent improvement: The more the attunement increased over the interview, the more favourable the course of the depression. The results are consistent with an interpersonal approach of depression.

A SOCIOBIOLOGICAL ANALYSIS OF CONFLICT IN BLENDED FAMILIES

Deborah Wilson Gilbert[1] and Carol Cronin Weisfeld[2]

[1]5445 El Camino, Columbia, Maryland 21044, USA
[2]University of Detroit Mercy, 8200 West Outer Drive Detroit, Michigan, USA, 48219; Fax: 313-993-6397

This study examines the impact of stepchildren and biologically-shared children on marital quality. Sixty remarried couples completed the Marriage and Relationship Questionnaire and listed all the stepchildren and biologically-shared children present in the family. The couples were divided into three groups based on the proportion of nonshared stepchildren present in the stepfamily. It was found that couples with higher proportions of stepchildren and lower proportions of biologically-shared children in the family reported greater levels of marital quality. This finding was the opposite of what was initially expected. Further analysis indicated that when mixes of children are present (stepchildren from previous marriages, and children born to the remarried couple), the marriage is more likely to be adversely affected than when stepchildren only are present. Significant correlations between net genetic relationship and marital quality corroborated the initial results. Comparisons of each spouse's gross genetic relationship was not found to be associated with marital quality. The data strongly imply that the introduction of a biologically-shared child into a stepfamily creates jealousies and tension due to the differential treatment of the children based on genetic relatedness. These nepotistic dynamics appear to affect the marriage.

THE ADAPTIVENESS OF RECOGNITION

Daniel G. Goldstein, Gerd Gigerenzer, and Geoffrey F. Miller

Center for Adaptive Behavior and Cognition, Max Planck Institute for Psychological Research, Leopoldstr. 24, 80802 Munich, Germany, Fax: 89/38 602 252; email: goldstein@mpipf-muenchen.mpg.de

Does an organism's preference for recognized foods, habitats, social partners, and mates make adaptive sense, or is it just a neophobic bias? We show, using mathematical analysis and computer simulation, that recognition can function as a powerful heuristic because it exploits correlations between the relative frequencies of events and their adaptively relevant features. For instance, in iterative game theory experiments, people who recognize each other are less likely to defect; and in food choice experiments in rats, recognized foods are rarely avoided as the probable causes of recent food poisoning. However, adherence to a recognition heuristic has some counter-intuitive, negative consequences. One is the "less-is-more effect" which predicts when organisms will make progressively less accurate inferences as their recognition knowledge increases. By carefully analyzing this heuristic and its relation to various environments, we have been able to make and test precise predictions about when and to what degree relying on recognition is adaptive or detrimental.

A NEURAL NETWORK APPROACH FOR THE CLASSIFICATION OF BODY MOVEMENTS

Karl Grammer and Martin Fieder

Ludwig-Boltzmann-Institute for Urban Ethology Althanstrasse 14 A-1090 Vienna/Austria; e-mail: karl.grammer@univie.ac.at

Humans are the only species that do not show any visually observable physical sign of ovulation, and human female sexual cripsis is thought to be complete. In this preliminary approach we tried to find out if female body movements change qualitatively during cycle. In order to accomplish this 150 video clips (duration four seconds each) of a female body movement were analysed with a digital motion energy detection system. The resulting data stream was fed to a time-delayed neural network (TDNN). In addition saliva oestrogen levels were determined by EIA. The results show that it is possible to train the network for the recognition of high oestrogen levels from behaviour data. We will demonstrate the validity of the patterns we found and discuss inhowfar males could be able to learn the connection between ovulation and qualitative changes in female behaviour.

BEHAVIORAL SOCIAL SUPPORT PREDICTS DEPRESSION OUTCOME

William Hale, Jaap Jansen and Netty Bouhuys

Dept. of Psychiatry, Academic Hospital, PO Box 30.001, 9700 RB Groningen, The Netherlands, Fax: +31-50-3696727

The interactions of depressed persons with others play an important role in the etiology of depression. According to Coyne (1976) depressed persons elicit support behaviors intermixed with rejection from others. Brown (1994) found that depression chronicity is related to a lack of social support. We investigated whether observable behaviors (which may reflect support-seeking and giving) in an interaction of depressed patients and others (i.e., partner and stranger) (1) predict depression outcome and (2) discriminate between patient and others interactional patterns.

25 depressed out-patients interacted with their partner and a sex-age matched stranger at hospital intake. Both interactions (patient-partner and patient-stranger) were video-taped for ethological analysis. It was found that (1) less speech of the partner predicted poor depression outcome and that (2) partners displayed less encouragement (yes-nodding and "um-hum"-ing during listening) and speech and more active listening (intense body touching during listening) than strangers. Data suggest that partners behave less supportive than strangers. In particular, low amounts of speech is predictive of unfavorable depression outcome.

"HEIMAT"-ATTACHMENTS AND RETURN TO THE NATIVE PLACE. ON THE BEHAVIORAL BIOLOGY OF MIGRANTS

Elieser G. Hammerstein

D-12105 Berlin, Kurfürstenstr. 97; Tel./Fax: +49-30-705 18 66

"Heimat" is an ambivalent concept; only few languages have a special substantive for it. "Heimat"-attachment develops like habitat imprinting; later emotional bonds to other places are conditioned by material success experiences. Return from emigration is sporadic; homesickness is neither the cause nor the reason for it. Among emigrants from Germany and among turkish "guest laborers" the biographical data of returners and non-returners were compared. The findings were: Regardless of the reasons for their emigration, their legal status, their self-definition and their ethnic affiliation—when migrants are allowed to stay in the new country and then live there in stable partnerships, build themselves a base of existence and have children, who grow up in the new country and continue to live there—they usually do not return to their country of origin. This is congruent with the general postulate that readiness for migration declines as biological goal attainement (i.e. securing livelihood and offspring) increases.

ROUGH-AND-TUMBLE PLAY AS A CONTACT STRATEGY

Gabriele Haug-Schnabel

Forschungsgruppe Verhaltensbiologie des Menschen (FVM), Albertstr. 21a, D-79104 Freiburg, Germany, Fax ++49-761-39579, eMail bensel@ruf.uni-freiburg.de

Long-term observations of behaviour in groups of 3–6 year old children demonstrate that rough-and-tumble play is an important element of behaviour in this age group. The basic conditions for its occurence as well as for its undisturbed course are in keeping with the prerequisites for playing. In a "relaxed field", to be defined by varying factors, the frequency of playful aggression increases, just as the likelihood of a playful reaction to it. The results indicate that rough-and-tumble play in the group serves different purposes: in stabilizing the playgroup, in making contact and in pacification following a serious conflict. Its significance in bridging the difference in order of precedence as well as differences in competency between group members and in order to set up mixed-sex playgroups is discussed.

WHEN DOES SELECTION FAVOR STUPIDITY, SELF-DECEPTION, BIASES?

Mario Heilmann

Dept. of Psychology, UCLA, Los Angeles, Los Angeles, CA 90095, USA, mheilman@ucla.edu, mheilman@a3.com

The increased intelligence of humans has some reproductively detrimental side effects. Special mechanisms are needed to counteract these, by selectively blurring our logic on some specific issues, such as impression management, social norm conformity, and existential conflicts. In spite of overwhelming evidence to the contrary, we tend to steadfastly maintain and project a rosy image of ourselves as honest, rational, and self-aware altruists living in a just society. Some of the mechanisms identified by psychological research are: positive illusions, self-serving and other biases, overconfidence, conformity, impression management, deception, and unconscious pursuit of self interest. Often well-being is positively correlated with these cognitive distortions. Social sanctions hit those who see the hypocrisy of their society, the arbitrariness of its norms, the emptiness of its initiation rites and gods. Galileo and many other creative thinkers were persecuted for espousing an objective truth. Uncritical belief in socially created reality, and unquestioning conformity usually prove to be much safer. Awareness of self deception is the first step towards overcoming it. To keep the system intact, we need to be unaware of our unawareness. Is it dangerous to do research that might upset these defensive mechanisms and illusions?

DATING LIKE IN THE AFRICAN SAVANNAH: AN EVOLUTIONARY PSYCHOLOGICAL CONTRIBUTION TO HUMAN MATING BEHAVIOR

Andreas Hejj

Inst. Psychology, University of Munich, Leopoldstraße 13, D-80802 Munich; Fax: 49-89-21805255; e-mail: Andreas@mip.paed.uni-muenchen.de

The theoretical background of the present paper is based on considerations of evolutionary psychology concerning the adaptations that men and women developed for the selection of an adequate partner who would optimise the reproductive rate of their own genetic information. How individuals of both genders present features that are relevant for their counterparts´ reproductive success on their first dates to enhance their own personal attractivity is traced from antique to present-day literature. Moreover, 136 heterosexual daters aged 20–30 were interviewed. The results are cross-validated by comparing the female cognitive scripts with what males think women would expect on a first date, and vice versa, in order to increase the reliability of the outcome. A similar behaviour pattern (optimizing self-marketing, relevant for reproductive success) emerges in diverse historical periods.

HUMAN AXILLARY ODOURS AND SEXUAL SELECTION: THEORY AND DATA

Klaus Jaffe and Asdrubal Briceño

Depto. Biologia de Organismos, Universidad Simon Bolivar, Apartado 89000, Caracas 1080A, Venezuela

Biodynamica, a complex simulation model of evolution, suggest that sexual selection is a basic component of natural selection and must be present in most sexual organisms, directing natural evolution. Odours are an important source for kin-recognition and sexual selection in a variety of vertebrates. Humans recognize conspecific individuals by their odour. Some of these odours may be involved in human kinship and sexual behaviour, playing a role as aphrodisiac and as kin-recognition signals. On the other hand, human odours differ in their power to attract arthropod parasites and hematophagous insect vectors. Individual variability in attractiveness to mosquitoes for example has been well documented. The present study attempts to characterize the chemical components of human axilary odours, analyzing the variability of the relative proportions of various volatile substances from axilary ethanol washings. Forty individuals were sampled repeatedly and each sample was analyzed with Gas Chromatography and GC-Mass Spectrometry. A multifactorial analysis of the GC spectra allowed to discern odours originating from specific foods and suggest the existence of species- sex- family- and individual-specific characteristics of the chemical blend of human axillary odour. The data suggests that correlations between specific chemical blends of human axillary odours and criteria for reproductive pair formation are feasible and would fulfill the criteria predicted by the theoretical model.

SELF-ESTEEM, FRIENDSHIP, AND VERBAL AND NONVERBAL INTERACTION

Gudberg Jonsson

Groupe de Recherche sur la Parole—Psychologie des Processus Cognitifs—University of Paris VIII and The Human Behavioral Laboratory—University of Iceland—Address (July): Masholar 6, 111 Reykjavik, Iceland; tel. home: (354)-557-43-55; Office: (354)-569-45-85, e-mail: bjalli@mmedia.is

Two studies are presented where the real-time structure and synchronization of verbal and non-verbal behavior are analyzed and related to self-esteem and friendship. Two different dyadic interaction situations have been analysed: a) sixteen 10 min interactions between therapist and drug addict (5 boys, 11 girls; aged 15–17 years, mean 16); b) five 6–10 min dyadic interactions between male students (aged 19–27, mean 22.3). A special software, THEME, was used to detect patterns in the real-time behavior records. The results indicate that communicational behavior is highly synchronized and structured. In research a) we found a strong correlation between self-esteem and number and frequency of behavioral patterns and the total number of recorded events. In research b) no correlation was found between self-esteem and number and frequency of patterns, but the variation in self-esteem scores was very small. Strong correlations were found between friendship (between the interactants) and number, frequency, level and length of patterns. In both studies, particular behavioral patterns were found to belong to a high self-esteem category and others to the friendship category.

COGNITIVE AND PHYSIOLOGICAL RESPONSES OF MEN TO FEMALE PHEROMONES

Astrid Jütte

Ludwig Boltzmann Institute for Urban Ethology, c/o Inst. Human Biology, Althanstrasse 14, A-1090 Vienna, Austria; email: a8111gbx@vm.univie.ac.at; fax: 0043-1-31336-788

The composition of human female vaginal pheromones changes during menstrual cycle. In order to investigate if men are able to perceive these cues and thus ovulation olfactorily and are able to respond in a biological senseful manner, human synthetic vaginal pheromones from three different cycle-days were tested against control (water) on male subjects. Assessments of personal attributes of women's photos and voices and subject's behavioural readinesss were measured by questionnaire. For testosterone-levels saliva was sampled before and after inhalation. Pheromones produced better judgements of female photos and voices than controls. Saliva-testosterone increased following the inhalation of the ovulation-sample but showed nearly no response to other odours. A differentiated model of signalling ovulation is outlined: women do not directly attract male interest by their smell, but provoke physiological reactions which affect male cognitive processing of female attractiveness.

NEW APPROACHES TO HUMAN AGGRESSION

Jan Kamaryt

Inst. of Philosophy, Acad. Science Czech Republic, Jilská 1, 110 00 Praha 1, Czech Republic; Fax: +422/242 202 57

'Aggression' is a polyvalent, equivocal, and elusive concept. Van der Dennen (1980, Problems in the Concepts and Definitions of Aggression, Violence, and Some Related Terms. Groningen, Polemological Institute) has listed 106 more ore less different definitions of aggression and related concepts such as 'agonistic behavior' and 'violence'. Aggression has been variously conceptualized. Controversies reign over many questions, e.g. whether aggression is a unitary construct or a multifactorial one, whether to include or exclude violence, etc. In psychology, Caprara (1994; Aggressive Behavior 20, 291) represents a change from the unitary conception of aggression towards specifying one including tolerance towards violence. A most representative survey concerning main theories of human aggression, violence and the predominant role of threat perception and fear in all kinds and forms of human aggression and violence are presented in many studies and books by van der Dennen (1995, The Origin of War. The Evolution of Male-Coalitional Reproductive Strategy, Groningen, Origin Press; including an immense bibliography of 150 pages length).

MODELLING NATURAL HUMAN MOVEMENT

Guido Kempter

Gerhard-Mercator-Universität Duisburg, Lab. Interaktionsforschung, Lotharstraße 65, D-47057 Duisburg, Tel. 0203/379-2815; fax 0203/379-2423; email hd253ke@rs1-hrz.uni-duisburg.de

Ethological research has shown that the perceptual filters that steer innate releaser mechanismes are particularily sensitive to *movement*-features (Lorenz 1964, Vergleichende Verhaltensforschung. In: Lorenz & Leyhausen. Antriebe tierischen und menschlichen Verhaltens, Piper, München). Likewise, human interaction studies suggest that the gestural and facial movements generated in face-to-face interaction strongly affect the process of person perception—that is, the way in which we form attitudes and opinions about other people (Frey 1996, Prejudice and the Theory of Inferential Communication. In: Eibl-Eibesfeldt & Salter: Indoctrinability, Warfare & Ideology, Berghahn, Oxford). However, due to the difficulties inherent in the simulation of complex movement patterns, most dummy-based investigations into perceptual filter routines have been working with static morphological features only. In an attempt to expand this research to the study of the specific *movement* characteristics that are salient within the process of person perception, a computer-animated system has been developed that permits high-resolution re-construction of natural human movement as well as its systematic variation. The data base for this movement animation is provided by a protocol which resolves—according to the logic of the "Berner System"—complex body movement into 55 dimensions. It is explained how this tool can be used for a systematic investigation into the perceptual mechanisms that govern the process of impression formation among humans.

THE SEMANTIC GESTURES IN THE CZECH POPULATION

Zdenek Klein

Psychiatric Center, 181 03 Prague 8, Czech Republic

The importance of nonverbal behavior rapidly increases in the cases of disturbances of verbal communication. We can observe such disturbances in pathological conditions (e.g. psychosis) but also in so called normal interaction. (e.g. transcultural differences). Our contribution is the first attempt to make an inventory of semantic gestures most frequently used in the Czech population. Schematic drawings of 143 gestures were presented to males (N=804) and females (N=916) and their interpretations of meaning were recorded. The pictures of gestures were presented in collections containing 20 drawings. Average age of males was 24.3 years (min. 16, max. 65) and females 24 years (17–59). The subjects were students of various Prague colleges, their friends and relatives. Conditions for being included in the samples were the following: 1. Czech origin (also forefathers), 2. absence of any kind of psychic disturbances. Each of the presented gestures was evaluated by 100 males and by 100 females at least. This collection—an ATLAS OF SEMANTIC GESTURES—can serve for the diagnosis of pathological interpretation of gestures and for transcultural comparisons.

QUASI-AGGRESSION IN APES AND THE ORIGINS OF HUMOR

Alexander G. Kozintsev[1] and Marina L. Butovskaya[2]

[1]Museum of Anthropology and Ethnography, Saint-Petersburg 199034, RUSSIA, e-mail: sasha@kozintsev.spb.su
[2]Inst. of Ethnol. & Anthropol., Leninsky prosp. 32-A, Moscow 117334, Russia, Fax:7-095-9380600, e-mail: recon@iea.msk.su

Cases of quasi-aggression (throwing feces, spiting) directed by apes (chimpanzees and orangutans) towards humans are investigated and considered as a possible analogy to rudimentary forms of human humor. Apes practiced quasi-aggression only at people who might be expected to react in a comparatively peaceful manner, bu not to whom they were afraid of. When the target was hit, the animals expressed joy by making a play face, hooting, clapping and stamping around. Such mock aggression is strikingly similar to certain patterns of human behavior used in archaic rituals of reversal. Throwing excrements and pouring urine at people was a prominent part of Western European medieval carnivals and featured very largely in Renaissance comic art. It has survived in Russian agricultural rites and the ritual clowning practices of North American Indians. Ritualized but no less laughter-provoking derivatives are even more common (demonstration of buttocks). Spitting may have been a prototype of more civilized forms of hostile humor like tongue-showing. Our findings support the idea that some important component of humor is rooted in the prehuman stage.

BEHAVIOURAL PATTERNS IN STRANGER'S TALK

Kirsten B. Kruck

Research Centre for Human Ethology in the Max-Planck-Society, D-82346 Andechs

Synchronization is necessary for pairbonding. This most complex form of interpersonal coordination, where multiple action units are geared together, was explored in an experimental setting with 48 mixed- and 12 same-sex dyads. Nonverbal behaviour during the initial 10 minutes of first-ever encounters was videotaped. To visualize the syntactical real-time pattern, a special research method (THEME) was used. Behavioural analysis with THEME allows the analyst to focus quantitatively on rhythms initially perceived to exist in the behavioural flow. THEME's results were correlated with self-reports. Synchronization was found not only in mixed-sex courting dyads, but in all cases. Neither its speed nor appearance was influenced by intensity of interest. However, form varied with female interest. When the female was more interested in the male, patterns became longer as more types of units appeared and existing types were repeated; also the hierarchical organization of interpersonal coordination became more complex. This study indicates that it is female choice that initiates pairbonding.

ETHOLOGICAL ANALYSIS OF SOCIAL INTERACTION IN TWIN CHILDREN

Colette M. Lay, Nancy Segal,[1] and Diane L. Christian

[1]California State University, Department of Psychology, Fullerton, California 92634, USA, fax: 714-449-7134, NSEGAL@FULLERTON.EDU

The Nearest Neighbor Technique was developed for studying social and spatial affinities of sex-age classes within hamadryas baboon troops (Kummer, 1968). This technique has been successfully adapted to analyses of play preferences among young school children. This method affords a viable tool for comparing these behaviors between monozygotic (MZ) and dizygotic (DZ) twin pairs. An ethological twin study of genetic influence on social interaction and spatial proximity was conducted. Thirteen MZ twin pairs were compared with 13 DZ twin pairs with reference to several categories of social closeness. It was found that MZ twins maintained closer proximity and engaged in more frequent interaction with their co-twin than DZ twins. Coordination of behavior ("matching") was also analyzed using several categories of matching behavior. MZ twins showed greater coordination in social relatedness than DZ twins. These findings are consistent with evolutionary-based studies which find that social affiliation may vary as a function of genetic relatedness.

ON THE BIOLOGY OF REASSURANCE

Antje Letzer

Forschungsgruppe Verhaltensbiologie des Menschen, Universität, Biologie 1, Albertstraße 21a, 79104 Freiburg, Germany; fax +49-761/39579

Since John Bowlby (1958) and Mary D. Salter Ainsworth (1969) perceived and described with their attachment theory the nature of the mother-child bonding as an extensive phylogenetically evolved protective program, the child's reassurance from his mother and the mother's function as a "secure base" have become generally accepted realities. Against this background the visual contact behavior of 60 children aged 10 to 24 months toward their caregiver was observed at a public playground in 30-minute focal samplings. The assignment of their visual behavior into five looking categories was only possible by taking into account the playing situation, not just by measuring such looking parameters as length or frequency of looking. No continued increase of the frequency of reassurance could be found with increasing self chosen distance to the caregiver. Pauses in the contact to the caregiver were significantly longer before looks of reassurance than before other kinds of looks. The results suggest that a low-level activated requirement for contact can be satisfied not only with looks of reassurance, but also with the other four looking categories.

ONTOGENETIC DEVELOPMENT OF PRESPEECH VOCALISATIONS

Katrin Lind and Kathleen Wermke

Department of Human Ethology and Chronobiology, Institute of Anthropology, Humboldt University (Charite), Berlin, Germany; Fax: +49-30-2802 6448, email: wermke@rz.charite.hu-berlin.de

Recordings of vocalisations from one child were made daily during the first year of life. Spectral and melody analyses were carried out with a Digital Speech Lab CSL (Kay Elemetrics Corp.). Special acoustic features of exspiratory cry vocalisations were selected for the description of ontogenetic changes in prespeech vocalisations. The signal parameters which occupy a key position in cries as well as in non-cry vocalisation seem to be fundamental frequency (F_0) and its variations in time (melody). Additionally we computed the mean F_0, maxima and minima values of F_0 and its range. For the description of micro variability we used "jitter" and several perturbation quotients. The intraindividual variability in sound features of infants is known to be very high. Our longitudinal study with daily recordings demonstrates this variability and its ontogenetic changes. Results of the analyses of vocalisations of the first eight weeks are presented. We describe the cry development with a high time resolution and discuss the appearance of so-called regression-phases.

CONFLICT RESOLUTION IN PRESCHOOL CHILDREN

Tomas Ljungberg, Anna J. Lindqvist Forsberg, and Karolina Westlund

Div. of Ethology, Dept. of Zoology, University of Stockholm, S-106 91, Stockholm, Sweden, FAX: +46-8-16 77 15; e-mail Tomas.Ljungberg@zoologi.su.se

Behaviours used in aggressive interactions between children have previously been described in great detail. However, much less is known about how children resolve conflicts and post-conflict behaviours. We have videotaped preschool children (3–5 years of age) at three different daycare centers in Stockholm by means of hidden videocameras during free play. More than 400 conflicts have been analysed in over 50 hours of observation. In the post-conflict period, we have found the occurence of several types of friendly or affiliative behaviours between former opponents—body contact, invitation to play or cooperate, giving objects or symbolic advantages or verbal apology being most common. More than half of the conflicts were followed by such behaviours. A number of non-friendly behaviours were also observed, most commonly revenge, redirected aggression and displacement activities. These behaviours occurred more often when friendly behaviours were not shown. Studies of conflict resolution in primates have shown that affiliative behaviours are used to reconcile opponents after aggressive interactions. Our interpretation is, that the friendly behaviours observed in the period after conflicts in preschool children also have such a reconciliatory role.

DEVELOPMENT OF SLEEP-WAKE AND FEEDING RHYTHM IN GERMAN INFANTS

Brita Löhr and Renate Siegmund

Dept of Human Ethology and Chronobiology, Inst. of Anthropol., Humboldt University (Charité), D-10098 Berlin, Germany

The aim was to investigate the different stages during the development of a diurnal pattern and its dependence on social influences. The behaviour of twenty-one fullterm and healthy infants during the first six months of life were recorded using parent diaries. Data of total time, number and length of sleep phases, wakefulness phases and feedings were calculated per day and in relation to day- and night-time. We found a change from a monomodal to a bimodal distribution of the length of sleep phases and a time stabilisation of sleep phases and feeding acts in relation to time of day. Interindividual differences were marked in the values of all parameters and in the time it takes to develop a diurnal pattern. It seems that first-born children develop longer sleep phases in a shorter time and show more free-running patterns than second-born children. It is worth to discuss the differences in the actual realisation of a self-demand feeding and sleeping and their effects on the shaping of the infant rhythm.

BEHAVIORAL REAL-TIME SYMMETRY, THE T-PATTERN FAMILY AND THEME SEARCH

Magnus S. Magnusson

Human Behavior Lab., Univ. of Iceland, IS-101 Reykjavik.
http://www.rhi.hi.is/~msm/behavior.html; Fax: (354) 5625219.

For an eventual structural and functional "grammar" of behavior many types of *hidden* patterns must be considered. Human behavior is here viewed as the performance of members of the proposed *T-pattern* family. These are *particular types* of repeated hierarchical/syntactic intra- and interindividual real-time patterns describing related aspects of the temporal organization of verbal and/or nonverbal behavior. Related terms are: a) "markers"; T-pattern components with particular pre-, post-, and/or "bi"-dictive powers b) "packets"; simultaneously sequential and non-sequential structures and c) pattern alternation groups describe behavioral streams as a set of alternating T-pattern family members. Examples from empirical human interaction studies are presented.

Factors making such patterns hidden to the naked eye are illustrated and a computer algorithm is described for pattern detection essentially through 1) "wild pattern growth" based on the detection of *critical interval* relations between pairs of time point series 2) pattern competition and 3) pattern selection. The corresponding theme software provides in addition multimedia techniques for data collection and interactive inspection of detected patterns in digitized video (avi).

CULTURE AS COURTSHIP: DARWINIAN DEMOGRAPHICS OF PUBLIC CULTURAL DISPLAYS

Geoffrey F. Miller

Center for Adaptive Behavior and Cognition, Max Planck Institute for Psychological Research, 24 Leopoldstr., Munich 80802, Germany (+49) (089) 342473 (fax); miller@mpipf-muenchen.mpg.de

The diverse forms of human culture (e.g. language, art, music) evolved largely through sexual selection as courtship displays. This hypothesis suggests strong sexual dimorphism and distinctive life-history patterns in rates of cultural production. I gathered data on age and sex of producer for over 15 000 works of human culture randomly sampled from literary, art, and music references sources, including over 2800 books, 3200 modern paintings, 2700 rock albums, and 2400 jazz albums. In each form of cultural display, males produced at vastly higher rates than females, and male productivity peaked in young adulthood, matching other peaks of sexual competitiveness (e.g. muscular strength, status-seeking, homicide rates). Also, like bird song, human cultural display functions as a reliable indicator of many heritable traits, shows clear reproductive advantages, and decreases rapidly after sexual pair-bonding. These data contradict other theories that culture evolved mainly to promote social learning, tribal solidarity, or sexual oppression. Cultural production is mostly sexual display by young males.

SYMMETRY, MATE SELECTION, GOOD GESTALT AND THE FRONTAL ANIMAL SCHEME

Ricarda Müssig

Neustadterstr. 7, D-71687 Karlsruhe, Germany

What do symmetry and mate selection have to do with each other? Besides the reasons recently discussed, the visual signal of the symmetric Frontal Animal Scheme (drawn by children as head-legs scheme) also plays a role. In phylogeny, it developed interdependently with the evolution of bilaterally symmetric animals with sufficiently good eyes. Its earliest function was to signal a predator, prey, rival or mate approaching. Only in primates with clinging youngs did it become a mother scheme and baby scheme. (bilateral symmetry with a vertical symmetry axis is favoured even by small babies.) The frontal mother (and baby) scheme, modulated by the perception of people around, explains why people with perfect symmetry are thought not only to be beautiful, healthy, and not mutilated, but also to be morally good and suitable mates. The frontal animal scheme is also the archetype of the good gestalt; the same rules are valid for both. Here too we find the connection between symmetry, beauty and good.

ORIGINS OF EMOTIONS

Emese Nagy and P. Molnar

Institute of Behavior Science, Semmelweis Medical University, Budapest, Nagyvarad sq. 1089, Hungary; Fax: 3612102955 email:nagyeme@net.sote.hu

The origins, i.e. phylogenesis and ontogenesis, as well as the mechanisms of basic emotions (happiness, fear, disgust, surprise, sadness, and anger) remain an unsolved and challenging problem for all disciplines concerned. Our complex approach (psychophysiological, ethological and developmental) revealed new aspects for an interdisciplinary integration. Newborns' invariant capacity to imitate all of facial expressions of emotions proves their inborn origin. Some emotional facial expressions have specific psychophysiological correlates from birth onward, providing a basis of our inborn interactive and empathic capacity. Most of them (smile, disgust, fear in our studies) designate a class of characteristic subtypes continuously developing during ontogenesis. We try to find and differentiate the root of every "tree" and to model their own characteristic mechanisms, referring to their phylo- and ontogenetic history and development.

FEMALE COURTSHIP BEHAVIOR IN SINGLES´ BARS

Barbara Niedner

Forschungsstelle für Humanethologie in der Max-Planck-Gesellschaft, Von-der-Tann-Str. 3-5, 82346 Andechs, Germany, Fax: 0049/8152/37370

When a woman wants to meet a stranger of the opposite sex for the first time in a singles´ bar, she is in a difficult high-risk situation. Neither person knows the intention of the other, and consequently non-verbal signalling becomes the major channel of communication. Because of their higher biological risks, females should prefer less obvious tactics in order to communicate interest in a potential partner. In this study I examine 300 randomly selected women (ages 18–35 years) who came alone or accompanied by a girl friend in a singles´ bar. The women were observed for 30 minutes starting from their entrance. The analysis centered on the following questions: Do females show self-presentation by entering the bar? In what way do single and accompanied females differ? How does the behavior of a woman change when a man passes by? Which tactics do women use to initiate verbal contact with men? The results suggest that the observation of women in field situations may provide clues to criteria used by females in initial selection of male partners.

GROUP EGOISM, GROUP SELECTION, AND NATIONALISM

Andrew Oldenquist

176 Walhalla Rd., Columbus OH 43202, fax: 614 784 9606; e-mail: aoldenqu@magnus.acs.ohio-state.edu

This is a conceptual/theoretical paper inviting relevant research. It is hypothesized and argued that humans are innately social, which includes the bases for language, pair bonding, and genetic predisposition to be socialized. Innate sociality in turn creates two underpinnings of nationalism and loyalty: social identities, necessary for a fully developed idea of the self, and group egoism. Group egoistic values, such as the value I attach to my family, my country, my company, and so on, are (a) like egoism, focussed on a particular thing and not on a kind of thing; (b) unlike egoism, social, creating the idea of our country, etc.; and (c) non-instrumental. Without group egoism, loyalties and the two main types of nationalism, assimilative and separatist, are impossible. Finally, the question is raised in the light of suggestions of Boyd and Richerson, whether loyalties and nationalism, as types of altruism, are best explained by group selection.

AUTISM: INSIDE AND OUT

John Richer

Paediatric Psychology, John Radcliffe Hospital, Oxford OX3 9DU, UK, Fax: Int+ 44 1491 826472.

A defining feature of autistic people is their difficulty in communication. Their subjective experience has therefore been glimpsed only darkly. Recently a few people who were autistic have recovered sufficiently to have written accounts of their experiences. The experience they describe is remarkably similar to what would be expected from an ethologically influenced account of autistic behaviour. The translation to and from these two views, the inside and the outside, supports the ethologically influenced description of autistic behaviour. It reminds us how ethologists try and enter the minds of the individuals they observe, and it provides insights into the relationship between subjective experience and publically observable behaviour: an essential aspect of intersubjectivity.

PHYSICAL CORRELATES OF FEMALE BEAUTY

Gudrun-Ingrid Ronzal

LBI for Urban Ethology, c/o Inst. For Human Biology, Althanstr. 14, A-1090 Vienna; email: a8750657@unet.univie.ac.at; Fax: 0043-1-31336-788

As there is still a lot of considerable controversy around attempts to identify the parameters that actually cause a woman to be judged attractive or unattractive, this study tried to find some universally valid single features of facial and bodily attractiveness. Common characters as big eyes, lips and a small nose for example, as well as anthropological indices, like androgynscore or body mass index were tested. Further interest concerned the question, whether attractivity of faces or bodies is more important in the general impression, and if the results are applicable to different cultures. Pictures of hundred different american as well as japanese females (in seven views each), photographed by Gomi Akira were used for this study. All were scanned and several predefined points were measured from every view. Any distance between points and areas was calculated out of the measured coordinates. The digitized pictures were presented to male subjects by an interactive computer-program, and rated for attractiveness.

FAMILY RESEMBLANCE AND MOTHER'S FACIAL BEAUTY

Frank K. Salter and Kirsten B. Kruck

Max Planck Research Center for Human Ethology, D-82346 Andechs, Germany

Salter (1996, in press, Ethology & Sociobiology) hypothesizes that female attractiveness partly consists of "carrier features". These are phenotypic features indicating a marked tendency to transmit the biological father's features to offspring. Carrier features were hypothesized to be ultimately attractive because they increase paternity confidence. We tested the hypothesis in the facial channel in a rating study of close-up photographs of a racially homogeneous sample of 89 families (fathers, mothers, and children photographed separately). Colouring was suppressed by using photocopies. Raters of mixed ages, sex and occupation rated parents' faces for several qualities including attractiveness. A separate group of raters judged various intra-family resemblances in a range of facial features. They also rated paternity. Results replicate previous studies that find high interspouse correlation in attractiveness. There was a weak but highly significant correlation between mothers' facial attractiveness and paternal resemblance. Furthermore, paternal resemblance increased in families where the father was relatively unattractive compared to the mother, as predicted. Control studies were inconclusive. Methodological improvements are suggested.

SEXUAL BEHAVIOURS IN MELANESIAN SOCIETIES—EVOLUTIONARY MODELS?

Wulf Schiefenhoevel

Forschungsstelle für Humanethologie in der Max-Planck-Gesellschaft, Von-der-Tann-Str. 3-5, D-82346 Andechs, Germany, FAX:0049/8152/37370

How are concepts of sexuality and reproduction shaped in traditional societies? Do their members know about biological paternity, do they live in promiscuous groups, do they "fall in love", how do they sleep with each other? The scientific literature on these topics often builds on wrong assumptions. Beyond voyeurism and seeking sensation or the sexual paradise, the sexual systems of traditional cultures deserve study (despite considerable methodological problems), especially as they may serve as models to understand human sexuality in general. Here, sexual concepts and practices of the two different types of populations in Melanesia, that of the archaic Eipo in the Highlands of West-New Guinea and that of the Trobriand Islanders, are presented. Whereas there are marked differences with regard to the role of women and men, sexual antagonism and preferences for place, time and position for sexual intercourse, there are also universal features. One of them is that members of traditional societies, too, fall in love and think in terms of romantic eroticism. Examples will be given and it will be examined whether sexuality in these societies can serve as models for the sexual past of our species.

SEGMENTATION IN BEHAVIOUR AND ITS CONNECTION WITH BRAIN FUNCTIONS

Margret Schleidt[1] and Jenny Kien[2]

[1]Forschungsstelle Humanethologie in der Max-Planck Gesellschaft, D-82346 Andechs
[2]Jerusalem

It can be shown (with the method of film analyses) that natural human behaviour is segmented into action units—that is, functionally related groups of movements—with durations of a few seconds (Schleidt 1992, Universeller Zeittakt im Wahrnehmen, Erleben und Verhalten. Spektrum der Wissenschaft 12: 111). This phenomenon can also be shown to exist in non-human primates and in other mammals. In humans, a similar segmentation can be found in planning and preparatory behaviour, as well as in speech and perception. This widespread temporal segmentation may be related to the functioning of short term memory and may thus be a central feature of neuronal integration. In the literature, segment length was thought to be determined by either capacity constraints or temporal factors. In contrast, we show that segment length depends on the interplay between capacity and temporal factors (Schleidt and Kien: Segmentation in behaviour and what it can tell us about brain functions. Human Nature submitted).

TONIC COMMUNICATION: A RE-VIEW OF HUMAN NONVERBAL BEHAVIOR

Wolfgang M. Schleidt

Robert-Hamerling-Gasse 1/22, A-1150 Wien, Austria (Universität Wien, retired)

Tonic communication is a specific form of communication that is characterized by more or less frequent repetitions of the same kind of signal, maintaining a link between sender and receiver (Schleidt 1973). This update of my paper emphasizes preverbal, nonverbal and metaverbal features of human communication, and proposes to distinguish between different kinds of communication: communion, manipulating, displaying, and ordering, as they relate to different kinds of bonds among communicants, and explores their evolution. Tonic communication is discussed here as an example of an originally nonverbal form of human communication that, using the stream of words as a carrier, has further evolved within the verbal domain, into what Malinowski (1923) has termed "phatic communication": a special case of human communication "in which ties of union are created by a mere exchange of words."

RECONCILIATION AMONG KINDERGARTEN CHILDREN

Alain Schmitt

Ludwig-Boltzmann-Institut für Stadtethologie, c/o Humanbiologie, Universität Wien, Althanstr.14, A-1090 Vienna, Austria, Fax: (+43-1) 31 336-788; alain.schmitt@univie.ac.at

Although reconciliation is very important e.g. in religious systems, peace educational programs and daily life, there are only few empirical data at hand. These say that children as young as two years make up spontaneously and that reconciliation ends in playing together. But many questions remain unanswered. Particularly, making up has never been linked to social structure and conflict intensity. This was done in an observational study of kindergarteners. Among 21 4–6 year olds and for 234 conflicts, 30% of which were reconciled, there was no covariation between presence of conciliatory gestures and social status (sociometric power, popularity, rejectedness; observed regard) or relationship status (sociometric friendship/enmity, observed familiarity). In contrast, likelihood of reconciliation was influenced by factors directly related to the conflict. It was highest if children played together before the conflict, if it ended in compromise and if it was hot tempered and long-lasting. Conciliatory gestures were almost always accepted by addressees (91.4%). Reconciliation relieved the distress originating from the conflict and enhanced post-conflict proximity and play.

TIME PATTERN IN ACTIVITY BEHAVIOUR OF TROBRIAND ISLANDERS

Renate Siegmund,[1] Matthias Tittel,[1] and Wulf Schiefenhoevel[2]

[1]Dept. of Human Ethology and Chronobiology, Institute of Anthropology, Humboldt University (Charité), Berlin, Germany, ax: +49-30-2802 6448; email: rsiegmd@rz.charite.hu-berlin.de
[2]Research Unit of Human Ethology in the Max-Planck Society, Andechs, Germany

The inhabitants of the village of Tauwema (Trobriand Islands, Papua New Guinea) are one of few traditional living societies where people are still living uninfluenced by technical civilisation of the modern world. 39 inhabitants from 7 families were included in this study. Activity patterns were registered continuously with microelectronic actometers over 7 days and analysed under the aspect of sleep duration and synchronisation in activity among family members. Comparison of the onset-times of the main activity period indicate the existence of a strong social synchronisation among the inhabitants in the morning. The mean sleep and rest duration of the younger infants was between 9 and 12 h per day (n=4) with age-dependent differences. That of the adults averaged at 8.4 ± 1.0 h per day. Sleep duration in married couples was gender-specific. Wives slept 58 min ± 20.6 min per day longer than their husbands (n=7).

THE INFLUENCE OF EMOTIONS ON THE PERSISTENCE OF CONSCIOUS PERCEPTUAL REPRESENTATIONS

Kurt Solokowski

Psychologie im F6 3, Bergische Universität, Gesamthochschule Wuppertal, Gaußstr. 20, D-42097 Wuppertal

Aroused emotions (re)configurate cognitive and physiological processes and serve to optimize the preparation of adaptive actions. This study focusses on the influence of emotions on the persistence of conscious perceptual representations during the presentation of an ambiguous figure, the Necker cube. The independent variable was the induced mood after viewing slides with either sad, happy or neutral looking human faces. After the mood induction, the participants viewed the Necker cube in a balanced repeated-measures design under two instructions: to „hold" the actual perspective as long as possible and to „change" it as fast as possible. The reversal-times were measured. After viewing happy faces in the „change" condition the mean time between reversals was 2.5 s and after viewing sad faces it was 3 s. In the „hold" condition after viewing glad faces the mean time beween reversals was 5 sec and after viewing sad faces it was 6 sec. The results are discussed in terms of the adaptive value of conscious representations for the sequencing of actions and their voluntary controlability.

HORMONAL DEVELOPMENT AND PHEROMONAL PERCEPTION IN MAN

Franca Ligabue Stricker[1] and Brunetto Chiarelli[2]

[1]Dip. Biologia Animale, via Accademia. Albertina 17, 10123 Torino—Italy, FAX + 39 11 -812 45 61; E-mail: stricker@inrete.alpcom.it
[2]Istituto di Antropologia, via Del Proconsolo 12, 50100 Firenze—Italy

Studies on olfactory perception of pheromone-like substances were conducted on 2800 subjects aged 3–22 to find connections between hormonal and olfactory development, thus demonstrating pheromonal phenomena also in humans. The results showed that the perception of biologically meaningful odours begins with a phase of very high olfactory neutrality, coinciding with neutrality both sexual and in personal smelling. Then it gradually increases from childhood through puberty up to adult age. Since puberty and gonadotropin development proceed at the same pace, these results confirmed the strong interactions between olfactory, endocrine and reproductive systems (which forms the basis of pheromonal communication in animals). This theory was further confirmed by the high percentage of anosmic responses to pheromonal substances among adults with abnormal variations in hematic concentrations of gonadotropins and sexual hormones.

ENVIRONMENTAL INFLUENCE ON SOCIALITY BETWEEN STRANGERS

Christa Sütterlin and Kirsten B. Kruck

Forschungsstelle fuer Humanethologie, D-82346 Andechs, Germany

Preference for savanna-like parks displaying depth of field and a rich diversity of plant life (phytophilia) may be an evolved, innate preference for an environment of evolutionary adaptedness. However, cultural evolution has placed humans in a man-made environment that is often maladaptive: urban settings increase symptoms of stress and social isolation. To measure the influence of environment on first-encounter sociality, we arranged two experimental settings: "New York", an urban-style interior design; and "Park", a floral interior depicting sweeping lawns and a pond. Interactions of undergraduate mixed- and same-sex dyads were videotaped through a one-way mirror for six minutes each. Subjects reported their frames of mind before and after this period, and their feelings for their partner and the environment at the end. Interactions in "New York" contained more expressive gestures and automanipulation than in "Park". Smile, laughter, and longer conversations were more prevalent in "Park". However, good company proved more conducive to these behaviours than environment. In good company any room becomes friendlier.

HUMAN LANDSCAPE PREFERENCES

Erich Synek

LBI for Urban Ethology, c/o Inst. for Human Biology, Althanstr. 14, A-1090 Vienna; email: a8825147@unet.univie.ac.at

The choice of habitat has a strong influence on survival. Thus the involved mechanisms have been under strong selection during human evolution. Humans lived the most time of their evolutionary past in the savannas of tropical Afrika. Therefore our aesthetic answers to landscapes should be influenced by environmental savanna-like key-features. The "savanna-hypothesis" was tested in an experiment under the assumption that the less experience an individual has with the environment it is actually living in, the more it should prefer savanna-like landscapes. Earlier studies used pictures or slides of natural environments, with the disadvantages of different daylight, weather situation and position of the camera. We avoided these by generating virtual landscapes, which only differ in three features: hilliness, treedensity and ground vegetation. Subjects were 11 and 17 year old pupils (before and after puberty), the landscapes presented as slides. Every subject rated three out of 18 slides on how much they would like to "live in" or "visit" them.

SOCIAL AND INDIVIDUAL BEHAVIOR OF CREW MEMBERS IN CONFINEMENT EXPERIMENTS

Carole Tafforin and Raymond Campan

Laboratoire d'Ethologie et Psychologie Animale—Universite Paul Sabatier, 118 route de Narbonne—31062 Toulouse Cedex—France, Fax : (33) 61 55 61 54 ; E.mail : tafforin@cict.fr

The aim of the study was to emphasize long term behavioral adaptation of small human groups (6, 4 and 3 subjects) living and working together in confined environment (i.e. close proximity, reduced space and closed habitat) over a long time (1, 2 and 4 months). Such conditions were simulated in three experimens conducted by the European Space Agency: ISEMSI '90 (Isolation Study of European Manned Space Infrastructure), EX-EMSI '92 (EXperimental campaign for European Manned Space Infrastructure) and HUBES '94 (HUman Behavior in Extended Space flight). From video recordings performed during these experiments, our ethological method consisted in the observation, description and quantification of the crew members' motor interactions and of the crew's spatial organization during working tasks and daily life activities. Data are correlated with psychological (e.g. communication, stress state, mood profile etc.) and physiological (e.g. hormonal, cardiovascular, anthropometric etc.) parameters measured in other investigations on the same crew. They are discussed for application to space crews (orbital station) and for further investigations on Antarctic teams (polar station).

„MORPHING" AS A RESEARCH TOOL TO STUDY THE PROCESS OF EMOTION RECOGNITION

Harald G. Walbott

Institut für Psychologie, Universität Salzburg, Hellbrunnerstraße 34, A-5020 Salzburg, Tel.: (0043) 0662-8044-5132 or -5104; Fax: (0043) 0662-8044-5126; harald.walbott@sbg.ac.at

Much research is available on recognition of emotion from facial expression, mostly concerned with the accuracy of emotion recognition, but not with the underlying processes. Here it is asked whether emotion recognition is a categorical process or a linear process. One way to study that is to use computer morphing techniques available now to manipulate photographic stimuli depicting emotional facial expression. When a facial expression of one person is morphed into another facial expression of the same person (cf. anger and fear) emotion judgements of the different morphing steps may be used to study categoriality vs. linearity of judgements. Results of two studies are reported in which judges were confronted with four morphing series (four different persons, emotions anger and fear, ten intermediate morphing steps each). Results indicate that judgements of anger and fear cannot be adequately described by linear models, but that „jumps" in the distribution of judgements indicate categorical processes. In the second study this result was replicated and reaction times of judges (by presenting the stimuli and recording the judgements via PC) in addition were analysed. Here, results indicate significantly shorter reaction times of judges the less ambiguous stimuli are with respect to the expressed emotions, i.e. at the ends of each morphing sequence.

SEX APPEAL: A USEFUL STRATEGY IN SOCIAL LIFE OF BARBARY MACAQUE FEMALES?

Bernard Wallner and John Dittami

Institute of Zoology, Althanstr. 14, A-1090 Vienna, AUSTRIA

A lot of primate species show secondary sex characteristics during infertile cycle phases. Yerkes indicated in 1943 (*Chimpanzees: A laboratory colony.* New Haven, Yale University Press) that swollen chimpanzee females gain a lot of privileges in the social group. The present investigation was made on a semi-free ranging group of barbary macaque females in Affenberg Salem (Germany) during non mating season. The physiological and behavioral influence of attractiveness was measured on 24 focus animals living in a multi male-female society. The dimension of infertile, postestrous anogenital swelling was used to define the parameter attraction. To investigate the effects of these swellings females were categorised as having a small, medium or large development of this character. The results show that swellings were independent of the dominance hierarchy in the social group. The behavioral data demonstrate that large swellings are mainly associated with grooming contact and spatial cohesion with males and a high degree of male social intervention. Attractive females are less targets of male aggression. Endocrinologically the fecal estrogen/progestin ratio reflected anogenital development. Higher estrogen components correlated with increased swelling. Adrenal activity, measured as corticosteroid output was lower in the „large" females. In general the results demonstrate the positive social and physiological effects of swellings and indicates attractiveness as a substrategy additional to dominance which allow one to speculate about their function during non-mating periods.

SOME CORRELATES OF MARITAL SATISFACTION

Carol C. Weisfeld[1] And Glenn Weisfeld[2]

[1]Dept. of Psychology, University of Detroit Mercy, 8200 W. Outer Dr., Detroit, MI 48219 USA, fax 1-312-993-6397
[2]Wayne State University, Detroit

By analogy to animals, the human pair bond probably evolved to facilitate biparentalism and reproductive success generally. Married couples are expected to evaluate their mates on qualities like those for which they were chosen, namely potential as progenitors and parents. Our data on US marriages, using the Russell-Wells questionnaire, can be interpreted in this light. Marital satisfaction was related to viewing the spouse as a good parent. Where free choice is reduced, marriages are less happy: where the bride was pregnant and, in our Turkish sample, in arranged marriages. The pair bond is threatened more by female infidelity than male; men in the four cultures studied (US, UK, British Pakistani, Turkish) expected the wife to tolerate infidelity more than women expected the husband to. Also, men felt more possessive than wives (US data only). In all four cultures, husbands made more of the important decisions, and (UK and US data only) this arrangement enhanced the wife's satisfaction even more than the husband's. His higher income (US only) and her greater attractiveness (US, UK) were also positive factors. Other correlates of marital success discussed include fecundity, homogamy, dominant nonverbal behavior by husband, sexual satisfaction, and wife's economic dependence on husband.

EVOLUTION STRATEGIES OF COMPLEXITY: DEVELOPMENT OF PRESPEECH SOUNDS

Kathleen Wermke[1] and Werner Mende[2]

[1]Dept of Human Ethology and Chronobiology, Institute of Anthropology, Humboldt University (Charite), Berlin, Germany; Fax: +49-30-2802 6448; email: wermke@rz.charite.hu-berlin.de
[2]Freie Universitaet Berlin, Faculty of Geophysics, Germany

Our longitudinal study in singletons revealed that the phases of prespeech development are good examples for the evolutionary principles of modular composition of complexity and the principles of repetition and specialisation of modules (building blocks). With the help of these principles, the ontogenesis of complex prespeech voice features becomes understandable, and the existence of innate universal developmental programs—acting from the very first cries until the appearance of speech sounds—can be traced. We have also been testing these hypothesised universals for prespeech development in twins. Here our preliminary results revealed a surprisingly high parallelism in identical twins concerning the onset of some voice characteristics, especially the fundamental frequency and its variation in time and vibrato episodes. Changes of voice (cry) characteristics of infants during the first year of life were investigated by signal analysis methods, using spectrograms, fundamental frequency (F_0) determination and several indicators of microvariability of F_0.

MOTHER-INFANT ACQUISITION OF SPEECH COMMUNICATION

Jeanette M. Van der Stelt

Inst. of Phonetic Sciences, University of Amsterdam, Herengracht 338, 1016 CG Amsterdam, The Netherlands, Tel. +31-(0)20-525 2183; Fax. +31-(0)20-5252197; email: j.vanderstelt@fon.let.uva.nl; Internet: http://fonsg3.let.uva.nl

Aim: Speech communicative problems in two year olds are often thought to originate from deviant earlier mother-infant interaction, especially when physical or mental causes for the problems are absent. This study attempts to find a basis for this idea. Method: Split-screen monthly videorecordings of two normal mother-girl pairs were made in naturalistic home settings during two years. Their vocal and behavioural development is described by means of a multi-channel ethological method. Mother and infant are seen as a unique sensori-motor system exchanging movements which selectively acquire communicative meanings. Three characteristics of human communicative systems are studied: visual and auditive intersubjective tuning, transmission of intentions, and vocal turn-taking by the mother upon speech motor landmarks of the infant. Result: The two pairs differed in their interaction styles during the first six months already, explaining further developmental processes. At the age of two, one girl started a speech therapy period of about three years, the other girl was a precocious talker. Discussion: Evaluation of interactions at risk is suggested and discussed.

SOME LINKS BETWEEN COGNITION, SITUATION AND EVOLUTION

J.P. van de Sande

Rijks Universiteit Groningen, Psychologisch Instituut Heymans, Grote Kruisstraat 2/1, 9712 TS Groningen, Netherlands, Fax: 050-3636304, E-mail: J.P.van de Sande@PPSW.RUG.NL

Human cognition, although presently by far the most popular subject in social psychology, in most cases has only weak links with human behaviour. In order to understand this galling inconsequence, many psychologists exert themselves to make better and better models of human cognition. In this paper I will try to show that by looking at cognition as an evolved art, some reasons for this discrepancy can easily be seen: Evolution is for a large part adaptation to the situation, so we should ask ourselves what function cognition has in this sense. An important, but relatively neglected aspect of our cognitive functioning, the perception of situations, will then be discussed. I will present a method to study situation perception together with some results. These results will then be discussed in their relation to social psychological research on the one hand, and to sociobiological insights on the other.

SOCIAL BEHAVIORS OF YOUNG CHILDREN IN A WADING POOL

J.P. Vieille and Jean-Louis Millot

Laboratoire de Psychophysiologie, U.F.R. Sciences et Techniques, route de Gray, 25030 BESANÇON, France

We studied the ontogenesis of social behaviors and motor and posture skills. 56 children (7 to 30 months, 28 girls) were filmed during free group activity in a wading pool or playroom, with or without teacher presence. Film analysis provided a description and quantification of 85 behavioral items grouped together according to morphological and functional characteristics into 10 behavioral categories. The number of occurrences and the duration of the behaviors differed according to age, sex, prior practice, environment and social situation. Most of the skills (e.g. walking, standing up) and social behaviors (approaching, offering an object) increased in number of occurrence and duration during development, independent of the situation, whereas others decreased (crawling, sitting; crying). Only aggressive behaviors (biting, hitting) were less frequent in the wading pool than in the playroom. Motor skills varied with practice, which was different in the playroom and wading pool. Finally, in the wading pool and in the presence of an adult, boys aged 13 to 24 months were more often isolated than girls, who more often made offers. The results indicate that an aquatic environment encourages the emergence, acquisition, expression and mastery of certain motor and posture skills and social behavior. The children were less often isolated and were less aggressive. The progressive familiarization with the wading pool environment was accompanied by a decrease in conflictual situations. Familiarization was expressed differently according to the "objective" or active presence of the teacher. Social behavior, with the exception of imitating, was more frequent in the wading pool.

SOCIAL STATUS, AFFECT, AND SEX IN HUMAN ADOLESCENTS

Leslie Ziegenhorn

Dept of Psychology, Wayne State University, Detroit, MI 48202 USA, Phone: (313) 577-2800; Fax: (313) 577-7636

The ethological underpinnings of behavior and affect were examined using interviews and self-report measures on 170 adolescents. This exploratory study emphasizes the utility of ethological theory for clarifying issues in clinical psychology. This work complements previous research suggesting relations between dominance-subordination, serotonin, and some forms of depression. Preliminary analyses of gender differences revealed that status was related to more variables in males than females. While status was inversely related to depression in both sexes, the correlation between low status and suicidal ideation was stronger for males than females. Status and number of sexual partners were positively correlated in males, but negatively correlated in females. The relation between sexual activity and affect, athleticism, and risky driving also varied as a function of gender. Differences in the emotional correlates of sexual activity may reflect adaptive functions of pride and shame in reinforcing different mating strategies for the sexes. Antisocial behavior positively correlated with sexual activity in both sexes, which raises the question of potential fitness-enhancing functions of some forms of adolescent rebellion. The theoretical implications of these preliminary results warrant further inspection.

CONTRIBUTORS

R. Robin Baker
School of Biological Sciences
University of Manchester
3.239 Stopford Bldg.
Manchester M13 9PT, England, UK

Thomas J. Bouchard, Jr.
Department of Psychology and Institute of
  Human Genetics
Univ. of Minnesota
Elliott Hall
75 E. River Rd.
Minneapolis, Minnesota MN 55455

C. Sue Carter
Department of Zoology
University of Maryland at College Park
1210 Zoology-Psychology Bldg.
College Park, Maryland 20742-4415

Robin I.M. Dunbar
ESCR Research Center for Economic
  Learning and Social Evolution
Department of Psychology
University of Liverpool
PO Box 147
Liverpool L69 3BX
England, UK

Irenäus Eibl-Eibesfeldt
Max-Planck-Institut für
  Verhaltensphysiologie / Humanethologie
von-der-Tannstr. 3-5, D-82346 Andechs,
  Germany

Martin Fieder
LBI Urban Ethology, c/o Institut für
  Humanbiologie
Universität Wien
Althanstr. 14, A-1090 Wien, Austria

Valentina Filova
Institute for Automation
Department for Pattern Recognition and
  Image Processing
Technical University Vienna
Treitlstr. 3/1832
A-1040 Vienna, Austria

Karl Grammer
LBI Urban Ethology, c/o Institut für
  Humanbiologie
Universität Wien
Althanstr. 14, A-1090 Wien, Austria

Karl Sigmund
Institut für Mathematik
Universität Wien
Strudlhofg. 4, A-1090 Wien, Austria

Peter K. Smith
Goldsmiths College
University of London
New Cross, London SE14 6NW, England, UK

Glenn Weisfeld
Department of Psychology
Wayne State University
71 W. Warren Av.
Detroit, Michigan 48202

CURRICULA VITAE OF EDITORS

Klaus Atzwanger, born 1965, studied biology in Vienna and made his MD and PhD at the Research Center for Human Ethology in the Max-Planck-Society (Andechs, Germany) on aggressive behaviour of kindergarteners and car-drivers. He is currently working on projects dealing with ethnocentrism and with the influence of public space on social behaviour.

Karl Grammer, born 1950, studied biology in Munich and worked for ten years at the Research Center for Human Ethology in the Max-Planck-Society (Andechs, Germany). Recently, he published a book on flirting behaviour. Currently, he is scientific director of the Ludwig-Boltzmann-Institute for Urban Ethology.

Katrin Schäfer, born 1967, studied biology in Vienna and made her MD on how public places constrain the behaviour of users. She currently makes her PhD and works on impressions public space makes on users and participates in projects dealing with ethnocentrism and with physical anthropology.

Alain Schmitt, born 1960, studied veterinary medicine and psychology at the University of Vienna and worked there for nine years on animal and human behaviour subjects. Currently, he is a psychologist at a child protection center in Vienna.

CONTACT INFORMATION

DDr. Alain Schmitt
affiliation: Ludwig-Boltzmann-Institute for Urban Ethology (Vienna)
LBI Urban Ethology, c/o Institut für Humanbiologie
Universität Wien, Althanstr. 14, A-1090 Wien, Austria, Europe
tel +43 1 31 336 1340, fax +43 1 31 336 788

Mag. Dr. Klaus Atzwanger
affiliation: Research Center for Human Ethology, Max-Planck-Society (Andechs, Germany)
LBI Urban Ethology, c/o Institut für Humanbiologie
Universität Wien, Althanstr. 14, A-1090 Wien, Austria, Europe
tel +43 1 31 336 1340, fax +43 1 31 336 788
tel +43 1 526 18 20 5, fax +43 1 526 18 20 9
e-mail: a8111GCA@vm.univie.ac.at

Doz. Dr. Karl Grammer
affiliation: Ludwig-Boltzmann-Institute for Urban Ethology (Vienna)
LBI Urban Ethology, c/o Institut für Humanbiologie
Universität Wien, Althanstr. 14, A-1090 Wien, Austria, Europe
tel +43 1 31 336 1253, fax +43 1 31 336 788, email karl.grammer@univie.ac.at

Mag. Katrin Schäfer
affiliation: Institute for Human Biology
Institut für Humanbiologie
Universität Wien, Althanstr. 14, A-1090 Wien, Austria, Europe
tel +43 1 31 336 1339, fax +43 1 31 336 788, email katrin.schaefer@univie.ac.at

NAME INDEX

Abstracts of papers or posters presented at the 13th ISHE conference 1996 are indicated by an "a" immediately following the page number.

Abraham, 131
Adelson, 102
Aiello, 78
Ainsworth, 209
Albers, 154
Albert, 149
Alberts, 149
Aldis, 48, 49
Alexander, 36, 106
Allport, 132
Altemus, 152, 157, 158
Apicella, 31
Arbib, 100
Archer, 148, 149, 154
Argyle, 43, 92
Aristotle, 141
Arvey, 131
Aschoff, 7
Asher, 54
Astington, 81
Aston, 53
Attili, 189a
Atzwanger, v–viii, 19, 189a, 190a, 227
Auger, 174
Axelrod, 58, 68

Baddeley, 198
Bagatell, 148
Bahrke, 149
Bailey, 178
Bainbridge, 181, 186
Bajcsy, 114
Baker E, 153
Baker R, viii, 34, 37, 106, 163–189, 225
Barber, 34
Bard, 31
Barling, 136

Baron, 99
Barret, 169, 177
Barrett, 127
Barton, 77–79
Bassili, 101
Bastiani, 10
Beach, 178
Beaudichon, 190a
Bechinie, 191a
Beck, 190, 193
Becker, 43, 143
Bekoff, 48
Bell-Krannhals, 15, 16
Bellis, 106, 164–165, 168–172, 174–178, 183–184, 186
Belsey, 165
Bensel, 191a
Benshoof, 106
Bernieri, 99
Berry, 101
Berthold, 142
Betsworth, 126
Bever, 79
Bichakjian, 192a
Bickerton, 192a
Billings, 35, 39
Birkhead, 177
Blatchford, 50
Blau, 134–135
Block, 136
Blurton Jones, 31, 48, 49, 51
Bluthe, 157
Boerlijst, 69, 71
Boero, 192a
Bogaert, 177
Boissou, 149
Boltzmann, 44
Boodoo, 127
Bouchard, vii, 121–140, 225

Bouhuys, 193a, 201a, 203a
Boulton, 31, 48, 49, 53, 54, 55, 57, 60, 61
Bowlby, 14, 209
Boyd, 70–71
Boykin, 127
Bramel, 37
Brand, 193a
Briceno, 205a
Bridges, 150
Brody, 127
Bronson, 141, 148
Brooks-Gunn, 134, 149, 150
Brothers, 77
Brown, 95, 142, 143
Brown-Sequard, 142
Bruins, 157
Brüne, 194a
Buck, 26, 29, 31
Budack, 15
Bujan, 174
Bujatti, 194a
Burks, 58, 134–135
Burris, 148
Burt, 127
Buss, 13, 137
Butovskaya, 195a, 208a
Byrne, 79–81

Camire, 105
Campan, 219a
Campbell, 123–124, 132
Cannon, 31
Caprara, 207a
Cardon, 123
Carey, 133
Carnap, 128
Carson, 58
Carte, 158

229

Carter, viii, 141–162, 225
Carvalho, 53
Carver, 107
Casagrande, 195a
Castell, 128
Cattell, 27
Ceci, 127
Chance, 194
Chance, 99
Charlesworth, 11, 13–14, 16, 36
Cheney, 81
Chiarelli, 218a
Chowduri, 61
Christian, 209a
Clark, 169,176
Clutton-Brock, 78
Cohen, 84, 136
Coie, 54
Collaer, 50
Connolly, 48
Cook, 176
Cooper, 136
Corter, 151
Cosmides, 97, 116
Costabile, 49, 52, 53
Coyne, 203
Cracknell, 186
Craig, 4
Crawford, 196a
Cunnigham, 107
Cutting, 101
Czerlinski, 196a

Dabbs, 149, 154
Dahlstrom, 132
Damasio, 38
Dantzer, 157
Darwin, 2, 14, 29, 31, 42, 67
Davenport, 178
Davis, 29
Dawis, 131
Dawkins, 95
de Waal, 42
Delville, 157
Dennett, 80
Depaulo, 95, 107, 116
Depinet, 127
Desjardins, 148
Diamond, 176,177,179
Diener, C., 130–131
Diener, E., 130–131
DiLalla, 133–134
Dimitrov, 197a
Dissayanake, 36
Dittami, 199a, 220a
Dixon, 151, 155
Dodge, 52
Donald, 200
Dong, 37, 40

Dorfman, 128
Döring, 169
Dunbar, vi–vii, 26, 77–89, 116, 225

Eaves, 132, 136
Ehrhardt, 35
Eibl-Eibesfeldt,
Eibl-Eibesfeldt, v, 1–24, 26, 29, 35, 36, 39, 43, 92, 96, 98–99, 101, 207a, 225a
Ekman, 29, 30, 34, 42, 92, 95, 97, 99
Ellis, 30, 39
Elworthy, 197a
Enquist, 86
Escher, 18
Essa, 102
Ewert, 12
Eysenck, 132

Fagen, 54
Fancher, 128
Fassnacht, 48
Feather, 132
Feingold, 132
Fenk, 198a
Fenk-Oczion, 198a
Ferris, 157
Fieder, vii, 91–120, 198a, 199a, 202a, 225
Field, 33
Filova, vii, 91–120, 199a, 225
Finkel, 129
Finnan, 62
Fisch, 190
Fischer, 35
Fischer, 87
Fischmann, 199a
Fisher, 43
Fleming, 150, 151
Ford, 178
Forgas, 97
Frame, 52,
Franz, 200a
Frean, 72
Freedman, 14,
Freeman, 127, 135
Frey, 99, 207
Fridlund, 28–29, 31, 38, 42
Friesen, 29, 34, 42, 92, 97, 99
Fromm, 10
Fry, 48–49, 51, 53, 60
Frykholm, 101
Fürlinger, 200a

Gage, 175
Galileo, 204
Galton, 11, 134
Gangestadt, 171, 179–180

Garner, 196
Gati, 124
Gaylord Simpson, 5
Geerts, 193a, 201a
Geoge, 151
Gershowitz, 169
Gibber, 151
Gibson, 135
Gigerenzer, 202a
Gilbert, 201a
Ginsberg, 170
Gjerde, 136
Glover, 170
Goffman, 107
Goldberg, 39
Goldstein, 196a, 202a
Gomathi, 174
Gomendio, 164
Good, 15
Goodall, 27, 30
Goodman, 10
Gordon, 136
Gottesman, 133
Gottfedson, 127
Gottman, 43
Grammer, v–viii, 15, 19, 43, 91–120, 171, 180, 191a, 198a, 199a, 202a, 225–228
Grant, 194
Gratzl, 6
Greidanus, 157
Groos, 48
Gruter, 16
Gubernick, 150

Hafner, 153–155
Haiderer, 6
Hale, 193a, 203a
Hamilton, 58
Hammerstein, 203a
Hampson, 152, 155
Hansen A, 100,
Hansen J, 124–125, 132
Happe, 79, 81
Harcourt, 79–80, 175–176
Harding, 153, 156
Harlow, 33
Harper, 96, 97
Hartl, 176
Harvey, 175, 176
Harvey, 78
Haskovec, 194
Hass, 14
Hassenstein, 19
Haug-Schnabel, 204a
Hayes,122, 137
Heath, 132, 136
Hedley, 186
Heeschen, 15

Name Index

Heideman, 141
Heilmann, 204a
Heinroth, 2, 3
Heinson, 74–75
Heinz, 15
Hejj, 205a
Herrnstein, 127
Herzog, 15
Herzog-Schröder, 15
Hess, 14, 29
Hewitt, 197a
Hines, 50
Hochstetter, 3
Hof, 16
Hofbauer, 66
Hokanson, 37
Hold, 15
Holenstein, 198
Holmes, 136
Holzinger, 127, 135
Horn, 128
Hrdy, 169
Huber, 12
Huck, 170
Hudgens, 136
Hudson, 35
Hufschmidt, 197
Hum-Fagen, 48
Hume, 67–68
Humphreys, 50–51, 53–54, 57, 60
Hunter, 53
Hutt, 14, 39
Huxley, 7

Iacono, 128
Inoff-Germain, 149–150
Insel, 157
Irvine, 74
Izard, 26

Jackson, 42, 124, 133
Jaffe, 205a
James, 1, 31
Jang, 133
Jankowiak, 35
Jansen, 203a
Jardine, 132
Jenkins, 34
Jensen, 134
Jerison, 77–78
Jöchle. 169
Johansson, 100, 165, 178
Johnston, 97
Jolly, 194
Jonsson, 206a
Jouannet, 174
Juel-Nielsen, 127
Jütte, 93, 206a

Kahn, 42
Kakadiaris, 114
Kalin, 37
Kamaryt, 207a
Kamin, 128
Kandel, 12
Kant, 4
Katz, 149
Keesing, 18
Kempter, 207a
Kendler, 136
Kessler, 136
Keverne, 86, 150
Kien, 216a
Kim, 174, 177
Kinderman, 79, 81
Kinsey, 148, 178
Kirkealdy, 136
Klebanov, 134
Kleiman, 156
Kleimann, 155
Klein, 208a
Klopfer, 151
Koehler, 7
König L, 6
König O, 5, 6
Koyama, 80
Kozintsev, 195a, 208a
Kraemer, 33
Krappmann, 61
Krauss, 95
Kraut, 97
Kravitz, 147
Krebs, 95
Kreuz, 149
Kruck, 19, 97, 189a, 209a, 215a, 218a
Kruijt, 11
Kudo, 77, 79
Kummer, 209
Kuo, 1, 9
Kwann, 148

Ladd, 50
Lamb, 11
Lange, 31
Lay, 209a
LeDoux, 26
Lee, 164, 174, 177
Lehrman, 8, 9, 14
Leimar, 86
Leland, 186
Leslie, 80–81
Letzer, 209a
Lever, 57
Levinson, 95
Levy, 197a
Lewin, 38
Lewis, 30, 40, 49–52, 57

Lewontin, 128
Leyhausen, 207
Lichtenstein, 128, 135–136
Ligabue Stricker, 218a
Light, 135
Lind, 210a
Lindgren, 69
Lindqvist Forsberg, 210a
Lindzey, 132
Liu, 102
Lively, 66
Livesley, 133–134
Ljungberg, 210a
Lloyd Morgan, 1
Loehlin, 128–129
Löhr, 211a
Loneragan, 194
Lorenz A, 2–3
Lorenz, K, 1–16, 20, 39, 92, 97, 116, 207
Luine, 153, 156
Lykken, 121, 123–126, 128–129, 136–137

MacArthur, 99
MacDonald, 48, 58, 62
Maes, 114
Magnusson, 190a, 195a, 211a
Malinowski, 216
Mallick, 37
Manning, 179
Markl, 92
Markov, 102
Marler, 10
Marshall, 169
Martin, 132–133
Masters, 16, 30
Masutani, 30, 31
Matheson, 53
Mattei-Muller, 15
Maynard Smith, 66, 68, 73
Mazur, 11
McCall, 131
McCandless, 37
McClearn, 128–129, 136
McCoy, 146
McDougall, 26–28, 34, 36, 39
McGrew, 14
McGue, 121, 124, 126, 128–129, 136–137
McGuire, 30
McLean, 27
Mealey, 35
Meehl, 134
Mehrabian, 95
Meisel, 156, 158–159
Mende, 221a
Menesini, 49, 56
Mersch, 193a

Metaxas, 114
Michael, 146–147
Michalson, 30, 40
Milgram, 197–198
Milinski, 72–74
Miller, 86, 202a, 212a
Millot, 193a, 223a
Mitchell, 135
Moghaddam, 102
Møller, 169, 176, 179
Molnar, 213a
Moloney, 125–126
Money, 35
Montagner, 195a
Montepare, 107
Moore, 98
Morell, 74
Morgenstern, 66
Morris, 16, 116
Morton, 170
Moster, 136
Moyer, 30–31
Murphy, 147
Murray, 127
Müssig, 212a

Nadel, 79
Nagy, 213a
Nash, 66, 68
Neale, 123, 136
Neill, 48, 55, 59
Neisser, 127
Nelson, 142
Nesselroade, 128, 136
Newman, 127
Newton, 151
Nichols, 192
Niedner, 213a
Nissen, 152
Niyogi, 102
Noonan, 106
Nowak, 69–70, 72

O'Connell, 81
O'Keefe, 79
Oatley, 34
Ockenden, 179
Oldenquist, 214a
Olmstead, 196a
Orwell, 21
Oswald, 61
Otis, 2

Packer, 74–75
Paikoff, 149, 150
Panksepp, 30
Paquette, 59
Parke, 58
Parker, 164, 175, 181

Patore, 37
Patterson, 132
Pavlov, 1, 26
Pawlowski, 80
Pearlstone, 136
Pedersen, 128–129, 136, 150, 151
Pellegrini, 48–60
Pentland, 101–102, 114
Perpler, 49
Pfaff, 146, 159
Pillard, 178
Pinker, 192
Plomin, 39, 128, 133, 136
Plutchik, 27–28, 31
Polivy, 196
Pomianowski, 179
Pool, 99
Pope, 149
Popenoe, 21
Pöppel, 198
Popper, 4
Porges, 158–159
Povinelli, 81
Prange, 150
Premack, 80
Proffitt, 101
Propping, 121
Provine, 29, 87, 92
Pryce, 151
Pugh, 28, 34, 36, 38–39
Pursel, 165
Purvis, 79

Rahe, 136
Ricci Bitti, 43
Richer, 214a
Riederer, 194
Riedl, 8
Riess, 9
Roberts, 156
Robinson, 131
Roca, 152–155
Roes, 41, 42
Rohe, 123
Roldan, 164
Ronzal, 198a, 214a
Rosch, 115
Rose, 128, 149
Rosenblatt, 150
Rosenblum, 151
Rosenthal, 95, 99, 116
Rosenzweig, 33, 35
Rostand, v
Rothman, 127
Rozin, 34
Rubinow, 153–154
Runeson, 101
Rushton, 177
Russel J, 29, 34, 43

Russell R, 136
Russell W, 99
Russell-Wells, 221

Sachdev, 194
Sachs, 156, 158–159
Sackin, 42
Salter, 11, 16–17, 20, 207a, 215a
Salzen, 11,
Sawaguchi, 77
Scarr, 128, 134–136
Schacht, 169
Schäfer K, v–viii, 19, 189a, 227–228
Schäfer M, 49–50, 53–54
Schaie, 37
Schanberg, 33
Scheich, 12
Scheier, 107
Schell, 43
Scherer, 30
Schiefenhoevel, 15, 215a, 217a
Schleidt M, 198a, 216a
Schleidt W, 92, 216a
Schmidt, 153–154, 159
Schmitt, v–viii, 190a, 217a, 227
Schwartz, 197
Scott, 27–28
Seeman, 153–154
Segal, 128, 131, 209a
Selten, 71
Senft, 15
Seyfarth, 81
Shapiro, 157
Shaver, 131
Shaw, 131
Sherwin, 146–147, 152–155
Shields, 127
Shimoda, 43
Short, 177
Siddiqi, 95
Siegel, 6
Siegmund, 211a, 217a
Sigmund, vi, 65–76, 225
Sillen-Tullberg, 169
Sinervo, 66
Singh, 171
Skinner, 2, 26
Sloman, 190
Sluckin, 49–50, 61
Smees, 49–50, 56
Smith, vi, 14, 31, 47–64, 225
Snyderman, 127
Sogon, 30–31
Solokowski, 218a
Sonka, 102, 103, 114
Sosa, 151
Spennemann, 197
Sperry, 10

Name Index

Stacey, 21
Stalin, 6
Stanley, 34
Starner, 102
Staw, 131
Steiner, 34
Stent, 12
Stresemann, 7
Sudgen, 67, 70, 73
Surbey, 196
Susman, 149–150
Sütterlin, 19, 189a, 218a
Svärd, 176
Symons, 59
Synek, 219a

Tafforin, 219a
Tannenbaum, 146
Taubman, 131
Taylor, 127
Tellegen, 121, 123–124, 126, 128–129, 137
Tesser, 133
Thelen, 42
Thorne, 62
Thornhill, 106, 167, 171, 180
Thorpe, 7
Tinbergen, 1–2, 4, 7–9, 12, 29, 43, 92
Tittel, 217a
Tobach, 11
Tooke, 105
Trivers, 37, 67, 72, 105
Trumler, 6

Uvnas-Moberg, 151–152, 156–159

van Bezooijen, 29
van de Sande, 222a
van den Berghe, 16
van der Dennen, 207
van der Stelt, 222a
van Goozen, 152, 154–155
van Hooff, 27
van Hooff, 98
van Ree, 157
van Schaik, 78
van Valen, 167, 179
Velle, 39
Vermigli, 189a
Vernon, 132–133
Vieille, 223a
Voland, 196
von Cranach, 6
von Frisch, 7
von Holst, 4, 7, 12
von Neumann, 66, 68
von Rombert, 7
von Salisch, 61
von Uexküll, 2

Walbott, 30, 95, 116, 220a
Walburton, 147
Waldman, 128
Wallen, 145
Waller, 123, 133, 135
Wallhäuser, 12
Wallner, 220a
Walters, 196a
Watson, 2, 26
Wedekind, 72–73
Weibull, 66
Weiche, 200a

Weinberg, 128, 134–135
Weiner, 136
Weisfeld C, vi, 43, 201a, 221a
Weisfeld G, 25–46, 221a, 225
Well, 43
Wells, 136
Welsh, 132
Wermke, 210a, 221a
Westlund, 210a
White, 134
Whiten, 79–80
Whitmore, 186
Wickler, 7, 11
Wiessner, 15–16
Wiklund, 76
Willerman, 128
Willows, 12
Wilson, 12, 132
Wimersma, 157
Wingfield, 155
Witmean, 2
Wolff, 38
Woodruffe, 80
Wright, 131
Wu, 70

Yerkes, 220
Yong, 87
Young, 92

Zahn-Waxler, 136
Zarrow, 169
Ziegenhorn, 223a
Zivin, 42
Zumpe, 146–147

SUBJECT INDEX

Academic success, 149
Action units, 15, 158, 216
Activity patterns, 192, 211, 217
Adaptation, 5, 9–10, 13, 25, 27, 33, 41
Adolescence, 148–149, 223
Adoption studies, 134; *see also* Twins
Affective disorders, 153; *see also* Akathasia; Alzheimer disease; Anorexia; Autism; Depressed mood; Phobia
Aggression, 5, 10–12, 15, 19, 29–31, 34, 148–149, 155, 158, 195, 207–208
Aging, 154
Akathisia, 194
Alliances, 79; *see also* Grooming cliques
Alzheimer disease, 152
Anger, 34, 37, 149
Anogenital swellings, 220
Anorexia, 196
Anxiety, 153
Apes: *see* Non-human primates
Appetitive behavior, 4
Art, vi, 18, 36, 197, 212; *see also* Environmental aesthetics
Assortative mating, 132
Attachment/bonding, 203, 209
Attractiveness, 36, 180, 196, 198, 205, 214–215, 220–221
Autism, 214
Automatic image analysis: *see* Digital image analysis

Bats, 79
Beauty appreciation, 36
Behavior sequences; *see* Berner System; THEME
Behaviorism, 1–2, 26
Behaviour
 analysis, 91, 102
 cartographic description of: *see* Berner System
 categorisation, 100, 109, 112, 114–115, 145
 recording methods, 91, 99, 115, 117
 pattern, 30, 49
Berner system, 99, 193, 207
Birth, 150
Bisexuality, viii, 178

Bodily states, 159
Body
 image, 196, 214
 movements, 92, 95, 100, 102, 107,114
 postures, 92–93, 95,
 size, 155
Boredom, 35
Brain, 77, 143, 197, 216
Breast feeding, 152

Carnivores, 79
Castration ,141, 148
Children, 47, 52, 54, 62, 81, 189–190, 195, 201, 204, 209–210, 217, 223; *see also* Infants
Clothing, 106–107, 111, 200
Cognitive
 capacities, 26
 dysfunction, 152
 processes, 158
 psychology, 26
 skills, vi, 80–81, 96–97, 190, 197, 202, 204, 217, 222
Communication, 8, 11, 18, 91–115, 189, 216, 222; *see also* Behavior sequences; Cognitive skills; Conflict resolution; Cooperation; Courtship; Dating; Digital image analysis; Face; Gesture; Language; Mother–infant interaction; Non-human primates; Non-verbal behavior; Odors; Pheromones; Reconciliation; Rough-and-tumble play; Social; Speech; Status; Synchronization; Time; Tonic
Companionship, 35
Computer simulation, 69, 197, 202, 207; *see also* Neural networks
Concealed ovulation, 106, 117, 169, 202, 206
Confinement, 219
Conflict resolution, 66, 210, 217; *see also* Reconciliation
Conformity, 197
Consciousness, 78
Consummatory behavior, 4, 11
Contempt, 34
Conversation, *see also* Gossip
 group size, 83–84
 topics, 84–85

235

Cooperation, vi, 7, 66–75, 190
Copulation, 170
Courtship: *see* Mate choice
Critical realism, 4
Cross-cultural analysis, 17, 31, 100, 103, 191
Crying, 191, 210, 221
Cuckoldry, 215
Cultural ethology, 18
Cultural evolution, 192
Czech gesture atlas, 208

Dating, 205
Deception, 80, 86, 95–97, 101, 105,
Depressed mood, 53, 149, 153, 190–191, 193, 201, 203
Deprivation experiment, 9
Development,
Development, 1, 5, 8–9, 39, 168, 191, 210–211, 213, 218, 221, 223
Digital image analysis, vii, 91, 99, 101–102, 105, 107, 110–112, 114, 191, 198–199, 202
Disgust, 34
Displacement activity, 194
Divorce, 136,
Domestication, 5
Dominance: *see* Status
Drowsiness, 34

Ejaculate, 164, 173, 175, 179
Emotions, vi, 19, 25–43, 71, 101, 149, 153, 218, 223; *see also* Affective disorders; Aggression; Anger; Anxiety; Attachment; Beauty appreciation; Boredom; Conformity; Contempt; Crying; Disgust; Empathy; Fear; Frustration–aggression hypothesis; Humor; Jealousy; Loneliness; Love; Lovesickness; Mother love; Motivation; Nationalism; Nausea; Pride; Reconciliation; Reciprocal altruism; Phytophylia; Self esteem; Shame
  basic, vi, 213, 220,
Empathy, 34
Environmental aesthetics, 189, 218–219
Environmental effects, 123, 125–126, 128
Erection, 148
Erotic stimuli, 148
Ethogram, vi, 25–26
Evolutionary psychology, 77–89, 192, 197, 200, 202, 204–205, 208, 212–213, 218–219, 222
Evolutionary stable strategy, 13
Exploration, 210
Extra-pair copulation, 106, 165, 181, 183

Face, 15, 29, 99–102, 117, 193, 214, 220
Facial expression: *see* Face
Facial feedback hypothesis, 31
Facial Action Coding System FACS, 99
Family resemblance, 189, 215
Fatherhood certainty, viii, 189
Fatigue, 34
Fear, 37

Fecundity, 221
Female
  aggression, 150
  choice, 107, 117, 168
  sexual behavior, 146
  competition, 196, 200
Fertile period, 169
Figthing, 58
  play, 47, 54
  real, 47, 50, 53–54, 62
Fixed action pattern, 4, 5, 26; *see also* Action unit
Flowback, 164
Fluctuating asymmetry, 167, 171, 179; *see also* Symmetry
Friendship, 35
Frontal animal scheme, 212
Frustration–aggression hypothesis, 37

Game theory, vi, 65–75, 190, 202; *see also* Prisoner's dilemma
  an-eye-for-an-eye, 37, 70
  animal examples, 66, 74–75
  chicken, 73
  contrite tit-for-tat, 70–71
  firm-but-fair, 72
  generous tit-for-tat, 69
  history, 65–66
  (non)zero-sum, 66, 68
  neutral drift, 71
  pavlov, 70–74
  perfect foresight and rationality assumption, 66, 68
  remorse, 71
  rock-scissors-paper, 66
  snowdrift, 73
  strategies, 66–75
  tit-for-tat, 69–74
  trial and error, 70
  weakling, 71
  win-stay-lose-shift, 70
Genetics, 136–137
  behavior, 121, 133
  factor, 1, 122, 124
  variance, 125
Gestalt, 18, 212
Gesture inventory/atlas, 208
Gossip, 84; *see also* Conversation
Grooming, vii, 83
  clique size, 79
Group behavior, 83, 214
Group size, 78, 82–83

Habitat choice, 203, 219
Hair Flip, 93–94, 103, 109
Handedness, 197
Happiness, 130
Head movements, 94, 100
Heritability, 122, 125, 128
Homology, 3

Subject Index

Hormones, viii, 29, 86–87, 141–160, 199, 206, 218, 220; *see also* Odors; Pheromones; Sex appeal
  adrenal steroids, 144
  androgenic hormones, 143–144, 147, 149, 155
  catecholamines, 145, 158
  endogeneous opiates, 86–87, 158
  estrogen, 106, 110–112, 114, 143, 146, 156
  gonadal steroids, 143, 153, 155–156
  melatonin, 145
  neuropeptides, 143, 145, 156
  oxytocin, 145, 147, 151, 157–158
  progestagenes, 143, 147, 150
  replacement therapy, 144
  sex hormones, vii, 220
  vasopressin, 145,158
Hot flash, 146
Human Ethology
  and sociobiology, v
  as social enterprise, v
  history-future, v, 1–18
  interdisciplinarity, v
  fundamental concepts, v
Humanities, 21
Humor, 36, 208
Hunger, 29, 33
Hunting, 200
Hygienic practices, 34

Ideologisation, 20
Imprinting, 2, 3
In-pair copulation, 165, 174
Individual variability, 195
Infantile research, 41
Infants, 191, 195, 210, 211, 221; *see also* Children, Mother–infant interaction
Infidelity, 170
Infidelity, viii, 221
Innateness: *see* Nature–nurture
Innate releaser mechanism, 207
Insemination, 170
Instinct/Instinktbewegung, 1–4, 26
Intelligence/IQ, vii, 127–129, 134
Intensionality, 80–81
Intentionality, 80; *see also* Intensionality
Intersubjectivity, 213–214

Jealousy, 40, 221

Key stimulus, 5
Kindchenschema, 5
Kolymbetics, 194

Lactation, 151
Landscape preferences, 218–219
Language
  evolution, vi, 77–89, 192, 198, 212
  function, vi–vii, 83–87
Laughter, 87, 92–93, 95–98,
Law, vi

Learning, 152; *see also* Game theory
Linguistics, 192, 198
Loneliness, 35
Long-term relationship, 175
Lordosis, 146
Love, 35, 191, 215
Lovesickness, 191
Lying, 96, 116; *see also* Deception

Manipulation, 96, 98–99, 116,
  physiological, 96,
  subluminal, 103
Marital satisfaction, 201, 221, 223
Masturbation, 164, 174, 183
Mate choice, viii, 4, 40, 80, 86, 103, 105, 110, 112, 196, 205, 209, 212–213, 215, 220
Mate-guarding, 158
Maternity, 150
Max-Planck-Institut für Verhaltensphysiologie, 2, 7
Motion Energy Detection MED: *see* Digital image analysis
Memory, 79, 152
Menopause, 154
Menstrual cycle, 105, 116–117, 146, 153
Migration, 203
Milgram experiments, 197
Milk production, 152
Mimetic skills, 200
Monkeys: *see* Non-human primates
Monogamy, 156
Morphing, 220
Mother infant interactions, 151, 209
Mother love 151
Motion/movement analysis: *see* Digital image analysis
Motivation, 4, 26, 30
Multidimensional Personality Questionnaire MPQ, 129, 131

Nationalism, 214
Nature–nurture problem, 3, 5, 8, 10, 14, 20, 192, 197, 209, 214, 222
Nausea, 34
Neo-Darwinism, 43
Neocortex size, 78–79
Neural mechanism, 29
Neural network, vii, 10, 41, 111–112, 117, 202
Neuroethology, 12
Neuroleptics, 194
New-age ethology, 194
Noise-annoyance, 36
Non-human primates, vii, 30, 77–89, 151, 200, 208, 216, 220
Nonverbal behav./communic., 95–96, 98, 116, 190,193, 195, 199, 202, 206–209, 211, 213, 216–217, 222

Obedience: *see* Conformity
Obesity, 33
Observation as method, 16, 43, 47–48, 51, 53, 55, 60, 62

Odours, 218; *see also* Pheromones
  human axillary, 205
  human vaginal, 206
Ontogenesis: *see* Development
Orgasm, 17, 164, 173, 180
Ovarian cycles, 146
Ovulation, 169

Parental authority, 150
Parental behavior, 150
Parental investment, 189, 201
Parturition, 150
Pattern analysis, 98–99, 101, 112; *see also* Digital image analysis, Neural networks, THEME
Perceptual capacities, 18, 26
Person perception, 207
Pheromones, 96, 205–206, 218; *see also* Odors
Phobias, 37
Phytophilia, 18, 218
Play face, 50; *see also* Rough-and-tumble-play
Play preferences, 209
Political communication, 192
Pregnancy, 150
Premenstrual syndrome, 153
Prespeech, 210, 221
Pride, 34, 36
Prisoner's dilemma game, 67–73; *see also* Game theory
  alternating 67–69
  repeated, 67–73
  simultaneous, 72–73
Prospect refuge quality, 189
Prototypes, 112,
Prozac, 145, 153, 158
Psycho-acoustics, 84–85

Race, 134,
Reciprocal altruism, 41, 68, 72, 74
Recognition, 202
Reconciliation/reassurance, 42, 209–210, 217
Refrigerator mother, 135; *see also* Mother love
Rejuvenation, 142
Religion, 132
Reproductive success, 12, 40
Resemblance, 189, 215; *see also* Family resemblance
Risk, 95–96, 114, 116
Risk of conception, 171
Ritualisation, 96
Rodents, 145
Rough-and-tumble play, vi, 47–62, 204; *see also* Play

Savanna hypothesis, 36, 219
Schizophrenia, 153
Self deception, 204
Self differentiation, 2, 10
Self esteem, 206
Self presentation, 105, 107, 199, 213
Sensory capacities, 26
Sensory motor behavior, 200

Serotonin, 145, 158
Sex appeal, 86, 220
Sex/gender differences, vi–vii, , 39, 47, 62, 85, 195, 217, 223
Sexual
  attractiveness, 40
  behavior, 146, 158
  crypsis, 169
  feeling, 34
  interest, 148
  liasons, 166
  proceptivity, 146
  receptivity, 147
  signals, 111
Sexual behavior, 215, 223
  selection, 163–185, 205, 212
Shame, 36
Signals, 91–93, 95–101,
  honesty, 96, 97
  decoding, 92, 95, 101, 117
Single's bars, 213
Skin exposure, 107,
Sleep patterns, 35, 146
Smelling, 34
Social
  attitudes, 131
  behavior, v, 189–190, 195, 214, 218–219, 223
  bonding, 33, 47, 57, 62, 155; *see also* Attachment
  isolation, 201, 203
  motives, 27
  relationships, 79, 200, 201, 203–204, 209, 216, 217
  releasers, 3
  skills, 58, 62, 79–81
  tool, 47
Sociobiology, 11–13, 16
Socioemotional content, 26
Space: *see* Confinement
Spatial orientation, 152
Speech, 84–85, 198–199
Sperm
  competition, viii, 106–107, 117, 164, 168, 175, 180, 182
  number, 173
  retention, 173
  warfare, 164
Status/dominance, vi–vii, 36, 59, 192, 194, 217, 221, 223
Stimulus–response psychology, 1–2
Stress, 136, 157, 217–218
Strong vocational interest blank SIVB, 123, 125
Symmetry, 167, 171, 179, 198, 212
Synchronization, 98, 209

Tactical deception: *see* Deception
Tactile pain, 31
Tactile pleasure, 31
Tasting, 33
Taxis, 4
Television, 192

Subject Index

Territoriality, 157
Testes, 141, 148, 176
THEME: *see* Time structure analysis
Theory of (other's) mind/ToM, 79–82
Thirst, 33
Time structure analysis, 190, 195, 206, 209, 211, 216
Tonic communication, 92, 216
Tracking method, 114–115,
Traditional societies, 215, 217
Tropism, 26
Twin studies, vii, 121–138, 209, 222
    monozygotic reared apart MZA, 122, 124, 127–128
    monozygotic reared together MZT, 123
    dizygotic reared together DZT, 123
    dizygotic reared apart DZA, 123–124
    Minnesota study of reard apart MISTRA, 124, 128, 131

Twin studies (*cont.*)
    Swedish adoptive study SATSA, 128
    Unrelated individuals reared together URT, 123, 128

Universality, 29
Universals, vi; *see also* Basic emotions
Urban environment, vi, 18, 19, 189
Urban ethology, 18–19

Vaginal lubrication, 146
Verbal behaviour, 96, 98, 152, 206
Violation, 37
Violence, 149; *see also* Aggression
Virtual reality, 199
Visceral experiences, 29, 159

Walking speed, 190
War, 207